T0139928

Studies in Systems, Decision and Control

Volume 180

Series editor

Janusz Kacprzyk, Polish Academy of Sciences, Warsaw, Poland
e-mail: kacprzyk@ibspan.waw.pl

The series "Studies in Systems, Decision and Control" (SSDC) covers both new developments and advances, as well as the state of the art, in the various areas of broadly perceived systems, decision making and control–quickly, up to date and with a high quality. The intent is to cover the theory, applications, and perspectives on the state of the art and future developments relevant to systems, decision making, control, complex processes and related areas, as embedded in the fields of engineering, computer science, physics, economics, social and life sciences, as well as the paradigms and methodologies behind them. The series contains monographs, textbooks, lecture notes and edited volumes in systems, decision making and control spanning the areas of Cyber-Physical Systems, Autonomous Systems, Sensor Networks, Control Systems, Energy Systems, Automotive Systems, Biological Systems, Vehicular Networking and Connected Vehicles, Aerospace Systems, Automation, Manufacturing, Smart Grids, Nonlinear Systems, Power Systems, Robotics, Social Systems, Economic Systems and other. Of particular value to both the contributors and the readership are the short publication timeframe and the world-wide distribution and exposure which enable both a wide and rapid dissemination of research output.

More information about this series at http://www.springer.com/series/13304

Jaime Gil-Lafuente · Domenico Marino
Francesco Carlo Morabito
Editors

Economy, Business and Uncertainty: New Ideas for a Euro-Mediterranean Industrial Policy

 Springer

Editors
Jaime Gil-Lafuente
Departament d'Empresa
Universitat de Barcelona
Barcelona, Spain

Francesco Carlo Morabito
Department of Civil, Energy, Environment
and Materials Engineering
Mediterranea University of Reggio Calabria
Reggio Calabria, Italy

Domenico Marino
Mediterranea University of Reggio Calabria
Reggio Calabria, Italy

ISSN 2198-4182   ISSN 2198-4190   (electronic)
Studies in Systems, Decision and Control
ISBN 978-3-030-13140-1   ISBN 978-3-030-00677-8   (eBook)
https://doi.org/10.1007/978-3-030-00677-8

This Springer imprint is published by the registered company Springer Nature Switzerland AG
The registered company address is: Gewerbestrasse 11, 6330 Cham, Switzerland

Introduction

On 4th and 5th of September 2017, the XXVI International Conference of the European Academy of Management and Business Economics was held at Reggio Calabria, a welcoming Mediterranean city in Italy. The main motto of the conference was *Economy, Business and Uncertainty: ideas for a European and Mediterranean industrial policy*.

The modern economy is made up of a «complex system in evolution» in which single individuals are joined by relational forces, that the dynamic characteristics cannot be represented by means of individual approaches, but through collective properties submitted to successive non-reversible scansions. Thus, it is imaginable that each economic system, in its evolution, manifests both a multiplicity of equilibrium points, each dependent on previous historical interrelations, and the presence of inefficiencies and lock-ins that may be selected during the evolutionary course of the system, to the detriment of the possible efficient solutions.

The government of the economy, read as a complex system in evolution, excludes, therefore, the possibility that commands can be expressed thinking of a prescribed type mechanism, as would happen if the system under analysis were substantially closed and characterized by a low level of interrelations between the agents.

Within this paradigm, the involvement of professors and researchers from different Spanish, Italian and Latin American universities allowed us to develop and enrich scientific knowledge in specializations such as finance, marketing, operational techniques, management, production, human resources, business organization or management computing, among other subjects.

The level of some of the works presented and discussed in the forum on teaching and research experiences which AEDEM makes up led us to consider the possibility of compiling them in a document so that they could be published by a world-renowned publishing house. The generosity and intense work of the professors Francesco Carlo Morabito and Domenico Marino have made possible that today this book becomes a reality.

Jaime Gil-Lafuente
Domenico Marino
Francesco Carlo Morabito

Contents

Bibliometric Analysis on Customer Dissatisfaction

Laura Pascual-Nebreda[✉], Alicia Blanco-González,
and Francisco Díez-Martín

Departament of Business Administration, Faculty of Social Science and Law,
University Rey Juan Carlos, P°. de los Artilleros, s/n, 28032 Madrid, Spain
{laura.pascual,alicia.blanco,francisco.diez}@urjc.es

Abstract. This research analyzes the consumer's dissatisfaction regarding any possible failure made by brands or companies. In order to do this, the behavior, feelings, and negative and even aggressive behavior that costumers can have towards a company, product or brand are thoroughly studied. It is a relatively unexplored research field within the academic literature, in spite of being a subject of great repercussion or relevance at the present time. For this purpose, a study on bibliometric analysis has been carried out to verify the degree of relevance of this topic in the scientific field, discovering who are the authorities of the same, the life cycle of the articles on this subject, the articles keywords or their connections. With all this, we can deepen the academic literature, establish the knowledge structure in this area and propose future empirical research lines.

Keywords: Dissatisfaction · Customer · Brand sabotage
Instrumental aggression · Bibliometric analysis

1 Introduction

This study analyzes the negative behavior of the consumer regarding any possible issue or problem in their relationships with a company, that is to say, the customer's dissatisfaction. Consumers do not always act or react the way they expect. Consumers' preferences, habits, feelings and behaviors are in a changing and complex environment, and therefore it is important to study their reactions and the causes that lead to customer dissatisfaction and discontent. To be successful, you have to know the consumers, know how they want to be treated, what can bother them or why they might be disappointed. As is well known, sometimes your best client can become your worst enemy.

There are internal and external factors that can encourage a negative reaction from the customer towards the company when an unfair situation is perceived, which will depend on the culture, personal influence, situational determinants, perception, experience and learning or attitudes, among others. Why are there consumers who simply decide to make a personal complaint to the company and others opt for the option of hanging a video on YouTube telling their bad experience with the intention of boycotting or damaging the company or the brand? Why are there other consumers who prefer to do nothing? If the relationship with the company is good, will this have a

© Springer Nature Switzerland AG 2019
J. Gil-Lafuente et al. (Eds.): AEDEM 2017, SSDC 180, pp. 1–11, 2019.
https://doi.org/10.1007/978-3-030-00677-8_1

damping effect, or will you feel more betrayed? In short, these are the questions that are answered in this research.

Although they are key questions for companies, they have not been deeply treated in the academic literature. Even though many investigations have been carried out regarding consumer purchase behavior, few have been the ones that study the post purchase behavior, the negative reactions of the consumers, and the types of behaviors that they can lead to dissatisfaction (Kähr et al. 2016; Johnson et al. 2017).

In short, the objective of this work is to achieve a deep knowledge about negative reactions carried out by consumers when they are not happy with a brand or a company and what are the consequences for the companies, using a bibliometric analysis for this purpose.

2   Methodology and Sample

Bibliometrics is part of scientometry which applies mathematical and statistical methods to the whole scientific literature and to the authors who produce it, with the aim of studying and analyzing the activity and scientific evolution (Ramos 2017). The instruments used to measure the aspects of this social phenomenon are the bibliometric indicators, measures that provide information on the results of scientific activity in any of its manifestations. The objective is to create a representation of the structure of the research area by dividing elements (documents, authors, journals, words) in different groups. The display is then used to create a visual representation of the classification that emerges.

The structure of bibliometric analysis provides insights into the influence of authors around this issue, as well as changes in their influence over time are as follows

- First they delimit the subfields that constitute the intellectual structure of management strategically.
- Subsequently, relations between the subfields are delimited.
- After the authors play a fundamental role in the construction of two identified or more conceptual domains of research.
- Finally, graphics map the intellectual structure in two-dimensional space to visualize spatial distances between intellectual subjects.

This bibliometric study was conducted using two keywords: Dissatisfaction and client (customer) which was introduced into the database Web of Science (WOS). Web of Science is an online service of scientific information, provided by Thomson Reuters, integrated into ISI Web of Knowledge, WoK. It provides access to a set of databases in cited in journals, articles, books and other printed material covering all fields of academic knowledge. Previous publications are available to see research published on a determined topic through access to their references cited, and also to the publications mentioning a particular document to discover the impact of scientific work on a current investigation. Finally, it can connect you to the full text of and other resources and access them using a search based on keywords. These databases contain nearly 10,000 sciences, technology, social sciences, arts, and humanities magazines and over 100,000 conference proceedings and conferences which are updated weekly.

After entering into the WOS these two keywords (dissatisfaction and consumer/customer), the following filters were applied: fields in the area of social sciences (Social Sciences); specific subfields the area of social sciences and business economics and psychology, behavioral sciences, etc., and related subfields were eliminated such as engineering or medicine; Finally, we selected exclusively scientific articles (not Congresses, thesis or others). The total sample of items incorporated into these filters was 380 items for a period of time from 1980 to the present.

Then we proceeded to delete the inconsistencies of the database and prepared a new file from a classification confirming the names of the authors.

3 Results

First, in Table 1 most cited documents are presented (in this case specifically articles). These are items that have had the greatest impact, influence or have contributed more to the development of the discipline being treated. The most cited articles represent the basis of knowledge of a discipline and they reflect a consensus among its protagonists (Ramos 2017), which allows knowing the list for required reading in a particular field or discipline. The statistical technique used was by counting frequency, having previously done the relevant deleting.

As the article shown in said Table 1 the article that has been mentioned 66 times, was developed by Bitner, M. and has been published in the Journal of Marketing under the name "Evaluating service encounters: the effects of physical surroundings and employee responses", followed by the article written by Oliver that has been cited 65 times in the Journal Marketing Research, which is entitled "A Cognitive Model of the Antecedents and Consequences of Satisfaction Decisions". Coincidentally the third most cited article belongs to the same author as the second most cited article, "Customer delight: foundations, findings, and managerial insight". Therefore, it is stated that the author with the most cited articles is Oliver, adding up to a total of 119 citations is this table. In addition, Parasuraman also has a high number in the Journal of Marketing and Journal of Retailing magazines, specifically with 94 citations overall.

Secondly, in Table 2 the most cited authors are presented, that is, those who have had greater impact, influence or have contributed more to the development of a particular discipline. The great utility in this point is to know what authors can be considered authorities on a particular subject, and if there is relation between the most productive and most cited authors. The statistical technique used is the counting of frequencies, although it has one limitation because only the first author of the article is considered.

Of the total items extracted in the sample (380), Oliver, R. has the most citations with 176 which reflect that he is an expert or an authority in analyzing dissatisfaction. Next is Fornell, C. with 102 quotes. As seen, this table relates to Table 3, since there are many authors that are repeated.

Thirdly, coauthors analysis (Fig. 1) which uses data coauthored performed to measure the collaboration. Quotations are used as a measure of influence, so if an article is cited it is considered important. This analysis is based on the assumption that the authors cite documents considered important for their work and examine the social

Table 1. Ranking of the most popular articles

Cites	Author	Year	Journal	Article
66	Bitner, M.	1990	*Journal of Marketing*	Evaluating service encounters: the effects of physical surroundings and employee responses
65	Oliver, R.	1980	*Journal Marketing Research*	A cognitive model of the antecedents and consequences of satisfaction decisions
54	Oliver, R. Rust, R Varki, V.	1997	*Journal of Retailing*	Customer delight: foundations, findings, and managerial insight
49	Keaveney, SM.	1995	*Journal of Marketing*	Customer switching behavior in service industries: an exploratory study
47	Parasuraman, A. Zeithaml, VA. Berry, L.	1985	*Journal of Marketing*	Problems and strategies in services marketing
47	Parasuraman, A. Zeithaml, VA. Berry, L.	1988	*Journal of Retailing*	Servqual: a multiple-item scale for measuring consumer perception of service quality
44	Parasuraman, A. Zeithaml, VA. Berry, L.	1996	*Journal of Marketing*	The behavioral consequences of service quality
42	Smith, AK.	1999	*Journal Psychology Marketing*	Buyer–seller relationships: similarity, relationship management, and quality

Table 2. Ranking of the most cited authors

176	Oliver, R.	76	Zeithaml, VA.	52	Smith, AK.
102	Fornell, C.	59	Singh, J.	52	Westbrook, RA.
92	Bitner, M.	59	Tax, SS.	51	Richins, ML.
84	Parasuraman, A.	53	Keaveney, SM.	50	Bolton, RN.

networks that scientists create in scientific articles (Acedo et al. 2006). A relationship between two authors is established when an article is co –published (Liu et al. 2013).

Co-authorship is a measure of collaboration that assumes that the creation of a publication is synonymous with being responsible for the work done. However, just because the name of a person appears as a co- author of a scientific paper, it does not necessarily mean that person contributed with any significant amount of work (Martin 1997). In addition, there could be scientists who contributed to the work but whose names do not appear on the authors list.

As seen in Fig. 1, small islands that reflect different positions within this field and the author that is featured in each can be observed. Interestingly in the field studied there are different views or aspects, since most of the figures are completely independent and do not keep relationship between them. In the figures of the group on the

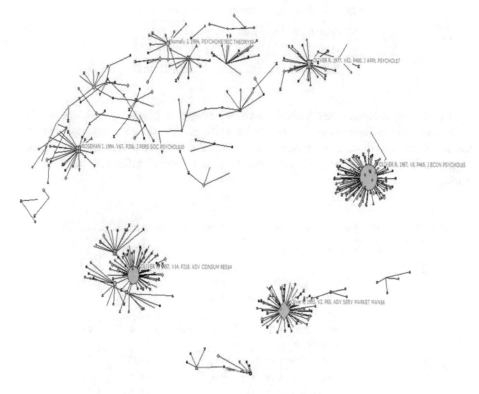

Fig. 1. Map of co-authors

right the author Oliver, R. prevails, and in the other three figures in the representation. Therefore Oliver, R. ranked first, he has been the most cited overall. Topics covered by Oliver shown on the map are economic psychology, marketing services or consumer behavior.

In addition to the other two figures that appear in the upper left part, the authors that predominate are Roseman, I. and Nunnally, J.; they keep some relationship between their researches, as they are joined by lines. Common themes that these authors treat are social psychology, and psychometric theories.

Fourth, in Table 3, the most productive authors in this research line appear, which are the experts or leaders in this scientific field and are the authors whose work should be known in a particular discipline. The statistical technique used is the frequency count of chain of text after making a good deletion.

This table represents the most productive authors, and therefore those who have written and published more articles in this field. Interestingly being the most fruitful or those that have more publications, are not the most cited. If Table 2 is examined, it is found that most of these authors do not appear. This is because these authors have not had enough impact and therefore do not match the authors most cited. The author with more number of papers published in this field is Mattila, AS., followed by Hyun, SS. and Bitner, M.

Table 3. Ranking of the most productive authors

5	Mattila, AS.	3	Pieters, R.	3	Lu, YB.
4	Hyun, SS.	3	Wang, LY.	3	Ostrom, AL.
4	Bitner, M.	3	Li, M.	3	Tronvoll, B.

Fifthly, we analyze which journals have had the greatest impact (Table 4), and how they have changed over time, in addition to which are the forum magazines that form the intellectual basis of a discipline. The statistical technique used was the frequency count, and the procedure used was extracting each journal reference and removing duplicates.

Table 4. Most cited magazines or forum of journals with impact index JCR year 2016

Number of Articles	Journals forum	Impact JCR	Quartil JCR
21	*Journal of Business Research*	**3.354**	**Q1**
17	*Total Quality Management & Business Excellence*	**1.368**	**Q3**
14	*International Journal of Service Industry Management*	***n.a.*** *	***n.a.****
12	*Journal of Service Research*	**6.847**	**Q1**
12	*Journal of Services Marketing*	***n.a.*** *	***n.a.****
8	*Tourism Management*	**4.707**	**Q1**
7	*African Journal of Business Management*	***n.a.*** *	***n.a.****
6	*Managing Service Quality*	***n.a.*** *	***n.a.****
6	*Journal of the Academy of Marketing Science*	**5.888**	**Q1**
6	*Journal of Service Management*	***n.a.*** *	***n.a.****
6	*Journal of Retailing*	**3.775**	**Q1**
6	*Service Industries Journal*	**3.012**	**Q2**

n.a.*: Not available. Magazines that in the year 2016 were no longer indexed in the JCR index, but in the year considered in the study.

As it is shown in Table 4, the magazine with the highest number of citations, 21 to be exact and leading the ranking is the Journal of Business Research, a JCR impact of 3,354 and belonging to quartile 1, which implies this it is a magazine of high importance and impact. Later the magazine Total Quality: Management & Business Excellence with a total of 17 citations with 1,386 JCR impact and belonging to quartile 3.

Sixthly, it is important to know when the investigation arises in a particular field, in which phase of the life cycle is this line of research, whether it is an emerging, mature or declining field and whether it is a fashion research (Ramos 2017). The set of basic documents can be divided into multiple time periods to capture the development of the field over time. Bibliometric data from each time period are analyzed separately and compared, to find changes in the structure of the field. This longitudinal analysis can reveal how particular groups within an intellectual structure emerge, grow or disappear. The statistical technique used is the frequency count, and the software called Bibexcel and Excel.

As shown in Fig. 2, in 1981 the first article on this subject was published, which reflects that it is a very fresh and new topic. As the years advance, investigations and therefore citations are more numerous. This reflects that the subject studied is an emerging field, and can be considered as an interesting and captivating discipline for many researchers.

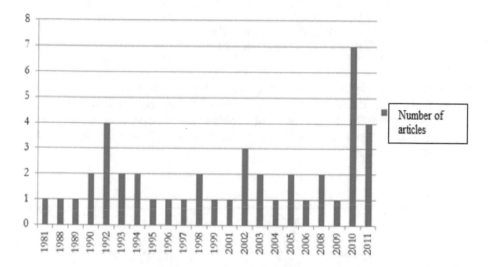

Fig. 2. Representation of the life cycle of the articles

Subsequently which terms are associated most frequently with a specific line of research are analyzed, when did these terms emerge and how they have evolved. While a co-word analysis has also been performed to cite publications, the analysis unit is a word, which means that thresholds for the appearance of words should be established. The statistical technique used is the count of the citations and the removal of duplicates. As shown in Table 5, the most important word that is repeated more times is satisfaction which makes reference to the feeling of comfort or pleasure you have when you have filled a want or covered a need, a term closely related to the field of study. Although in this research the study has been namely customer dissatisfaction, the word appears 235 times in the satisfaction articles of this field. The next word with more hits is client/customers, followed by the word service/services. The words that are repeated fewer times are analysis, information and complaint. All these terms relate in one way or another with this study.

Finally, a map of co- key words, co-words or associated words have been produced (Fig. 3), reflecting on how is the structure and dynamics of a conceptual field, what terms have contributed to the development of discipline, if they can identify "topics" associated with a research and how these concepts have evolved. It is interesting how

Table 5. Analysis of word relationships or keyword ranking

Frequency	Keywords	Frequency	Keywords
253	*Satisfaction*	82	*Behavior*
290	*Customer*[*]	79	*Responses*
264	*Service*[**]	76	*Performance*
157	*Quality*	72	*Management*
149	*Dissatisfact*	66	*Loyalty*
120	*Model*	63	*Recovery*
112	*Field*	57	*Perceptions*
111	*Consumer*	54	*Impact*

[*]Total result of terms customer/customers
[**]Total result of terms service/services

the building of density maps and strategic centrality of correlate (Callon et al. 1995). The purpose of this correlation matrix is to identify the structure underlying data, and tools used for processing have been multivariable analysis, social network analysis and mapping or SNA. The representation of the factor solution is made through the Zhao procedure and Stormantt.

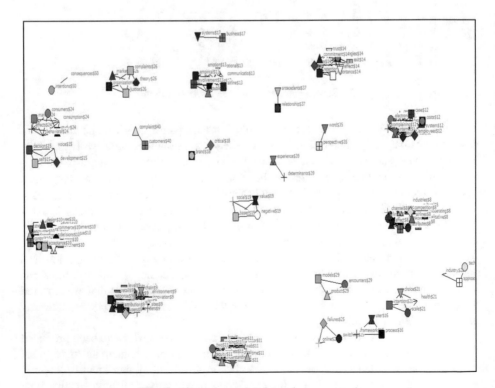

Fig. 3. Map of associated words or co-words

Output co-word analysis is a network of themes and relationships that represent conceptual space of a field. This semantic map helps to understand their cognitive structure (Bohrner et al. 2003). The co-word analysis uses text titles, keywords designated by the author, abstracts or full text to build a semantic map of the field. This method can be used to find links between subjects in a field of research and track its development (He 1998).

As seen in the Fig. 3 there are various relationships between keywords, such as: complaint and customer, critic and brand, experience and determinants, word and perspective, history and relationship, models, encounters and products, social, based, negative and value, consequences and intentions, complaints, theory, justice, marketing and organization, consumption, effects, and consumer behavior. In short, it is verified that all these relationships of words are drawn from articles are that are closely linked to this field of research.

4 Implications

In literature on consumer behavior it special attention has been paid to the study of conceptualization, antecedents and consequences of satisfaction as a phenomenon previous to a shopping experience or consumption. However, a minor concern is seen from the analysis of processes and variables accompanying dissatisfaction because the theoretical body commonly uses the same approach that was applied in the area of satisfaction. Achieving full "customer satisfaction" is a must in earning a spot in the "mind" of customers and the target market. Therefore, in order to maintain the objective of "customer satisfaction" it has crossed the borders of the marketing department to become one of the main objectives of all functional areas (production, finance, human resources, etc.) of successful companies. Because this is essential to analyze the issue of customer dissatisfaction, the causes, consequences, and the process it leads them to act in one way or another. It is considered that the study of customer dissatisfaction is in an evolutionary stage as there is growing interest in this field today, and although it has not been very much studied so far and it is a relatively new field which is captivating and transcendental. "Your most unhappy customer is your best source of learning." Bill Gates, Founder of Microsoft.

Bibliometric methods are used to describe the evolution of the activity of publication, most representative partners, the methodologies and the content of articles in order to explore the main research topics. These analyses identify potential avenues for future research that could be significant to advance the consolidation of the discipline. Therefore with bibliometric analysis contributions have been made for this line, using different types of measures and indicators to describe the evolution of production and the structure of the field. Bibliometric methods show great potential for quantitative confirmation of derived categories subjectively published reviews as well as exploration of the research field and identifying categories. Mapping science with bibliometric methods are useful for two main reasons: (a) help new investigators to a field to quickly understand the structure of that field and (b) introduce quantitative rigor in traditional literature reviews.

Magazines include more quality; therefore the synthesis presented in this research represents the state of the art research bibliometric management and organization. Bibliometric methods are not a substitute for reading and extensive synthesis. Bibliometrics can connect reliably publications, authors or magazines; Identify research streams; and produce, but research published maps depends on the researcher and his knowledge of the field to interpret the findings, which is the hard part.

It reveals that the study of customer dissatisfaction is a relatively recent issue, and it is increasingly studied and researched. It also has much relevance in the business world and marketing. The usefulness of this research has allowed us to know the different customer behaviors that have before experienced a service failure or brand failure such as reactions which are processed psychologically, and the consequences that these negative behaviors can have for businesses. Empowered by new technological possibilities, consumers can now wreak havoc on a brand with relatively little effort, and this is a very worrying issue and which is on the agenda. Many news stories show how public figures or anonymous people tell about their negative experience trying and often getting to damage the brand or company, uploading a simple video to any social network telling the terrible experience gained is already a danger to any company and therefore an alarming issue for them.

As for future researches, it is interesting to consider this issue to come to understand more deeply these complex behaviors and their consequences involved.

With the bibliometric analysis we can see what the global situation on this interesting subject is currently, and knowing the most expert authors in the field, items with the most impacting keywords which have greater relevance, the life cycle of articles that we have had throughout history and relationships between them, plus analysis of co-words offers new currents links between research and emerging issues. Thanks to this we can further research on benchmarks with great wisdom on this subject so fascinating, and has been so little studied until now.

Some limitations of this study are acknowledged. First, analysis of co-words provides a small number of significant clusters taking into account the number of keywords considered.

In addition, this paper focuses on a set of indicators and bibliometric techniques to examine the content of articles published in selected journals. Other objective analysis techniques data with different objectives are useful mapping methods to identify basic sets of articles, authors or magazines in particular disciplines, the possibility of using other bibliometric techniques that complement this study and provide a systematic description of the structure of the field of customer dissatisfaction before a service failure, since it is likely that several new bibliometric methods become prominent in the future.

This study opens up new possibilities for discovering important research areas. It provides theoretical and methodical suggestions that can improve the development of this subject as a discipline.

References

Acedo González, F.J., Barroso Castro, C., Casanova, C., Galán González, J.L.: Co-authorship in management and organizational studies: an empirical and network analysis. J. Manag. Stud. **43**, 957–983 (2006)

Börner, K., Chen, C., Boyack, K.W.: Visualizing knowledge domains. Annu. Rev. Inf. Sci. Technol. **37**, 179–255 (2003)

Callon, M., Courtial, J.P., Penan, H.: Cienciometría. El estudio cuantitativo de la actividad científica: de la bibliometría a la vigilancia tecnológica. Ed. Trea SL., Gijón (1995)

He, T., et al.: A simplified system for generating recombinant adenoviruses. Proc. Natl. Acad. Sci. U.S.A. **95**, 2509–2514 (1998)

Johnson, M.D., Herrmann, A., Huber, F.: The evolution of loyalty intentions. J. Mark. **70**, 122–132 (2017)

Kähr, A., Nyffenegger, B., Krohmer, H., Hoyer, W.D.: When hostile consumers wreak havoc on your brand: the phenomenon of consumer brand sabotage. J. Mark. **80**(3), 25–41 (2016)

Liu, J.S., Lu, L.Y., Lu, W.-M., Lin, B.J.: Data envelopment analysis 1978–2010: a citation-based literature survey. Omega **41**(1), 3–15 (2013)

Martín, B.R.: The use of multiple indicators in the assesment of basic research. Sci. **36**(3), 343–362 (1997)

Ramos, F.J.: Manual sobre Introducción a los métodos bibliométricos. Seminario Fundación Camilo Prado, Universidad Rey Juan Carlos (2017)

Holistic Learning Evidences in the Supervised Teaching Practice Reports

María de la Cruz del Río-Rama[1]([⊠]), Cristina Mesquita[2],
Maria José Rodrigues[2], and Rui Pedro Lopes[3]

[1] Department of Business Organisation and Marketing,
University of Vigo, As Lagoas s/n, 32004 Ourense, Galicia, Spain
delrio@uvigo.es
[2] School of Education, Polytechnic Institute of Bragança,
Campus de St. Apolónia, 5300-253 Bragança, Portugal
{cmmgp,mrodrigues}@ipb.pt
[3] Department of Informatics and Communications,
Politechnic Institute of Bragança,
Campus de St. Apolónia, 5300-253 Bragança, Portugal
rlopes@ipb.pt

Abstract. The curricular unit of Supervised Teaching Practice (STP), of the
Master programme in Pre-School Teacher Education and of the Master programme in Pre-school and Primary School Teacher Education, is organized in
two main phases: the teaching practice in the adequate education levels, and the
development of a final report. The final report is a document each student writes
and that reflects the course of the training, the critical and reflexive attitude
assumed in response to the challenges that were faced, and the processes and
performance of the professional routine experiences. The document should also
include a research component concerning the teaching action developed by the
student. The final report is, thus, an instrument that allows multiple dimensions
of analysis. The work described in this chapter resulted from the study of 62
reports, developed between 2012 and 2015, with the objective of analysing the
type of teaching-learning experiences, the content areas and the pedagogical
approach developed by students. A quantitative and qualitative approach was
followed, using text mining to complement the analysis, due to the large
quantity of text. This was complemented with content analysis to interpret the
resulting data. The results reveal that students describe and analyse teaching-
learning experiences in different content areas, valuing differentiated pedagogical strategies and that the reflective processes are centred in the description of
the practicum, without revealing a substantial critical thinking. The final report,
regardless of the weaknesses, is a valuable instrument in the training process,
because it allows the development of reflection and research tools about the
education practice, essential for the professional development of future teachers.

Keywords: Supervised Teaching Practice · Higher education
Teaching-learning experiences · Holistic learning

J. Gil-Lafuente et al. (Eds.): AEDEM 2017, SSDC 180, pp. 12–22, 2019.
https://doi.org/10.1007/978-3-030-00677-8_2

1   Introduction

Students of the curricular unit of Supervised Teaching Practice (STP) of the Preschool and Primary School Teacher Education master program contact with the professional context (kindergartens and primary schools) in a progressive way. This requires a supervision process, essential to help students improve reflexion about the contexts and their reality, calling for the "creation and sustainability of environments that promote the construction and the professional development, following a sustainable path of progressive development and professional autonomy" (Alarcão and Roldão 2010).

According to Vieira (2009), there should exist a close relation between pedagogy, comprising a conceptual dimension, and supervision, which integrates an experiential component, and whose integration results in praxis. The supervision practice appeals to the cooperation between all the actors as a life-long professional development process that places students, practitioners and supervisors together, sharing knowledge, functions and achievements.

However, supervision is not enough for future teachers to understand the professional reality they are experiencing. To be substantial, supervision has to focus on the multidimensionality of the pedagogical process. It should focus on the students' performance and in the reflexion that leads them to a sustainable and holistic approach about the profession. Practitioners and supervisors should stimulate students to think critically about what they are doing, why, and the impact of their actions. This dialogical process should be supported by documenting, questioning and inquiring, so that theory and practice come together in a joint development of educational action that should be constantly evolving.

In this perspective, supervision is a process that can contribute to make meaning about professional reality and, in this way, induce pedagogical change. According to Freire (1979), praxis is where theory and practice (or practice and theory) meet. It flavours the interpretation of the experience that leads us to a better understanding of the action, and to pedagogical intentionality. This complex process is where critical pedagogy develops.

This study assumes a holistic approach of the professional practices that connects the supervision and pedagogy. This approach is supported by a concept of practice that assumes the intentionality towards change and an ethical commitment with children, their families and the community.

2   The Final Report of the STP in the Development of Critical Thinking

As referred above, the STP should ensure the development of specific, multidimensional, knowledge, that are described, in writing, in a document that helps the students to understand and reflect on the path they are experiencing. The final report describes the teaching-learning activities developed through the STP duration, comprising several educational levels, subjects within the teaching domain, and the critical reflexion about them. The reflexion should be supported by pedagogical and scientific literature, as well as on information gathered from the practice, highlighting the critical analysis and the results obtained.

Considering the meaning of pedagogy discussed above, students should reveal, in their writing, a critical and reflexive attitude that allows them to unveil the research performed on the content areas and on the teaching-learning process, associating the content of the curricular areas and the way they articulate with the self-control, attitudes, representations, beliefs, preferences and styles, purpose and priorities, learning strategies and techniques, and the didactical process.

The written document should reflect the result of a constant reflexion that helps the student, in his effort towards autonomy, to perform changes concerning his concepts and practices. According to Alarcão and Tavares (2016), current trends are leading to a democratic supervision process and to strategies that value reflexion, cooperative learning, and to self-supervision and self-learning mechanisms.

The development of reflexion on the practice helps future teachers to identify coherent pedagogical approaches, to substantiate their options, to understand the value of some pedagogical strategies, to recognize their difficulties and to overcome them, and to develop the attitude for active and innovative pedagogical-didactical practice. This is further strengthened with the role of "facilitator of reflexion, raising awareness of his situation, helping him identifying problems and planning strategies for their resolution" (Amaral et al. 1996, p. 97).

2.1   Experiential and Holistic Pedagogical Experiences: How Students of the First Cycle of Education Learn

The basis for this study is the connection between the educational model and the pedagogical perspective. Several studies and international reports highlight the holistic, experiential and integrative pedagogical approach that teachers need to have in their practice for the concept of praxis to flourish. The teachers' actions targets the child as a whole, body, mind, emotions, creativity, history and social identity (Pires 2013).

Scientific literature has been revealing that holistic and integrative pedagogical approaches, that recognize the competence of children and listen to their voices, have greater impact on their learning and their future life (Eurydice 2009; OECD 2004, 2012; Siraj-Blatchford et al. 2002). Regarding this, the Starting Strong III report (OECD 2012) highlights that children learn better: (i) with integrative pedagogical approaches, where social and cognitive learning are regarded as complementary and equally important; (ii) when they are active and involved; (iii) when the interactions are frequent and meaningful; (iv) when the curricula is based on previous learning.

In particular, the afore mentioned report also highlights that curricula that value the self-initiated activities by children are more beneficial at a long term, they drive their participation in community services and motivates them to proceed studying. The quality of learning environments is considered, in some reports and scientific papers (Elliott 2006; Evangelou et al. 2009), as one of the most important factors in the process of development and learning of children. The evidences refer, as meaningful elements in this process: (i) the diversity of opportunities the learning experiences have; (ii) the intentional organization of the environment and the quality and diversity of the available materials; (iii) the experiences that value the contact with the nature (forest, field); (iv) learning by playing; (v) the existence of a structured, although flexible, daily and weekly routine; (vi) the appreciation of the children culture as a pedagogical resource; (vii) the involvement of parents in the school life of their children; (viii) the respect of the children's voice and the recognition of their participative competence; and (xix) the existence of qualified teachers, deeply involved in professional development processes that support their own research and learning.

In this context, the Starting Strong reports (OECD 2004, 2006, 2012) have been highlighting the need to ensure professional development opportunities, considering that the teaching procession should embrace a wide range of social responsibilities. Teachers should demonstrate a consistent knowledge of how children learn and develop, creating rich learning experiences for all, including the most vulnerable, involving children of diverse social and ethnical origins, in different levels of development. The functions of the professionals assume a complex nature, that implies the reflexion and development of critical thinking concerning the pedagogies developed in context.

3 Methodology

The study presented in the article assumes an exploratory approach that intends to verify, through the analysis of several STP reports, the areas of content that future teacher value and the pedagogical strategies used during the professional training. Considering the importance of a holistic perspective of the pre-school and primary school curricula, it is also intended the analysis of the degree of integration of the teaching activities. Finally, a relation is established between the main strategies and the content.

The study included all the reports from the last four years, in a total of 62 (17 from 2012, 12 from 2013, 22 from 2014 and 11 from 2015). All of them are available in the digital repository (https://bibliotecadigital.ipb.pt) in the PDF format, allowing a digital analysis.

Due to the large quantity of text (a total of 6723 pages, with 1832566 words), text-mining tools were used. Text organizes letters in words and these in phrases, gathering the information that can be stored, transmitted and read. Large quantities of text can make the interpretation of content and pattern discovery a difficult task. Using information processing algorithms, such as lexical analysis, pattern recognition, syntactical function annotation and natural language processing, among others, allows highlighting potentially useful information, difficult to assess otherwise. In its simpler form, it allows identifying the documents that satisfy given criteria in a large collection.

In this work, a classification process was used, to sort documents according to its content, looking for terms that define content areas and the teaching-learning strategies that were used. Text was initially pre-processed, to eliminate repeated forms and irrelevant words, as well as to minimize the different forms a word has by reducing flexed and derived words (for example, the word Didactic, DIDACTIC or didactic were converted to the common term didactic, as well as collaborate, collaboration, collaborative to the root collator).

Based on a significant number of reports, a dictionary of terms, representing content areas (Table 1) and teaching-learning strategies (Table 2) was built. Since all the reports are written in the Portuguese language, from now on the terms are kept in the original. The tables also present a translation, for reference.

Table 1. Dictionary of content area terms

Área de conteúdo	Content Area
Matemática	Math
Estudo do meio social	Social mean
Estudo do meio físico	Sciences
Musical Expression	Musical expression
Expressão Plástica	Plastic expression
Drama Expression	Drama
Linguagem Oral e Abordagem à Escrita/Língua Portuguesa	Spoken language and introduction to writing
Expressão Físico-Motora	Physical expression
Formação Pessoal e Social	Social and personal training

Source: Authors' elaboration

Table 2. Dictionary of teaching-learning strategies terms

Estratégias de ensino-aprendizagem	Teaching-learning strategies
Apresentação PowerPoint	Powerpoint presentation
Exposição	Expositive
Uso da narrativa	Use of the narrative
Visualização e discussão de vídeos	Viewing and exploring videos
Visitas de estudo	Field trips
Atividades práticas e experienciais	Practical and experiential activities
Investigação	Research
Leitura e exploração de textos	Reading and exploring texts
Fichas de trabalho	Forms
Jogo dramático	Drama and role play
Laboratório gramatical	Grammatical laboratory
Jogo fonético	Phonetic game
Resolução de problemas	Problem solving
Atividade integradora	Integrative activity
Trabalho colaborativo	Cooperative work

Source: Authors' elaboration

Each of these entries is characterized by words and sentences. For example, the sentences "memory game", "puppets", "theatre", "drama", "drama game", "dance", characterize the Drama Expression area, and the sentences "musical instruments", "songs", "beat", "rhythm forms", "song rhythm", "sound creations", "rhythm and sounds", are all associated to the Musical Expression area.

After this initial step, a histogram of different terms was built, both for the content area and for the teaching-learning strategies, to check the frequency of each term in all the STP reports. The analysis continued by grouping related terms in each report, resulting in several multiconnected charts. Finally, a heat map crossing the teaching-learning strategies and the content was built, to check their intersections and interdependencies. The next section present and discuss the data from each of this instruments.

4 Results and Discussion

The histograms of the different terms, related to the content (Fig. 1) and to the teaching-learning strategies (Fig. 2) allows sorting the frequency of each term in all the reports.

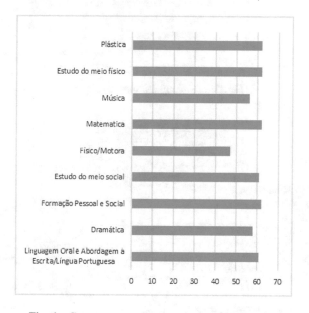

Fig. 1. Content areas. *Source:* Authors' elaboration

Most of the reports refer all the content areas related to pre-school and primary school education, meaning that students consider the development of all areas. However, mention to *Expressão Musical* (Musical Expression) (56 reports) and *Expressão Físico-Motora* (Physical Expression) (47 reports) are less referred. This fact may be associated with the offer of activities in extracurricular regime, existent in most institutions and schools.

The histogram related to the teaching-learning strategies developed during the teacher training period has higher incidence in the *Resolução de Problemas* (Problem Resolution) and in the *Atividades Experimentais* (Experimental Activities). Students use a diversity of active strategies, although their use is not transversal in the reports.

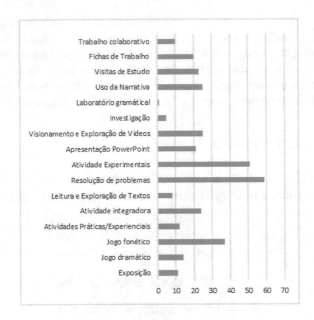

Fig. 2. Teaching-learning strategies.

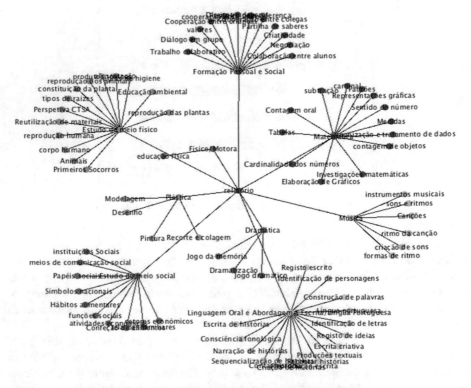

Fig. 3. Content area distribution. *Source:* Authors' elaboration

For better visualization, multilinked graphs were built, with the content area distribution (Fig. 3) and teaching-learning strategies (Fig. 4) in each report. The graphical representation is very similar in all the reports and therefore the pictures are an example of the global. From the analysis of the graphs, each report (represented in the centre of the picture) has several connections to the terms that characterizes it. The dispersion is remarkable, and it describes the representativeness of several areas and strategies in each report.

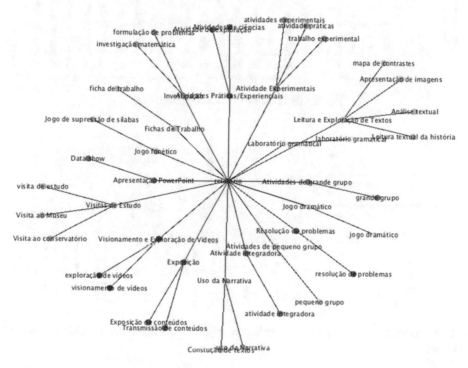

Fig. 4. Teaching-learning strategies distribution. *Source:* Authors' elaboration

Beyond references to the areas and strategies, students frequently mention the organization of classes around small group activities (60 reports) and large group activities (62 reports), which means that students choose the type of group organization that ensures the success of the activity, considering the pedagogical resources available. It may also mean a diversity of options, considering the educational levels in which they are developing activities (these data needs more research in future work so that the differences between the pre-school options and primary school can be assessed).

Usually, the report is structure in two main parts. The first is related to the pre-school education and the second to the primary school. It is possible to estimate the if a term is related to the first or to the second, according to the initial page it firstly appears. Although not infallible, it provides a reasonable heuristic. In this context, the initial position of the term *Atividade Integradora* (Integrative Activity) was retrieved in the 24

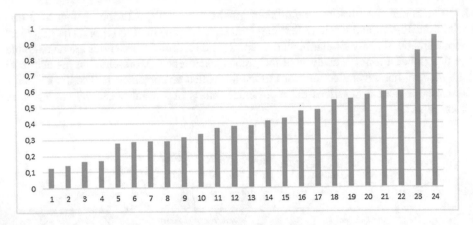

Fig. 5. Reference to "atividade integradora"

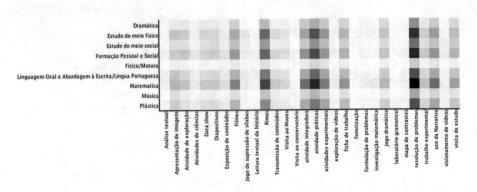

Fig. 6. Relation between teaching-learning strategies and content areas. *Source:* Authors' elaboration

reports in which this term is mentioned (Fig. 5). In 17 of the 24 reports, the term is initially mentioned in the first half, indicating that the integrative activities are more valued in pre-school. This requires further clarification, in future work.

Finally, a heat map was built, crossing the teaching-learning strategies with the content areas to verify the intersection and interdependencies (Fig. 6).

As expected, the crossing is darker (and also transversal) in the *Resolução de Problemas* (Problem Resolution) and in the *Atividades Práticas* (Practical Activities). From the chart, it is possible to see that both teaching-learning strategies have more expression in the area of Math, in the *Formação Pessoal e Social* (Social and Self Training) and in the *Estudo do Meio Físico* (Sciences). In relation to Math, this may indicate a large number of exercises solved by students, highlighting a more technical approach to teaching and less of building critical thinking. However, it is important to assess the context of the practical activities to understand if they are translated into moments of student implication or experientialism.

5 Conclusions

The STP reports are documents that allow studying the options students make during the professional training. Both the content areas and the teaching-learning strategies are described in them, allowing to know the frequency each approach or technique is used. The study presented in this document highlights the importance that the contact with the contexts has to provide students with the professional reality they will face in the future, allowing them to plan and implement different teaching-learning strategies in all content areas.

This exploratory study reflects on multiple possibilities of analysis, based on the data collected. It is clear that students develop teaching-learning experiences in several content areas, valuing some over others. However, a deeper perspective on the curricular integration is needed. The integrative activities, which describe a holistic of the teaching-learning process, are not sufficiently stressed in the first cycle of education (6 to 10 years old), suggesting a more vertical perspective on this education level than on the pre-school education.

It is still necessary to understand the pedagogical approaches that students follow in their practice, their concepts about the teaching-learning in both levels and in the role they attribute to the children and the teachers, in the context.

From the analysis performed, although no described on these pages, it is also evident the reflexion and research on the teaching practice, essential to the professional development of future teachers. These, and other aspects, will give continuity to the research, to better understand the training of future teachers within the STP.

References

Alarcão, I., Roldão, M.C.: Supervisão - um contexto de desenvolvimento profissional dos professores. Edições pedago, Mangualde; Ramada (2010)

Alarcão, I., Tavares, J.: Supervisão da Prática Pedagógica - Uma Perspectiva de Desenvolvimento e Aprendizagem. Almedina, Coimbra (2016)

Amaral, M.J., Moreira, M.A., Ribeiro, D.: O papel do supervisor no desenvolvimento do professor reflexivo: estratégias de supervisão. In: Alarcão, I. Formação reflexiva de professores: estratégias de supervisão, pp. 89–122. Porto (1996)

Elliott, A.: Early Childhood Education: Pathways to Quality and Equity for all Children. ACER Press, Camberwell (2006)

Eurydice: Early childhood education and care in Europe: tackling social and cultural inequalities. European Comission (2009). http://europa.eu/rapid/press-release_MEMO-09-66_en.pdf

Evangelou, M., Sylva, K., Kyriacou, M., Wild, M., Glenny, G.: Early years learning and development: literature review. Department for Children, Schools and Families, London (2009). http://www.foundationyears.org.uk/files/2012/08/DCSF-RR1761.pdf

Freire, P.: Conscientização: Teoria e Prática da Libertação. Cortez & Moraes, São Paulo (1979)

OECD: Starting Strong: Curricula and Pedagogies in Early Childhood Education and Care. Organisation for Economic Co-operation and Development (2004)

OECD: Starting Strong II: Curricula and Pedagogies in Early Childhood Education and Care. Organisation for Economic Co-operation and Development (2006)

OECD: Starting Strong III: Curricula and Pedagogies in Early Childhood Education and Care. Organisation for Economic Co-operation and Development (2012)

Pires, C.M.M.G.: A voz da criança sobre a inovação pedagógica (30 July 2013). http://hdl.handle.net/1822/25441

Siraj-Blatchford, I., Sylva, K., Muttock, S., Gilden, R., Bell, D. (eds.): Researching Effective Pedagogy in the Early Years. Department for Education and Skills, Nottingham (2002)

Vieira, F.: Towards a transformative vision of pedagogical supervision. Educação & Sociedade 30(106), 197–217 (2009). https://doi.org/10.1590/S0101-73302009000100010

Assessing Overall Fit and Invariance in a PLS Model of PIGS and V4 Countries' Financial Systems

Juan José García-Machado[1]([✉]) and Agnieszka Jachowicz[2]

[1] University of Huelva, Huelva, Spain
machado@uhu.es
[2] The University of Dabrowa Gornicza, Dąbrowa Górnicza, Poland

Abstract. This research focuses on the financial systems of the countries that are in the tail of the Eurozone (contemptuously named PIGS), that take part in the ESCB, and are subjected to the discipline of the ECB and the troika, versus the systems of the countries which are at the top of the New Member States (NMS) and do not have the Euro currency. In this paper we apply some PLS-SEM-based model fit measures and testing measurement model invariance as a prior study to make sure that we can compare and find significant differences between PIGS and V4 countries financial systems using a Multigroup Analysis. In our opinion, this study provides important and reliable information to the ECB, PIGS and V4 countries' policy makers.

1 Introduction

The financial markets of most European Union (EU) Member States, especially after World War II, and until the 1980s were dominated by banks, which were mainly owned by states. That is why domestic banking were main players in the economy. In Western Europe the situation changed at the end of the 1970s, after the oil price shocks, which were connected with the downfall of the Bretton Woods' system and the beginning of European integration. In the Visegrád Group (Czech Republic, Hungary, Poland and Slovakia) the situation was different and this alliance of four countries was established in 1991 (see García-Machado and Jachowicz 2017). A more detailed description and explanation about the V4 countries financial systems can be found in García-Machadoand Jachowicz (2016). But, in short we can describe the financial sector in the Visegrád group as follows:

- The dominating role of banks, including foreign ones, as main financial intermediaries (in terms of asset size);
- The second largest type of financial institutions are insurance companies and after the financial crisis of 2008 their asset size became bigger as a result of conversions;
- The depth of the financial markets is still unrewarding as measured by total assets to GDP ratio;
- The role and size of the stock exchanges is still low as a source of capital.

J. Gil-Lafuente et al. (Eds.): AEDEM 2017, SSDC 180, pp. 23–34, 2019.
https://doi.org/10.1007/978-3-030-00677-8_3

With regards to the PIGS countries, this term is an acronym used to refer to the economies of Portugal, Italy, Greece and Spain, four economies of the Southern Europe with problems after the financial and economic crisis, the burst of the real estate bubble, and the onset of the European sovereign – debt crisis. Ireland became associated with the term, either replacing Italy or as a second I (PIIGS). Sometimes a second G (PIGGS or PIIGGS), for Great Britain, was also added. Finally, another third I is added at times to include Iceland (PIIIGGS). The term is considered pejorative by affected countries and it is widely considered disparaging. The term was first used by the Financial Times and Barclays Capital in 2010 in comparison with the BRIC or BRICS countries (Brazil, Russia, India, China and South Africa), the G7 developed economies or another predominately economic – groupings of countries. From now on we will consider in this paper the first of the meanings. That is, we will include the countries of Portugal, Italy, Greece and Spain (PIGS) in comparison with Czech Republic, Hungary, Poland and Slovakia (V4 Group). On the other hand, as it is well-known, the financial crisis has motivated the PIGS countries Central banks' and Government interventions in several forms, including liquidity injections, direct public finance, publicly loan guarantee schemes and, even interest rates subsidization. However, the effectiveness of these interventions may be put in doubt.

The common points in the financial systems shared by the V4 and the PIGS group are as follows:

– In both groups a nationalized financial system dominates. In West Europe the changes began in the 1970s, whereas in Central Europe – in the 1990s;
– The main role played was still played by banks – initially by national banks but after that by foreign banks especially by German ones;
– Weak developed stock markets The best situation is in the Madrid Stock Exchange, where the Latin South American businesses are noticed and in Warsaw, which is the biggest market in this part of Europe;
– An unsatisfactory level of pension funds.

Governments, mainly in the Eurozone, and affected by the last financial crisis, have developed special programs to save their financial systems and restore their economies to growth. However, this would have been impossible without the participation of the largest central banks that have used unconventional tools aimed at restoring balance in the markets. The Eurozone partly caused the crisis to spread more quickly, but governments moved to rescue the currency. Countries from the V4 are not members of the Eurozone and they are not financially integrated with the Euro, and perhaps that is why the crisis did not hit them as strongly as it did the PIGS group of countries.

In this study, we apply some PLS-SEM-based model fit measures and testing measurement model invariance as a prior study to make sure that we can compare and find significant differences between PIGS and V4 countries financial systems using a Multigroup Analysis.

The paper is organized as it follows. We start reviewing the conceptual framework as theoretical background and giving an overview about model fit measures and techniques to assess the invariance in PLS-SEM context. Next, we describe the methodology, as well as the sample, data collection, component and data analysis. Following that we present empirical results. Finally, we provided a summary and conclusions.

2 Conceptual Framework

According to García-Machado and Jachowicz (2016), there is a large literature that attempts to analyze the transformation of the financial system and its efficiency as well as many interesting methods for assessing banking sectors in order to improve it and predict its development as well as the financial system as a whole. They estimate a PLS path model to study how the V4 countries' Financial Systems work in benefit of the economic welfare of its citizens. This study is also important because it show how PLS path modeling can be used to successfully assess complex model in macro finance and, in this case, provide some explanation of the relationships between the selected factors and the latent dependent variable economic welfare. It provides important and usable information to the V4 countries' policy makers.

Afterwards, García-Machado and Jachowicz (2017) carried out a PLS-SEM Multigroup analysis to study if making a partition of data in two separate groups for PIGS and V4 countries, could provide some useful knowledge about the differences in their financial systems, identifying observed heterogeneity, and checking if they are statistically significant. However, as Hair et al. (2018) point out, a primary concern before comparing group-specific parameter estimates for significant differences using a multigroup analysis is ensuring measurement invariance to be confident that group differences in model estimates do not result from the distinctive content and/or meanings of the latent variables across groups.

There is a heated argument about the overall goodness of fit in PLS path modeling and its suitability to apply it at the beginning or at the end of stages for applying PLS-SEM systematic procedure. For instance, as Hair et al. (2017) state, measures such as the chi – square ($\chi2$) statistic or the various fit indexes associated with CB – SEM (Covariance – Based Structural Equation Modeling) are based on the difference between the two covariance matrices. The notion of fit is therefore not fully transferable to PLS-SEM as the method seeks a solution based on a different statistical objective when estimating model parameters (i.e., maximizing the explained variance instead of minimizing the differences between covariance matrices). Nevertheless, research has proposed several PLS-SEM-based model fit measures, which are still in development and are not exempt from criticism.

One of the earliest proposed indexes was by Tenenhaus et al. (2004 and 2005), who proposed the Goodness-of-Fit index (GoF). However, for Henseler and Sarstedt (2013) it is not a good measure because it does not represent a goodness-of-fit criterion for PLS-SEM, unable to separate valid models from invalid ones, and the recommend avoid its use.

Later, Henseler et al. (2014) proposed the Standardized Root Mean Square Residual (SRMR), a model fit measure well know from CB-SEM, which has previously not been applied in a PLS-SEM context. It is the square root of the sum of the squared differences between the model-implied and the empirical correlation matrix (i.e. the Euclidean distance between the two matrices). A value of 0 for SRMR would indicate a perfect fit and generally, an SRMR value less than 0.05 indicates an acceptable fit (Byrne 2008). A recent simulation study shows that even entirely correctly specified model can yield SRMR values of 0.06 and higher (Henseler et al. 2014). Therefore, a cut-off value of 0.08 as proposed by Hu and Bentler (1999) appears to be more adequate for PLS path models. Ringle (2016) proposes a more flexible option of SRMR < 0.10.

Another useful approximate model fit criterion could be the Bentler-Bonett index or normed fit index (NFI) (Bentler and Bonett 1980). For factor models, NFI values above 0.90 are considered as acceptable (Byrne 2008). For composite models, thresholds for the NFI are still to be determined. Because the NFI does not penalize for adding parameters, it should be used with caution for model comparisons. In general, the usage of the NFI is still rare. Another promising approximate model fit criterion is the Root Mean Square error correlation (RMS$_{theta}$), which follows the same logic as SRMR but relies on covariances (Lohmöller 1989; Henseler et al. 2014). A recent simulation study (Henseler et al. 2014) provides evidence that the RMS$_{theta}$ can indeed distinguish well-specified from ill-specified models. However, thresholds for the RMS$_{theta}$ are yet to be determined, and PLS software still needs to implement this approximate model fit criterion. Initially, values below 0.12, are considered as acceptable.

Following Henseler et al. (2016) and Henseler (2017), the global model fit can be assessed in two non-exclusive ways: by means of inference statistics (i.e. so-called tests of model fit), or through the use of fit indices (i.e. an assessment of approximate model fit). In order to have some frame of reference, it has become customary to determine the model fit both for the estimated model and for the saturated model. Saturation refers to the structural model, which means that in the saturated model all constructs correlate freely. PLS path modeling's tests of model fit rely on the bootstrap to determine the likelihood of obtaining a discrepancy between the empirical and the model – implied correlation matrix that is as high as the one obtained for the sample at hand if the hypothesized model was indeed correct (Dijkstra and Henseler 2015). Bootstrap samples are drawn from modified sample data. If more than 5 percent (or a different percentage if an α-level different from 0.05 is chosen) of the bootstrap samples yield discrepancy values above the ones of the actual model, it is not that unlikely that the sample data stems from a population that functions according to the hypothesized model. The model thus cannot be rejected. There is more than one way to quantify the discrepancy between two matrices, for instance the maximum likelihood discrepancy, the geodesic discrepancy d_G, or the unweighted least squares discrepancy d_{ULS} (Dijkstra and Henseler 2015), and so there are several tests of model fit. Next to conducting the tests of model fit it is also possible to determine the approximate model fit. Approximate model fit criteria help answer the question how substantial the discrepancy between the model-implied and the empirical correlation matrix is.

With regard to the suitability to apply the model fit measures at the beginning or at the end of stages for applying PLS-SEM systematic procedure, Henseler et al. (2016), opposite to Hair et al. (2017), propose that the overall goodness of the model fit should be the starting point of model assessment. If the model does not fit the data, the data contains more information than the model conveys. The obtained estimates may be meaningless, and the conclusions drawn from them become questionable.

On a separate issue, researchers are increasingly interested in identifying and understand significant differences across two or more groups of data as a way to recognize that heterogeneity is often present. In fact, failure to consider such heterogeneity can be a threat to the validity of PLS-SEM results (Becker et al. 2013; Hair et al. 2012).

In García-Machado and Jachowicz (2017), they apply a PLS-SEM Multigroup Analysis to study if making a partition of data into separate groups for PIGS and V4 countries, we can gain knowledge about the differences in their financial systems, but researchers recommend to establish measurement invariance (also referred as measurement equivalence), as a primary concern before comparing groups of data, to be confident that group differences in model estimates result from neither distinctive content and/or meanings of the latent variables across nor measurement scale. Hult et al. (2008) point out that failure to establish data equivalence is a potential source of measurement error and when measurement invariance is not demonstrated, any conclusions about model relationships and questionable. Hence, multigroup comparisons require establishing measurement invariance to ensure the validity of outcomes and conclusions (Hair et al. 2018).

To assess measurement invariance, in a PLS context, researchers usually apply the Measurement Invariance of the Composite Models (MICOM) procedure developed by Henseler et al. (2016). The MICOM procedure builds on the scores of the latent variables, which are represented as composites, that is, linear combinations of indicators and the indicator weights as estimated by the PLS-SEM algorithm. As Fig. 1 shows, the MICOM procedure involves three steps: (1) configural invariance, (2) compositional invariance, and (3) equality of composite mean and variances. The three steps are hierarchically interrelated (Hair et al. 2018).

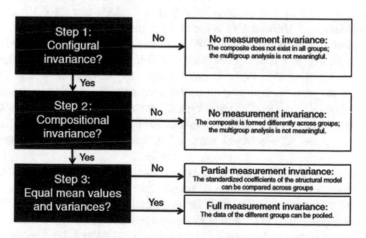

Fig. 1. The MICOM procedure Source: Henseler et al. (2016), p. 412

3   Sample, Data Collection, and Methodology

Our analysis covers a sample of 836 financial institutions from PIGS and V4 countries' financial systems: Czech Republic (25), Greece (12), Hungary (28), Italy (552), Poland (41), Portugal (31), Slovakia (13), Spain (134). The study period is 2000–2015. The empirical bank-level data was obtained from a sample of financial institutions that include commercial banks, savings banks, cooperative banks, real estate and mortgage banks, investment banks and securities firms. Most of these data were obtained from Bankscope International Bank Database[1]. Macroeconomic-level aggregate and financial data for each country were collected from FMI's International Financial Statistics and World Economic Outlook Database, World Bank's World Development Indicators, and Country economy website[2]. Stock exchange data was collected from PIGS and V4 stock markets (Athens Stock Exchange, Bratislava Stock Exchange, Madrid Stock Exchange, Milan Stock Exchange, Oporto Stock Exchanges, Prague Stock Exchange, Budapest Stock Exchange, Warsaw Stock Exchange and) in a daily basis and then annualized.

Data collection was performed from December 2016 to March of 2017. We built a very complete data set, which initially included 44 indicators or manifest variables and a sample size of 112 observations (for PIGS and V4 countries and 15 years). They total 4.928 data. Indicators include manifest variables at macroeconomic, banks and other financial intermediaries and stock markets levels, financial sector structural indicators and indicators for development and happiness. After debugging data set, finally, our sample includes 112 observations, with 37 indicators and 4.144 data. It includes less than a 5% of missing values.

We use our data set with 112 observations for our empirical PLS Model of PIGS and V4 Countries' Financial Systems analyses. The data set is from research that attempts to predict financial efficiency and, ultimately, economic welfare. Following Cohen's (1992, p. 158) recommendations for multiple OLS regression analysis, we would need 92 or 113 observations to detect R^2 values around 0.10, assuming a significance level of 10% or 5%, respectively, and a statistical power of 80%. In addition, following Nitzl's (2016, p. 26) recommendations, we would need 69 or 85 observations to detect a medium effect size of 0.15, assuming the same significance level and statistical power. Because our sample size in this study was 112, there appears to be no problem with respect to the necessary sample size.

Following the former studies for economic stability and welfare in V4 and PIGS countries, carried out by García-Machado and Jachowicz (2016 and 2017), and based on their results, we have selected some of indicators and chosen the same latent variables or constructs which are shown in Table 1.

In addition to the indicators for measurement models, and in order to a better assessment of the Economic Welfare construct, they had taken into consideration two new manifest variables or indicators (García-Machado and Jachowicz 2017): Human development index (HDI) value and the Average happiness. The Human Development Index (HDI) is a summary measure of human development and the Average Happines (Veenhoven 2017) shows how much people enjoy their life-as-a-whole on scale 0 to 10. Life-satisfaction is assessed by means of surveys in general population samples.

---

[1] The Bankscope International Database is a detailed database provided by Bankscope which contains information on over 30.000 international banks for a period up to 16 years of detailed accounts for each bank.

[2] www.countryeconomy.com.

Table 1. Indicators and latent variables or constructs

Latent variables	Indicators
Banking System Quality (BANK_QUAL)	Bank_capital/assets, Deposits/Assets, and Net_Int_Margin
Economic Welfare (ECON_WELF)	Deposits/GDP and GDPpc_US$
Financial System Efficiency (FIN_EFFIC)	Credit_Private_Sector and Domestic_credit
Investors Behaviour (INVEST_BEHAV)	Gross_Savings
Macroeconomic Framework (MACRO_FRMW)	GDP_Growth and Inflation
Stock Market Performance (STOCK_PERF)	Capitalisation, Listed_companies, Stocks Traded value, and Stocks_turnover

Source: García-Machadoand Jachowicz (2016), pp. 50–51.

In this study, we used two PLS software: ADANCO (v. 2.0.1), developed by Henseler and Dijkstra (2015) and SmartPLS 3 software (v. 3.2.6) developed by Ringle et al. (2015). ADANCO (ADvanced ANalysis of COmposites) is a software for confirmatory composite analysis and variance – based structural equation modeling.

The proposed model for economic stability and welfare is framed with respect to latent constructs as given in the diagrammatic design in Fig. 2.

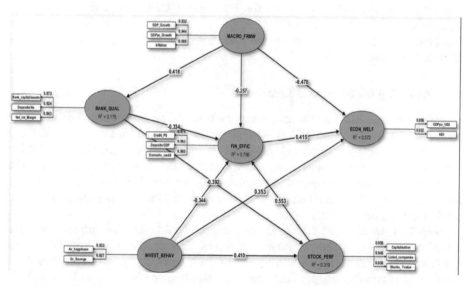

Fig. 2. PLS path model estimation results for all countries Source: Authors' own research

4   Results

4.1   Testing Goodness of Overall Model Fit

After running the PLS algorithm and the bootstrap procedure for the estimation of the model, we can provide the following overall goodness of fit measures to our PLS Model for PIGS and V4 countries' financial systems, as it is shown in Table 2.

Table 2. Goodness of overall model fit measures

	SmartPLS			ADANCO		
	Original Sample (O)	95%	99%	Value	HI95	HI99
SRMR						
Saturated Model	0.111		0.058	0.063	0.1052	0.0556	0.0597
Estimated Model	**0.113**		**0.073**	**0.082**	**0.1098**	**0.0714**	**0.0806**
d$_{ULS}$						
Saturated Model	1.673		0.459	0.536	1.5064	0.4209	0.4842
Estimated Model	**1.747**		**0.724**	**0.909**	**1.6404**	**0.6924**	**0.8839**
d$_G$						
Saturated Model	3.528		1.088	1.230	2.1542	0.4423	0.5167
Estimated Model	**3.463**		**1.946**	**3.027**	**2.1774**	**0.8166**	**1.5042**

The SRMR quantifies how strongly the empirical correlation matrix differs from the implied correlation matrix, therefore the lower the SRMR, the better the fit of the theoretical model (Henseler 2017). The SRMR original value is higher than the threshold of 0.08 suggested by Hu and Bentler (1999) and almost nearly of 0.10 proposed by Ringle (2016).

4.2   Assessing Measurement Invariance

To assess measurement invariance, we run the MICOM procedure which draws on the permutation testing instead of bootstrapping. As we told previously, the MICOM procedure involves three steps: (1) configural invariance, (2) compositional invariance, and (3) equality of composite mean and variances. Firstly, in Step 1, we ensure that configural invariance exist across the groups. As we can check, they have identical indicators, identical data treatment, and they draw on the identical algorithm setting, so configural invariance is established.

Step 2 of the MICOM addresses the compositional invariance and applies a statistical test to assess whether the composite scores differ significantly across the groups. For this, it examines the correlation c between the composite scores, and requires that c equals 1. Technically, the procedure tests the null hypothesis that c is 1. In order to establish compositional invariance, we must not reject this null hypothesis. That is, if the test yields a p-value largen than 0.05 (in case we assume a significance level of 5%), we can assume compositional invariance (Hair et al. 2018). Table 3 shows the results for the compositional invariance assessment.

Table 3. Compositional invariance assessment (MICOM Step 2)

	Original correlation	Correlation permutation mean	5.0%	Permutation p – values	Compositional invariance?
BANK_QUAL	0.666	0.993	0.978		No
ECON_WELF	1.000	1.000	0.999	0.354	Yes
FIN_EFFIC	1.000	1.000	1.000	0.355	Yes
INVEST_BEHAV	0.994	0.873	0.554	0.906	Yes
MACRO_FRMW	0.998	0.994	0.978	0.538	Yes
STOCK_PERF	0.355	0.999	0.997		No

Comparing the correlations with the 5% quantile reveals that the quantile is smaller than (or equal to) the correlation c for ECON_WELF, FIN_EFFIC, INVEST_BEHAV, and MACRO_FRMW. This result is also supported by the p-values that are higher than 0.05, indicating the correlation is not significantly lower than 1 for these constructs. Therefore, we can conclude that compositional invariance has been established for the constructs mentioned above, but not for all latent variables in the model, thus partial measurement invariance is not confirmed and multigroup analyses is not meaningful.

In case that Step 2 support partial measurement invariance, we can continue with Step 3 to check whether even the model holds full measurement invariance. Step 3 of the MICOM addresses the equality assessment of the composites' mean values and variances. Table 4 shows the mean differences between the scores as resulting from the original model estimation and the permutation procedure, respectively. Table 5 shows the analogous results for the composite variances.

Table 4. Step 3a of the MICOM procedure

	Mean – original difference	Mean – permutation mean difference	2.5%	97.5%	Permutation p – values	Equal mean values?
BANK_QUAL	−1.547	0.002	−0.399	0.354		No
ECON_WELF	1.157	−0.003	−0.367	0.340		No
FIN_EFFIC	1.587	−0.004	−0.378	0.372		No
INVEST_BEHAV	−0.331	0.009	−0.364	0.387	0.082	Yes
MACRO_FRMW	−0.918	0.013	−0.365	0.397		No
STOCK_PERF	1.147	0.005	−0.400	0.370		No

Finally, as we can see, not all mean values and variances are equal, not supporting full measurement invariance. In our case, this last step does not necessary because the MICOM procedure is hierarchical, and partial measurement invariance was not supported in Step 2.

Table 5. Step 3b of the MICOM procedure

	Variance – Original difference	Variance – Permutation mean difference	2.5%	97.5%	Permutation p – values	Equal variances?
BANK_QUAL	0.689	0.002	−0.482	0.469		No
ECON_WELF	0.409	0.001	−0.411	0.417	0.054	Yes
FIN_EFFIC	2.826	−0.002	−0.536	0.504		No
INVEST_BEHAV	0.842	−0.004	−0.517	0.499		No
MACRO_FRMW	−0.066	−0.004	−0.519	0.581	0.831	Yes
STOCK_PERF	3.188	0.011	−0.937	0.871		No

5 Discussion, Conclusion and Implications

Our study has implemented some PLS-SEM-based model fit measures and testing measurement model invariance as a prior study to make sure that we can compare and find significant differences between PIGS and V4 countries financial systems using a Multi-group Analysis.

Goodness of overall model fit measures showed that the model was unfit for d_{ULS} and d_G discrepancy and almost nearly to the threshold value for SRMR given by Ringle (2016). However, aside from model fit assessment criteria, it is important to note that they may often not be useful for PLS-SEM and must be used with caution. In addition, these criteria are in their very early stage of research and not fully understood (e.g., the critical threshold values). More specifically, Lohmöller (1989) states that some fit measures imply restrictive assumptions on the residual covariances, which PLS-SEM does not imply when estimating the model. In addition, we should consider the note of caution presented by Hair et al. (2017). While PLS-SEM was originally designed for prediction purposes, research has sought to extend its capabilities for theory testing by developing model fit measures. Model fit indices enable judging how well a hypothesized model structure fits the empirical data and, thus, help to identify model misspecifications. [...] Initial simulation results suggest that the SRMR, RMS$_{theta}$, and exact fit test are capable of identifying a range of model misspecifications (Dijkstra and Henseler 2015; Henseler et al. 2014). At this time, however, too little is known about these measures' behavior across a range of data and model constellations, so more research is needed. PLS-SEM focuses on prediction rather than on explanatory modeling and therefore requires a different type of validation. More precisely, validation PLS-SEM is concerned with generalization, which is the ability to predict sample data, or, preferably, out-of-sample data—see Shmueli (2010) for details. Against this background researchers increasingly call for the development of evaluation criteria that better support the prediction-oriented nature of PLS-SEM (e.g., Rigdon 2012, 2014) and for an emancipation of PLS-SEM from its CB-SEM sibling (Sarstedt et al. 2014). In this context, fit (as put into effect by SRMS), RMS$_{theta}$, and the exact fit test offer little value. In fact, their use can even be harmful as researchers may be tempted to sacrifice predictive power to achieve better fit. Future research must provide detailed explanations and recommendations on the computation, usage and interpretation of these outcomes.

With regard to the assessment of invariance measurement, we conclude that compositional invariance has not been established for all the constructs included in the model, thus partial measurement invariance is not confirmed and multigroup analyses is not meaningful. If partial measurement invariance cannot be established for a latent variable (e.g. BANK_QUAL and STOCK_PERF), then group – specific comparisons using multigroup analysis on any relationships involving these constructs are not feasible. In this case, following Hair et al. (2018) recommendations, researchers may refrain from running a multigroup analysis altogether and analyse each group separately without attempting to compare the results for the two groups. Alternatively, one can eliminate the latent variables that did not achieve compositional invariance, provided theory that supports this step. But, it is important remark that any changes in the model, PLS algorithm, bootstrapping, permutation, and MICOM procedures need to be run again and check all model quality criteria.

Finally, in future studies it would be interesting because this type of analysis will provide important and reliable information to the ECB, PIGS and V4 countries' policy makers which would allow them make better decisions.

References

Becker, J.M., Rai, A., Ringle, C.M., Völckner, F.: Discovering unobserved heterogeneity in structural equation models to avert validity threats. MIS Q. **37**, 665–694 (2013)

Bentler, P.M., Bonett, D.G.: Significance tests and goodness of fit in the analysis of covariance structures. Psychol. Bull. **88**(3), 588–606 (1980)

Byrne, B.M.: Structural Equation Modeling with EQS: Basic Concepts, Applications, and Programming. Psychology Press, New York (2008)

Cohen, J.A.: A Power Primer. Psychol. Bull. **112**(1), 155–519 (1992)

Dijkstra, T.K., Henseler, J.: Consistent and asymptotically normal PLS estimators for linear structural equations. Comput. Stat. Data Anal. **81**, 10–23 (2015)

García-Machado, J.J., Jachowicz, A.: Using PLS path modeling to investigate stability of financial system in V4 countries' welfare. Int. Res. J. Financ. Econ. **157**, 37–63 (2016)

García-Machado, J.J., Jachowicz, A.: A Comparison using PLS-MGA between PIGS and V4 Countries' Financial Systems. In: Laguna–Sánchez, Mᵃ.P., Blanco-González, A. (eds.) Business and Society: Responsible Research and Innovation, XXXI AEDEM Annual Meeting, pp. 387–401. European Academic Publisher, Madrid (2017)

Hair, J.F., Sarstedt, M., Ringle, C.M., Mena, J.A.: An assessment of the use of partial least square structural equation modeling in marketing research. J. Acad. Mark. Sci. **40**, 414–433 (2012)

Hair, J.F., Sarstedt, M., Hopkins, L., Kuppelwieser, V.G.: Partial least square structural equation modeling (PLS – SEM); an emerging tool in business research. Eur. Bus. Rev. **26**(2), 106–121 (2014)

Hair, J.F., Hult, G.T., Ringle, C.M., Sarstedt, M.: A Primer on Partial Least Squares Structural Equation Modeling (PLS–SEM), 2nd edn. Sage Publications Inc., Los Angeles (2017)

Hair, J.F., Sarstedt, M., Ringle, C.M., Gudergan, S.P.: Advanced Issues in Partial Least Squares Structural Equation Modeling. Sage Publications Inc., Los Angeles (2018)

Henseler, J.: Adanco 2.0.1. User Manual, KG, Composite Modeling GmbH & Co. Kleve, Germany (2017)

Henseler, J., Dijkstra, T.K.: ADANCO 2.0. Composite Modeling, Kleve, Germany (2015)

Henseler, J., Sarstedt, M.: Goodness-of-fit indices for partial least squares path modeling. Comput. Statistics **28**, 565–580 (2013)

Henseler, J., Dijkstra, T.K., Sarstedt, M., Ringle, C.M., Diamantopoulus, A., Straub, D.W.: Common beliefs and reality about PLS: comments on Ronnko and Evermann (2013). Organ. Res. Methods **17**(2), 182–209 (2014)

Henseler, J., Hubona, G., Ray, P.A.: Using PLS path modeling in new technology research: updated guidelines. Ind. Manag Data Syst. **116**(1), 2–20 (2016a)

Henseler, J., Ringle, C.M., Sarstedt, M.: Testing measurement invariance of composites using partial least squares. Int. Mark. Rev. **33**, 405–431 (2016b)

Henseler, J., Ringle, C.M., Sinkovics, R.R.: The use of partial least squares path modeling in international marketing. Adv. Int. Mark. **20**, 277–320 (2009)

Hu, L.T., Bentler, P.M.: Fit indices in covariance structure modeling: sensitivity to underparameterized model misspecification. Psychol. Methods **3**(4), 424–453 (1998)

Hu, L.T., Bentler, P.M.: Cut-off criteria for fit indexes in covariance structure analysis: conventional criteria versus new alternatives. Struct. Equ. Model. **6**(1), 1–55 (1999)

Hult, G.T.M., Ketchen, D.J., Griffith, D.A., Finnegan, C.A., González-Padrón, T., Harmancioglu, N., Huang, Y., Talay, M.B., Cavusgil, S.T.: Data equivalence in cross-cultural international business research: assessment and guidelines. J. Int. Bus. Stud. **39**, 1027–1044 (2008)

Lohmoller, J.B.: Latent Variable Path Modeling with Partial Least Squares. Physica, Heidelberg (1989)

Nitzl, C.: The use of partial least squares structural equation modelling (PLS-SEM) in management accounting research: directions for future theory development. J. Account. Lit. **37**, 19–35 (2016)

Ringle, C.M., Wende, S., Becker, J.M. SmartPLS 3.0, Boenningstedt: SmartPLS GmbH. http://www.smartpls.com (2015)

Ringle, C.M. Advanced PLS-SEM Topics: PLS Multigroup Analysis, Working paper, University of Seville, November (2016)

Tenenhaus, M., Amato, S., Esposito Vinzi, V.: A global goodness-of-fit index for PLS structural equation modelling. In: Proceedings of the XLII SIS Scientific Meeting, pp. 739–742. CLEUP, Padova (2004)

Tenenhaus, M., Esposito Vinzi, V., Chatelin, Y.-M., Lauro, C.: PLS path modelling. Comput. Stat. Data Anal. **48**, 159–205 (2005)

Veenhoven, R.: Happiness in Nations, World Database of Happiness, Erasmus University Rotterdam, The Netherlands (2017). http://worlddatabaseofhappiness.eur.nl. Assessed 17 Mar 2017

To Assess Collective Well-Being
with a Synthetic and Autocorrelate Index
Tourism of Italian Provinces

Domenico Tebala[1]([⊠]) [ID] and Domenico Marino[2] [ID]

[1] National Institute of Statistics, Catanzaro 88100, Italy
tebala@istat.it
[2] Mediterranea University of Reggio Calabria, Reggio Calabria 89100, Italy
dmarino@unirc.it

Abstract. Tourism is a phenomenon that is particularly important in the contribution to collective wellbeing. Moreover, data measured in a specific area (or province) can be influenced by what happens in nearby areas, generating what is commonly called "spatial autocorrelation" or "spatial interdependence". This study is aimed at identifying a composite "systemic" index to measure the impact of tourist goods and others determinants can make to collective well-being in the provincial context through a composite index through the BES methodology (Equitable and Sustainable Well-being) and to analyze possible spatial autocorrelations between Italian provinces.

The method of construction of the index followed these steps:

(1) description of the theoretical framework, methodology used and indicators;

(2) description of the descriptive analysis results: in order to assess the robustness of the identified method and, therefore, improve decision-making, we also completed an influence analysis in order to analyze the most significant indicators (software COMIC - COMposite Indices Creator);

(3) a summary of the conclusions through a georeferenced map of the synthetic index of Italian tourism provinces and a Cluster Map LISA which shows the provinces with statistically significant values of the LISA index, classified by five categories: (A) Not significant (white); (B) High-High (red); (C) Low-Low (Blue); (D) Low-High (light blue); (E) High-Low (Light Red) (software GeoDa).

Keywords: Tourism · Index · Autocorrelation · Provinces

1 Introduction

Tourism is a phenomenon that is particularly important in the contribution to collective wellbeing. In recent years, tourism has changed a lot, transforming itself into a very complex social and economic phenomenon, capable of having positive effects on the socio-economic well-being of a territory [1].

It is therefore an important added value to the economy of a city, of a region and hence of a whole nation. As a result, the development of this sector can be crucial to

© Springer Nature Switzerland AG 2019
J. Gil-Lafuente et al. (Eds.): AEDEM 2017, SSDC 180, pp. 35–42, 2019.
https://doi.org/10.1007/978-3-030-00677-8_4

improving existing conditions in a place. Moreover, data measured in a specific area (or province) can be influenced by what happens in nearby areas, generating what is commonly called "spatial auto-correlation" or "spatial interdependence". In this regard, the LISA indicators (Local Indicator of Spatial Association) provide a local dimension to the measure of autocorrelation, enabling each spatial unit (e.g. the province) to assess the degree of spatial association and similarity with the elements surrounding it. In the case of positive autocorrelation these associations can be of the High-High type (high values observed in a territorial unit and high values even in its vicinity) or Low-Low (low values observed in a territorial unit and low values even in its own around). Conversely, in the case of negative self-correlation, the associations will be of the High-Low or Low-High type. In all other cases, there will be no autocorrelation or non-significant autocorrelation.

However, despite you talk a lot of this importance, it is difficult to measure the extent of their contribution. So this study is aimed at identifying a composite "systemic" index to measure the impact of tourist goods and others determinants can make to collective well-being in the provincial context through a composite index through the BES methodology (Equitable and Sustainable Well-being) and to analyze possible spatial auto-correlations between Italian provinces. The province, that is the territorial sphere where the man is placed at the center of a wide range of environmental, cultural and artistic stresses, is the ideal place to build a synthetic and autocorrelate BES index.

The index construction will take place through the following steps:

(1) description of the theoretical framework, methodology used and indicators;
(2) description of the descriptive analysis results: in order to assess the robustness of the identified method and, therefore, improve decision-making, we also completed an influence analysis in order to analyze the most significant indicators (software COMIC - COMposite Indices Creator);
(3) a summary of the conclusions through a georeferenced map of the synthetic index of Italian tourism provinces and a Cluster Map LISA which shows the provinces with statistically significant values of the LISA index, classified by five categories: (A) Not significant (white); (B) High-High (red); (C) Low-Low (blue); (D) Low-High (light blue); (E) High-Low (light red)

2   Process to Calculate the Indicator

2.1   Description of the Theoretical Framework

Tourism is one of the main sectors influencing socio-economic development of the territories and is influenced by various determinants. The approach used involves the construction of macro areas (pillars) by aggregating elementary indicators. Both pillars and elementary indicators have been considered non-replaceable. To construct synthetic index, we adopted the following indicators and polarity [2] (Table 1):

Table 1. Indicators and polarity

Macro areas	Indicators	Polarity
Infrastructure [3]	Number of tourist ports, airports and railway stations per 100,000 inhabitants	+
Tourist facilities [4]	Number of hotels and number of places bed, number hotel extra exercises and sets bed per 100,000 inhabitants	+
Tourism demand [4]	Number days of presence (Italian and foreigners) in the complex of the receptive exercises per inhabitant	+
	Number of presences (Italian and foreigners) in the complex of the receptive exercises in the non-summer months (days per inhabitant) Total spends in goods and services sustained by a traveller	
	Per-capita GDP [5]	

2.2 Methodology

The matrix relating to data on Italian provinces was divided into four progressive steps:

(a) Selection of a set of basic indicators on the basis of an ad hoc evaluation model hinging upon the existence of quality requirements;
(b) Further selection aimed at balancing the set of indicators within the theoretical framework of the structure. Outcome indicators are impact indicators as the ultimate result of an action as a result of a stakeholder activity or process;
(c) Calculation of synthetic indices (pillars), by making use of the methodology proved more appropriate to obtain usable analytical information on the tourism of Italian provinces;
(d) Processing of a final synthetic index as a rapid empirical reference concerning the degree of tourism of Italian provinces.

Missing values were attributed via the *hot-deck* imputation and, where not possible, with Italy's average value.

The choice of the synthesis method is based on the assumption of a formative measurement model, in which it is believed that the elementary indicators are not replaceable, which is to say, cannot compensate each other.

The exploratory analysis of input data was performed by calculating the mean, average standard deviation and frequency, as well as correlation matrix and principal component analysis. Since this is a non-compensatory approach, the simple aggregation of elementary indicators was carried out using the correct arithmetic average with a penalty proportional to the "horizontal" variability.

Normalization of primary indicators took place by conversion into relative indexes compared to the variation range (min-max).

Attribution of weights to each elementary indicator has followed a subjective approach, opting for the same weight for each of them. Since, in some cases, the elementary indicators showed different polarity, it was necessary to reverse the sign of negative polarities by linear transformation.

For the synthetic indicator calculation, we used the *Adjusted Mazziotta-Pareto Index* (AMPI), which is used for the min-max standardization of elementary indicators and aggregate with the mathematical average penalized by the "horizontal" variability of the indicators themselves. In practice, the compensatory effect of the arithmetic mean (average effect) is corrected by adding a factor to the average (penalty coefficient) which depends on the variability of the normalized values of each unit (called horizontal variability), or by the variability of the indicators compared to the values of reference used for the normalization.

The synthetic index of the i-th unit, which varies between 70 and 130, is obtained by applying, with negative penalty, the correct version of the penalty method for variation coefficient (AMPI ±), where:

$$AMPIi - = Mri - Sricvi \qquad (1)$$

where *Mri* e *Sri* are, respectively, the arithmetic mean and the standard deviation of the normalized values of the indicators of the i unit, and $cvi=Sri/Mri$ is the coefficient of variation of the normalized values of the indicators of the i unit.

The correction factor is a direct function of the variation coefficient of the normalized values of the indicators for each unit and, having the same arithmetic mean, it is possible to penalize units that have an increased imbalance between the indicators, pushing down the index value (the lower the index value, the lower the level of health).

This method satisfies all requirements for the wellbeing synthesis:

- Spatial and temporal comparison
- Irreplaceability of elementary indicators
- Simplicity and transparency of computation
- Immediate use and interpretation of the obtained results
- Strength of the obtained results

An influence analysis was also performed to assess the robustness of the method and to verify if and with which intensity the composite index rankings change following elimination from the starting set of a primary indicator. This process has also permitted us to analyze the most significant indicators.

The analysis was conducted using the COMIC (Composite Indices Creator) software, developed by ISTAT. The software allows calculating synthetic indices and building rankings, as well as easily comparing different synthesis methods to select the most suitable among them, and write an effective report based upon results.

3   Description of the Results

Table 2 reveals a moderate variability, especially for the infrastructure index (σ = 7.653) while Table 3 shows significant correlations between the analyzed indicators of the macro areas: direct correlation between the tourist facilities index and tourism demand index (r = 0.779), between the tourist facilities index and infrastructure index (r = 0.515) and between infrastructure index and tourism demand (r = 0.407).

Table 2. Mean, σ and frequency – tourism macro areas

	Infrastructure	Tourist facilities	Tourism demand
Mean	99.339	99.828	99.643
σ	7.653	6.165	5.989
Frequency	110	110	110

Table 3. Correlation matrix of the macro areas

Macro areas	Infrastructure	Tourist facilities	Tourism demand
Infrastructure	1.000		
Tourist facilities	0.515	1.000	
Tourism demand	0.407	0.779	1.000

The influence analysis describes the indicators that most influence the composition of rosters in tourism of provinces. In analyzing Table 4, we can see that most significant macro area is infrastructure ($\sigma = 13.791$).

Table 4. Influence analysis: mean and σ of the shifts for basis indicator of macro areas

Macro areas	Mean	σ
Infrastructure	14.027	13.791
Tourist facilities	5.091	4.546
Tourism demand	8.264	5.874
Mean	**9.127**	**8.071**
σ	**3.699**	**4.081**

4   Discussion and Conclusions

The cartographic representation of the final composite index value, alongside with the descriptive analysis of data, yields the usual dualistic pattern South/Center-North of Italy as in other domains of BES, except for the case of Naples (Fig. 1).

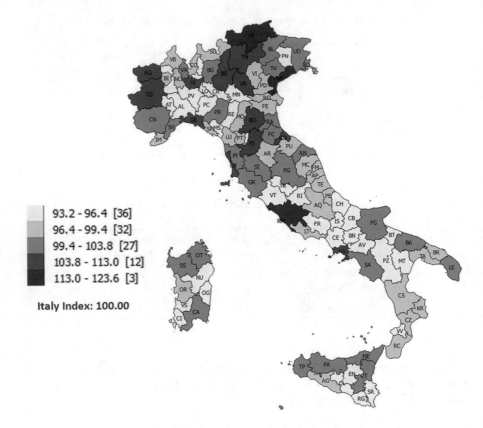

Fig. 1. Territorial distribution of the synthetic index of tourist provinces

In particular, the best performances are grouped in the Lazio (Rome), Trentino-Alto Adige (Bolzano), Lombardy (Milan), Tuscany (Florence and Leghorn) regions, but the most "touristic" province is Venice (123.61 Index – Italia Index 100), thanks, primarily, to the tourist facilities index (136.6) (27439 hotel extra exercises and 268105 sets bed) and the tourism demand index (116.9) (about 40 days of Italian and foreign presences in the total of resorts per inhabitant).

The province of Medio Campidano (Sardinia) occupies the last place in the ranking (index 93.1), preceded by Enna (Sicily) (93.4) and Barletta-Andria-Trani (Puglia) (93.6). However all the South is heavily penalized (60% of the provinces below the average are southern), except some isolated cases as Naples as already mentioned previously (index 109.6).

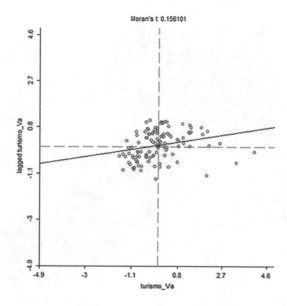

Fig. 2. Moran's index

What is the role of autocorrelation [6] or space interdependence? In other terms, is the index of a province influenced by what happens in neighboring provinces? Our results (Fig. 2) show a sufficient spatial interaction described by Moran's index (0.2) that is an index that varies between −1 and +1 and records the level of dissimilarity (negative autocorrelation) or similarity (positive autocorrelation) between contiguous territorial units. Negative values of autocorrelation are indicators of a process spatial repulsive type, positive values for the presence of an aggregative process that can be understood as a "mechanism" of territorial propagation of productivity.

In our case in the South there is a certain positive autocorrelation between 9 provinces, with low values observed in a province and low values even in the surrounding areas (Low-Low Blue Color) and in the North between 7 provinces with high values observed in a province and high values even in the surrounding areas (High-High Red Color) (Fig. 3).

Therefore, it was possible to identify spatial patterns able to describe areas of "multidirectional dependence", where contiguous areas show similar levels of the same phenomenon.

This again confirms the gap between the well-being of the South and the North with regard to the tourist sector.

Our systemic index would be useful to have an idea of the general status of tourism and to guide government actions.

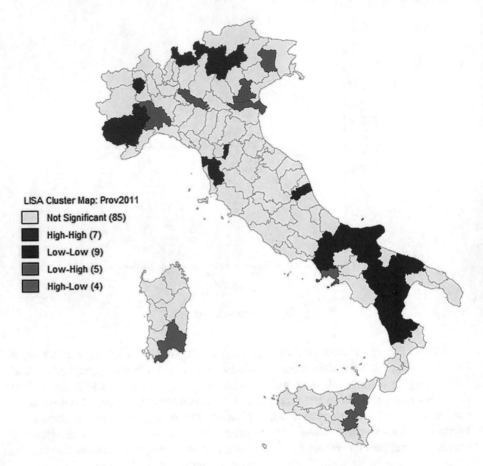

Fig. 3. Lisa Cluster Map

References

1. Pernicola, C.: Il Community Empowerment sulle Comunità Locali a vocazione turistica (2007)
2. Istat: UrBes 2015 – Il benessere equo e sostenibile in Italia (2015)
3. http://datiopen.it
4. http://dati.istat.it
5. http://ec.europa.eu/eurostat/data/database
6. http://www.istat.it/it/files/2014/10/M.-Mucciardi-E.-Otranto_Presentazione.pdf

Cooperation for External Knowledge Acquisition from Inter-organizational Relationships as Antecedent of Product Innovation: The Mediating Role of Absorptive Capacity

Beatriz Ortiz García Navas[1](✉), Joachim Bach[2],
Mario Javier Donate Manzanares[3], and Fátima Guadamillas Gómez[1]

[1] Castilla-La Mancha University, Toledo, Spain
{Beatriz.Ortiz,Fatima.Guadamillas}@uclm.es
[2] Schmalkalden University, Schmalkalden, Germany
j.bach@hs-sm.de
[3] Castilla-La Mancha University, Ciudad Real, Spain
MarioJavier.Donate@uclm.es

Abstract. This paper analyses the antecedents of company's product innovation based on knowledge acquisition from inter-organizational relationships and absorptive capacity. Focusing on Spanish biotechnology and pharmaceutics industries, it examines the mediating effects of the explorative and exploitative absorptive capacity sides on the relationships between social capital, knowledge acquisition made through strategic alliances, and product innovation. The study results partially support these mediating effects, and allow to conclude that an appropriate management of firm's inter-organizational networks to identify and acquire knowledge-based resources from cooperation partners, besides encouraging the development of internal absorptive capabilities, could be used for developing new products, and generating/maintaining competitive advantages related with these processes.

1 Introduction

In a growing complexity and high uncertainty environment, innovation has a strong role on firms' survival and success, being a key aspect from a strategic point of view [7]. Similarly, firm's ability to manage knowledge is currently essential for achieving profitable positions in technological-intensive industries [13]. When these companies cannot exclusively depend on their capabilities and internal competencies to develop new technologies [39], they need to access external knowledge for complementing their resources and capabilities using different strategies, such as the development of cooperation agreements or alliances [3, 44]. On this regard, strategic alliances play a critical role in innovation, because it eases firms into overcoming resource constraints and achieve superior innovative performance, not only by using internal resources, but also by acquiring knowledge-based capabilities from cooperative partners [47].

© Springer Nature Switzerland AG 2019
J. Gil-Lafuente et al. (Eds.): AEDEM 2017, SSDC 180, pp. 43–57, 2019.
https://doi.org/10.1007/978-3-030-00677-8_5

On the other hand, firm's absorptive capacity is an essential requirement for successful internal and external knowledge integration. This dynamic capacity can be defined as the ability that allow firms to recognise the value of external knowledge, so that it can be acquired, assimilated, transformed and exploited according to their objectives [5, 24, 43, 46]. Despite the importance of absorptive capacity for value generation, there are many ambiguities in the literature about its process operation [25, 37]. From this point of view, this research will try to analyse individually the explorative an exploitative aspects of this concept, in order to formulate their specific connexion with external knowledge acquisition.

In addition, the inter-organizational perspective of social capital establishes that the way a firm manages its relationships with external agents is an important source for both the identification and exploitation of new opportunities to improve company innovativeness [42]. Nevertheless, to date these relationships and their distinctive effects on knowledge acquisition, in this case to carry out by means of cooperation, has not been analysed by any scholar in the field [30].

Taking into account the aforementioned points, this study examines the mediating effects of the explorative and exploitative absorptive capacity sides on the relationships between social capital, knowledge acquisition made through strategic alliances, and product innovation.

The structure of the paper is as follows. First, we introduce the conceptual aspects and the hypotheses of the research. Next, we describe the research methodology and the results obtained from the statistical analyses applied by the authors to test the proposed hypotheses. Finally, we present the main conclusions of the paper and future research avenues for knowledge acquisition in firms.

2 Conceptual Framework

Knowledge acquisition is a mechanism by which a firm intentionally incorporates new technologies, ideas and know-how into its existing knowledge base from the external environment. Such acquisition is especially important in dynamic and innovative environments where organizations need to continuously access to a wide range of highly specialized technologies, expertise and capabilities that are difficult to be developed internally by a single firm [20]. Although not all the knowledge from outside of an organization is likely to be acquired by a firm [41], Organizational Knowledge Management literature has proposed different options through which firms can carry out knowledge acquisition, which can be grouped in two large blocks of acquisition strategies: (1st) direct purchases of external knowledge and (2nd) cooperation agreements. This paper is focused on the latter considering its strategic relevance for innovation and its relationship with inter-organizational social capital as antecedent. In this sense, cooperation is a flexible means to acquire certain tacit knowledge [8] and an opportunity to get access to abilities, technologies, markets, core competences and even network agents' strategic information [15]. In general, alliances can facilitate the access and acquisition of tacit, complex and specialize technological knowledge, although they frequently need to learn in some extent [38].

Moreover, in the recent years the external or inter-organizational perspective of social capital has focused on companies' external links as being determinant factors to explore and exploit new opportunities and competitive advantages [42]. Social capital is a collection of assets that derive from, are embedded in, and are accessible from a firm's networks of relationships [31]. This definition of social capital includes different aspects of the social context such as strong interactions and social links –structural social capital–, trusted relationships –relational social capital–, and systems of shared values that facilitate interactions between individuals located in a specific social context – cognitive social capital–, which are closely interrelated [31]. Consequently, their joint investigation is essential for understanding both exploiting on knowledge through interfirm relations and explaining innovation results [27].

Several studies identify knowledge acquisition as a direct benefit of social capital [1, 31]. Specifically for cooperation agreements, strong ties between partners because of prior relationships and repeated transactions are necessary for knowledge acquisition [21]. These strong ties also promote and enhance trust, reciprocity, long-term perspectives and mutual comprehension, enhance the interfirm collaboration [21].

However, to analyse the main benefits of social interaction and facilitate knowledge acquisition through a repeated interaction in cooperation agreements, firms must have the ability to identify and evaluate the strategic value of their partners' knowledge [45], –i.e. explorative side of absorptive capacity. In that sense, not every network agent should necessarily has valuable knowledge – particularly complex and tacit–, and can be a potential cooperation partner.

On the other hand, there is an implicit assumption that knowledge acquisition automatically starts once a firm identifies valuable knowledge; but, in fact, it rarely happens this way in practice [43]. In our view, the aspects related to the design and implementation of cooperation strategies are ignored (i.e., post-hoc aspects of knowledge identification; previous aspects of knowledge acquisition). A main argument of this study is that through its knowledge identification capabilities a firm will amplify its possibilities to access to a wider range of previously recognised tacit knowledge from its inter-organizational relationships. This potential access will allow the firm to assess knowledge value, and formulate and implement suitable strategies for its acquisition by means of cooperation agreements, depending on its current and/or future knowledge needs. We thus establish the first hypothesis:

H_1: A company's explorative absorptive capacity will have a mediating effect in the relationship between its inter-organizational social capital and external knowledge acquisition through cooperation.

Companies also increasingly perceive cooperation strategies as a crucial aspect that sustains its innovation performance [12]. Some of the reasons that justify the significance of external knowledge for innovation are its integration into firm's knowledge pool in order to create new knowledge [26], to avoid the over-confidence risk in internal knowledge, the appearance of learning traps, or the not invented here syndrome [36]. However, the access to a great amount of partner's knowledge in a cooperation agreement does not ensure its appropriate understanding, absorption and creation of new products [19]. Consequently, in alliances with more than two participants, companies can be exposed to the same quantity of external knowledge and, at the same time, they can obtain different innovation profits [14].

This is especially important for tacit knowledge acquisition, which must be processed into a collective understanding so organizations can turn this external knowledge from its collaborative partners into innovative capabilities. Consequently, firms should implement organizational mechanisms that allow the internal transfer of the externally acquired knowledge so that it can be integrated and combined with their prior knowledge base [6] –i.e. exploitative side of absorptive capacity. In that sense, very few researches consider how each different aspects of a firm's global absorptive capacity effects on innovation capability (e.g., [4]). From these arguments, this paper suggests that acquired tacit knowledge by means of cooperation will be more valuable if it is properly assimilated and integrated into a firm's prior knowledge base. The second hypothesis is as follows:

H_2: A company's exploitative absorptive capacity will have a mediating effect in the relationship between external knowledge acquisition through cooperation and its product innovation capability.

3   Sample and Methodology

This study selected Spanish Biotechnological and pharmaceutical industries to carry out the empirical analysis. An on-line survey was launched in March-May 2015 to 735 firms from these industries. The data on the number and information about firms were obtained from the SABI database (System for accounting information analysis). Table 1 shows the research specifications.

Regarding the measures of variables, we used validated scales from previous studies, which were conveniently adapted to our research's context (see appendix for

Table 1. Research specifications

Population	735
Geographical scope	Spain
Sample size	87 firms[a]
Unit of analysis	Firm or business unit
Data collection method	Online survey
Response rate	11.84%
Sampling error	9.87%; p = q = 0.5

[a]In order to verify the sample representability a T-test was applied, which did not show significant differences neither in relation to size (t = 1.119; p = 0.263) or age (t = 1.159; p = 0.247). In addition, the Harman Test was used to verify the non-existence of common variance for our data set. The results show the existence of four factors with eigenvalues higher than one that explained the 67.5% of total variance. Because the first factor explains only a 37.4%, common variance does not appear as a problem.

the list of items). Additionally, age, size, industry, and R&D effort were included as control variables, given that all of them have extensively been utilized by studies on knowledge acquisition strategies and innovation.

4   Statistical Analysis and Results

The Partial Least Squares (PLS) method was applied as statistical technique in order to test the hypotheses of our study. PLS is a multivariate analysis technique [17] based on variance analysis, used to model latent constructs under non-normality conditions for data and small sample sizes. The PLS path method is typically applied in two stages: (1th) analysis of the measurement model; and (2nd) analysis of the structural model).

Regarding the first stage, the reliability of reflective and formative variables is differently measured. In addition, the proposed research model includes two second-order constructs [23], a Type II –social capital– (reflective-formative), and a Type I – exploitative absorptive capacity– (reflective-reflective). In that sense, an exploratory factor analysis (EFA) was applied in order to validate the formative nature of the Type II second order construct. The EFA solution reveals a three-dimensional structure of social capital [16], according to the three types of social capital that are considered – structural, relational and cognitive–.

Table 2. EFA results

Items	Factors		
	1	2	3
ST_SC1	0.727	0.242	0.322
ST_SC2	0.868	0.073	0.148
ST_SC3	0.904	0.234	0.174
ST_SC4	0.855	0.276	0.069
ST_SC5	0.616	0.337	0.299
REL_SC1	0.312	0.183	0.836
REL_SC2	0.140	0.271	0.885
COG_SC1	0.425	0.651	0.327
COG_SC2	0.356	0.742	0.302
COG_SC3	0.190	0.881	0.103
COG_SC4	0.117	0.885	0.168

Varimax rotation with Kaisser
normalization was performed to
retrieve factor loadings. Variables
are listed in appendix

Formative variables should be interpreted according to weights (ß). In that case, the reliability is determined by the variance inflation index (VIF), which what allows corroborating that multicollinearity between independent variables in a regression

Table 3. Measurement model of formative second-order construct

Second order construct	Dimensions	VIF	Weights	Loadings
SC	ST_SC	1.709	0.240	0.750
	REL-SC	1.975	0.621*	0.915
	COG_SC	1.899	0.316	0.797

*$p < 0.001$ ($t_{(0.001;\ 499)} = 3.3101$)

model does not exist [17]. For this purpose, some authors advise VIF values lower than 3.3 [e.g. 35]. Furthermore, weights signification should be also considered (Table 3). Using bootstrap, the results show that structural and cognitive social capital weights are not significant ($p > 0.05$). However, their standardized loadings (λ) are higher than 0.5, and these two formative items should be kept [16].

On the other hand, the measurement model of reflective variables was estimated using confirmatory factor analysis in order to assess reliability −individual items, constructs− and convergent and discriminant validity of measures. Results (shown in Table 4) confirm that the measurement model is reliable and valid.

Concerning the analysis of the structural model, it allows the researcher to test the proposed mediation hypotheses by the analysis of path coefficients –β– and the determination coefficients (R^2).

To test mediating effects PLS uses bootstrapping as it provides indicators for both direct and indirect effects [18]. The results should fulfil two conditions [32] –Tables 5 and 6–. Specifically, for the mediating effect proposed by hypothesis 1, the indirect effect between social capital and cooperation, when the mediating variable explorative absorptive capacity is introduced in research model, is significant ($\beta = 0.153$, $p < 0.05$). Likewise, the direct path coefficient in the relationship between social capital and acquisition through cooperation is significant when the mediating variable is introduced into the research model ($\beta = 0.426$, $p < 0.000$). Consequently, the mediating effect of explorative absorptive capacity in the relationship between social capital and external knowledge acquisition through cooperation is commentary partial.

Regarding the mediating effect proposed in hypothesis 2, the indirect effect between external knowledge acquisition through cooperation and product innovation, when exploitative absorptive capacity is added as a mediator, is strong and highly significant ($\beta = 0.234$, $p < 0.001$). Moreover, the direct effect of knowledge acquisition by means of cooperation on product innovation is also positive and significant ($\beta = 0.202$, $p < 0.05$). Therefore, the mediating effect is considered as complementary partial.

R^2 coefficients indicate the amount of variance explained by the relationships in the model. Figure 1 shows that the model explains 49.4% of the variance of product innovation, 31.2% of the variance of exploitative absorptive capacity, 36.2% of the variance of external knowledge acquisition through cooperation, and 46.3% of the variance of explorative absorptive capacity. A model has enough predictive power if R^2 is at least 10% [9], condition fulfilled by our study model.

Table 4. Measurement model of reflective variables

Constructs	Range of loadings	CR[3]	AVE[4]	Correlations								
				ST_SC	REL_SC	COG_SC	EXPLR_AC	COOP	INN	EXPLT_AC	ASI/TRF	EXPLT
ST_SC	0.798–0.932	0.932	0.733	*0.839*								
REL_SC	0.924–0.930	0.924	0.859	0.513	*0.927*							
COG_SC	0.845–0.910	0.925	0.755	0.604	0.542	*0.869*						
EXPLR_AC	0.700–0.817	0.894	0.586	0.450	0.645	0.545	*0.765*					
COOP	0.743–0.840	0.848	0.651	0.509	0.501	0.457	0.513	*0.807*				
INNP	0.708–0.853	0.898	0.639	0.297	0.311	0.371	0.459	0.550	*0.799*			
EXPLT_AC[1]		0.830	0.712	0.408	0.552	0.502	0.688	0.558	0.596	*0.844*		
ASI/TRF	0.774–0.855	0.917	0.689	0.465	0.585	0.557	0.713	0.589	0.591		*0.830**	
EXPLT	0.744–0.794	0.808	0.585	0.157	0.271	0.208	0.387	0.285	0.383		0.447	*0.765**

[1]Second-order construct; [2]Acceptance level ≥ 0.707 [17]; [3]Acceptance level ≥ 0.8 [33]; [4]Acceptance level ≥ 0.5 [11].

*Root of the variance shared between the first- order constructs and their measures.

Diagonal elements (in italics) are the square root of the variance shared between the constructs and their measures; off-diagonal elements are the correlations among the items/constructs; for discriminant validity, diagonal elements should be higher than off-diagonal elements in the same row and column.

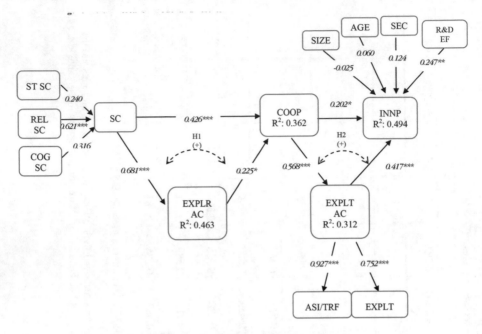

Fig. 1. Research model and results

Table 5. Mediating hypothesis 1 test

Effect on dependent variable	Path coefficient (β)	t
Direct	0.426***	3.759
Indirect	0.153*	1.792

*p < 0.05 ($t_{(0.05;\ 4999)}$ = 1.6479); **p < 0.01
($t_{(0.01;\ 4999)}$ = 2.3338); ***p < 0.001
($t_{(0.001;\ 4999)}$ = 3.1066)

Table 6. Mediating hypothesis 1 test

Effect on dependent variable	Path coefficient (β)	t
Direct	0.202*	2.056
Indirect	0.234***	3.256

*p < 0.05 ($t_{(0.05;\ 4999)}$ = 1.6479); **p < 0.01
($t_{(0.01;\ 4999)}$ = 2.3338); ***p < 0.001
($t_{(0.001;\ 4999)}$ = 3.1066)

Finally, regarding control variables, only R&D effort has a significant effect on product innovation (β = 0.245, p < 0.01). This is a logical result if we consider that those firms that make higher innovation efforts get a greater development of their ability to create new products.

5   Discussion, Conclusion and Implications

This research proposes that, on the one hand, the explorative side of absorptive capacity as mediating variable on the relationship between inter-organizational social capital and external knowledge acquisition made through cooperation. On the other hand, the exploitative side of absorptive capacity is a mediator in the relationship between acquired external knowledge by means of cooperation and product innovation capability. Nevertheless, the obtained results only partially support these hypothesized mediating effects. For hypothesis one, the results support there being both a direct and an indirect effect –via explorative absorptive capacity– of inter-organizational social capital on external knowledge acquisition. Consequently, the development of a high level of structural, relational and cognitive social capital with alliance partners will directly allow organizations to acquire a greater amount of tacit knowledge from them (than those firms that have a low level of social capital). This idea is consistent with the findings of Yli-Renko et al. [45] or Maurer [29]. As a result, we can state that the stronger interactions and social links, more reliable relationships, and the better systems of shared values are, the higher the level of available tacit knowledge from the collaborative partner will be. These kinds of links thus increase the likelihood that agents will carry out and complete cooperative agreements to acquire knowledge.

Additionally, the study findings emphasise the importance of valuable knowledge identification as the first stage of the absorption process, in accordance with Patterson and Ambrosini [37]. The fact that a company is able to develop capabilities to identify and assess the value of knowledge of their network partners in a greater extent than others can allows it to choose for the best alternative of knowledge acquisition and optimize such process (e.g., in terms of knowledge quantity and quality), a key issue for innovative firms.

Regarding hypothesis two, the greater or lesser importance of the exploitative side of absorptive capacity to product innovation capability could be justified by the different nature of the acquire knowledge (tacit vs. explicit) from a cooperation agreement. In that sense, the acquisition of explicit knowledge could have direct and positive effects on firms' product innovation. However, the development of an appropriate capacity to integrate and combine prior and new knowledge is essential for the acquisition success of tacit knowledge (i.e. the creation of new products and technologies).

The study results have interesting prescriptive implications for company managers in high-tech industries. Managers should understand that "good" management of inter-organizational social capital might allow their companies to develop dynamic capabilities related to the absorptive capacity of unique and complex knowledge. Moreover, the encouragement of the explorative side of this capacity is an essential aspect for firm's consideration, because can enable to verify which potential cooperation partners have the required knowledge, and could become the best choice for reaching a successful acquisition process, depending on company needs, timing, and its particular circumstances. Additionally, the executive directors should promote those abilities related to the exploitative side of absorptive capacity, using for example firm's human resource policies or tools for managing both internal and external knowledge, so that any obstacle that difficult its operation can be eliminated.

Among the limitations of this study, we can point out, firstly, the cross-sectional nature of the empirical analysis. Furthermore, the study does not consider if there are dependent relationships between external knowledge acquisition made through strategic alliances and the components of absorptive capacity in its "exploitative" side, just as their influence on product innovation capability. Future studies may focus on that analysing. Lastly, the distinction between types of knowledge (e.g. tacit vs. explicit; individual vs. collective) has not been considered, which it can be relevant for knowledge transfer between organizations.

To deal with the previous limitations, future research can concentrate on analysing the effect of specific network characteristics for acquisition and transfer of knowledge in alliances, deeming knowledge types. Additionally, the study could be replicated for validation in other different contexts, such as other countries or low-tech industries. Finally, a longitudinal study could focus on examining the effect of changes and evolution of inter-organizational networks and absorptive capacity over time.

Appendix. Research Items

Constr.	Dimensions	My firm usually… (From 1 – strongly disagree to 7-strongly agree)	Source
COOPERATION (COOP)		Develops alliances and/or cooperation agreements with universities	Fey and Birkinshaw [10]; Díaz et al. [8]; Valmaseda and Hernández [44]
		Develops alliances and/or cooperation agreements with customers and suppliers	
		Develops alliances and/or cooperation agreements with participants in the development of joint research projects promoted by government institutions	
SOCIAL CAPITAL (SC)	STRUCTURAL SOCIAL CAPITAL (ST_SC)	Acquires knowledge from our inter-organizational contacts' network	Yli-Renko et al. [45]; Maula, Autio and Murray [28]; Inkpen and Tsang [21]; Ambos and Ambos [2]; Pérez-Luño, Cabello and Carmona [34]
		Personally meets contacts who acquire external knowledge	
		Maintains narrow inter-relationships with contacts who acquire external knowledge	
		Maintains frequent inter-relationships with contacts who acquire external knowledge	
		(In general) Has contacts who acquire knowledge from among themselves	

(*continued*)

(*continued*)

Constr.	Dimensions	My firm usually... (From 1 – strongly disagree to 7-strongly agree)	Source
	RELATIONAL SOCIAL CAPITAL (REL_SC)	Has external relationships based on cooperation and mutual trust	
		Has external relationships with a high grade of commitment	
	COGNITIVE SOCIAL CAPITAL (COG_SC)	Shares goals and joint-project interests with its external relationships	
		Shares a common vision regarding environment and successful key factors whit its external relationships	
		Uses similar working techniques that external agents whom it is related	
		Has a similar business culture and management style that external agents whom it is related	
EXPLORATIVE ABSORPTIVE CAPACITY (EXPLR AC)	IDENTIFICATION	Has the ability to seek information within its environment	Segarra *et al.* [39]; Jansen, Van Den Bosch and Volberda [22]
		Has the ability to anticipate competitors movements	
		Has the ability to anticipate customers necessities	
		Has the ability to keep in touch with external institutions and specialised sources	
		Has personal, equipment and specialised services for environment monitoring	
		Has problems to recognise changes in our market/products*	
		Understands new opportunities to satisfy our customers quickly	
		Interprets changes in market pull quickly	
		Knows intuitively which areas can use acquired technology or external knowledge	
EXPLOITATIVE ABSORPTIVE CAPACITY (EXPLT AC)	ASIMILATION/ TRANSFORMATION	Considers the consequences of changes in the market in order to create new products and services	Jansen *et al.* [22]
		Uses TIC for registration and storage newly acquired external knowledge for future reference	

(*continued*)

(*continued*)

Constr.	Dimensions	My firm usually... (From 1 – strongly disagree to 7-strongly agree)	Source
		Recognises the usefulness of newly acquired external knowledge to incorporate it with existing knowledge quickly	
		Has employees who hardly share practical experiences*	
		Hardly takes advantage of newly acquired knowledge*	
		Establishes regular meetings to discuss the consequences of market trends and the development of new services	
		Has tools/techniques for distributing and sharing acquired knowledge	
		Spreads acquired knowledge to those areas in which it can be more effective quickly	
	EXPLOTATION	Considers which is the best way to exploit knowledge	
		Knows which area (department, employees) can best exploit new knowledge	
		Hardly uses new knowledge in new products*	
PRODUCT INNOVATION		Has introduced more innovative products/services as compared to its competitors	Škerlavaj Hoon and Lee [40]
		Has frequently emphasised the development of new patented products	
		Has satisfied the market through the quick development of its products	
		Has continuously changed the product design to quickly enter in new emerging markets	
		Has continuously improved the components and quality of its products	

*Item with inverse coding

References

1. Adler, P.S., Kwon, S.W.: Social capital. prospect for a new concept. Acad. Manag. Rev. **27**, 17–40 (2002)
2. Ambos, T.C., Ambos, B.: The impact of distance on knowledge transfer effectiveness in multinational corporations. J. Int. Manag. **15**, 1–14 (2009)
3. Cassiman, B., Veugelers, R.: In search of complementarity in innovation strategy. Internal R&D and external knowledge acquisition. Manag. Sci. **52**, 68–82 (2006)
4. Cepeda, G., Cegarra, J.C., Jiménez, D.: The effect of absorptive capacity on innovativeness. Context and information systems capability as catalysts. Br. J. Manag. **23**, 110–129 (2012)
5. Cohen, W.M., Levinthal, D.A.: Absorptive capacity. A new perspective on learning and innovation. Adm. Sci. Q. **35**(1), 128–152 (1990)
6. Cruz, J., López, P., Navas, J.E.: Absorbing knowledge from supply-chain, industry and science. The distinct moderating role of formal liaison devices on new product development and novelty. Ind. Mark. Manag. **47**, 75–85 (2015)
7. Delgado, M., Martín, G., Navas, J.E., Cruz, J.: Capital Social, Capital Relacional e Innovación Tecnológica. Una Aplicación al Sector Manufacturero Español de Alta y Media Tecnología. Cuadernos de Economía y Dirección de la Empresa **14**, 207–221 (2011)
8. Díaz, N.L., Aguilar, I., De Saá, P.: El Conocimiento Organizativo y la Capacidad de Innovación. Evidencia para la Empresa Industrial Española. Cuadernos de Economía y Dirección de la Empresa **27**, 33–60 (2006)
9. Falk, R.F., Miller, N.B.: A Primer for Soft Modelling. University of Akron, Ohio (1992)
10. Fey, C., Birkinshaw, J.: External sources of knowledge, governance mode, and R&D performance. J. Manag. **31**(4), 597–621 (2005)
11. Fornell, C., Larcker, D.F.: Evaluating structural equation models with unobservablevariables and measurement error. J. of Mark. Res. 39–50 (1981)
12. Gallego, J., Rubalcaba, L., Suárez, C.: Knowledge for innovation in Europe. The role of external knowledge on firms' cooperation strategies. J. Bus. Res. **66**(10), 2034–2041 (2013)
13. Galliers, R.D., Leidner, D.E.: Strategic Information Management Challenges and Strategies in Managing Information Systems. Routledge, NewYork (2014)
14. Giuliani, E., Bell, M.: The micro-determinants of meso-level learning and innovation. Evidence from a Chilean wine cluster. Res. Policy **34**(1), 47–68 (2005)
15. Hagedoorn, J., Duysters, G.: External sources of innovative capabilities: the preferences for strategic alliances or mergers and acquisitions. J. Manag. Stud. **39**(2), 167–188 (2002)
16. Hair, J.F., Hult, G.T.M., Ringle, C.M., Sarstedt, M.: A Primer on Partial Least Squares Structural Equation Modelling (PLS-SEM). Sage, Thousand Oaks (2016)
17. Hair, J.F., Ringle, C.M., Sarstedt, M.: Editorial-partial least squares structural equation modelling. Rigorous applications, better results and higher acceptance. Long Range Plan. **46**(1–2), 1–12 (2013)
18. Hayes, A.F., Scharkow, M.: The relative trustworthiness of inferential tests of the indirect effect in statistical mediation analysis does method really matter. Psychol. Sci. **24**(10), 1918–1927 (2013)
19. Hughes, M., Morgan, R.E., Ireland, R.D., Hughes, P.: Social capital and learning advantages. A problem of absorptive capacity. Strateg. Entrep. J. **8**(3), 214–233 (2014)
20. Iansiti, M.: From technological potential to product performance. An empirical analysis. Res. Policy **26**(3), 345–365 (1997)
21. Inkpen, A.C., Tsang, E.W.K.: Social capital, networks and knowledge transfer. Acad. Manag. Rev. **30**(1), 146–165 (2005)

22. Jansen, J., Van Den Bosch, F., Volberda, H.: Managing potential and realized absorptive capacity. How do organizational antecedents matter? Acad. of Manag. J. **48**(6), 999–1015 (2005)
23. Jarvin, ChB, Mackenzie, S.B., Podsakoff, P.M.: A critical review of construct indicators and measurement model misspecification in marketing research. J. of Consum. Res. **30**, 199–218 (2003)
24. Jiménez, M.M., García, V.J., Molina, L.M.: Validation of an instrument to measure absorptive capacity. Technovation **31**, 190–202 (2011)
25. Lewin, A.Y., Massini, S., Peeters, C.: Microfoundations of internal and external absorptive capacity routines. Organ. Sci. **22**(1), 81–98 (2011)
26. Liao, Y., Marsillac, E.: External knowledge acquisition and innovation. The role of supply chain network-oriented flexibility and organisational awareness. Int. J. Prod. Res. **53**(18), 5437–5455 (2015)
27. Martínez, R., Sáez, F.J., Rúiz, P.: Knowledge acquisition's mediation of social capital-firm innovation. J. of Knowl. Manag. **16**(1), 61–76 (2012)
28. Maula, M., Autio, E., Murray, G.: Prerequisites for the creation of social capital and subsequent knowledge acquisition in corporate venture capital venture capital. An Int. J. Entrep. Financ. **5**(2), 117–134 (2003)
29. Maurer, I.: How to build trust in inter-organizational projects. The impact of project staffing and project rewards on the formation of trust, knowledge acquisition and product innovation. Int. J. Project Manag. **28**(7), 629–637 (2010)
30. Mura, M., Radaelli, G., Spiller, N., Lettieri, E., Longo, M.: The effect of social capital on exploration and exploitation. Modelling the moderating effect of environmental dynamism. J. Intellect. Cap. **15**(3), 430–450 (2014)
31. Nahapiet, J., Ghoshal, S.: Social capital, intellectual capital and the organizational advantage. Acad. Manag. Rev. **23**(2), 242–266 (1998)
32. Nitzl, C., Roldan, J.L., Cepeda, G.: Mediation analysis in partial least squares path modelling. Helping researchers discuss more sophisticated models. Ind. Manag. & Data Syst. **116**(9), 1849–1864 (2016)
33. Nunnally, J.: Psychometric Methods. McGraw-Hill, New York (1978)
34. Pérez-Luño, A., Cabello, C., Carmona, A.: How social capital and knowledge affect innovation. J. Bus. Res. **64**(2), 1369–1376 (2011)
35. Petter, S., Straub, D., Rai, A.: Specifying formative constructs in information systems research. *MIS Q.* 623–656 (2007)
36. Purcell, R., McGrath, F.: The search for external knowledge. Electron. J. Knowl. Manag. **11**(2), 158–167 (2013)
37. Patterson, W., Ambrosini, V.: Configuring absorptive capacity as a key process for research intensive firms. Technovation **36**, 77–89 (2015)
38. Savino, T., Messeni, A., Albino, V.: Search and recombination process to innovate: a review of the empirical evidence and a research agenda. Int. J. Manag. Rev. **19**(1), 54–75 (2017)
39. Segarra, M., Roca, V., Bou, J.C.: External knowledge acquisition and innovation output: An analysis of the moderating effect of internal knowledge transfer. Knowl. Manag. Res. & Pract. **12**(2), 203–214 (2014)
40. Škerlavaj, M., Hoon, J., Lee, Y.: Organizational learning culture, innovative culture and innovations in South Korean firms. Expert Syst. Appl. **37**, 6390–6403 (2010)
41. Teece, D.J.: Capturing value from knowledge assets. The new economy, markets for know-how, and intangible assets. Calif. Manag. Rev. **40**(3), 55–79 (1998)
42. Teng, B.: Corporate entrepreneurship activities through strategic alliances. J. Manag. Stud. **44**(1), 119–142 (2007)

43. Todorova, G., Durisin, B.: Absorptive capacity on business unit innovation and performance. Acad. Manag. J. **44**(5), 996–1004 (2007)
44. Valmaseda, O., Hernández, N. Fuentes de Conocimiento en los Procesos de Innovación Empresarial. Las Spin-Off Universitarias en Andalucía. *ARBOR Cienc., Pensam. Y Cult.* 188 (enero-febrero), 211–228 (2012)
45. Yli-Renko, H., Autio, E., Sapienza, H.J.: Social capital, knowledge acquisition, and knowledge exploitation in young technology-based firms. Strateg. Manag. J. **22**, 587–613 (2001)
46. Zahra, S.A., George, G.: Absorptive capacity. A review, reconceptualization and extension. Acad. Manag. Rev. **27**(2), 185–203 (2002)
47. Zhang, H., Shu, C., Jiang, X., Malter, A.J.: Managing knowledge for innovation. The role of cooperation, competition, and alliance nationality. J. Int. Mark. **18**(4), 74–94 (2010)

Recent Trends in Volunteerism: A Comparison Between European and North/South American Countries

Domenico Marino[1](✉) ⓘ and Marina Schenkel[2]

[1] Mediterranea University of Reggio Calabria, Reggio Calabria, Italy
dmarino@unirc.it
[2] Udine University, Udine, Italy
marina.schenkel@uniud.it

Abstract. In a preceding work on European data it has been found that during the crisis volunteers have increased in the whole Europe, with some exceptions (Schenkel et al. 2016). The most popular fields of volunteer activity are sport and artistic/cultural associations. Sport events are also a main occasion of "episodic volunteerism", which has gained popularity in the last years. The participation to community and neighbourhood organizations is also increasing.

The same trend is not to be recognized in the other Western countries, across their different stage of development and institutional settings. The reductions in spending on public services, the increase in unemployment and the decrease in employment have happened also across the Atlantic, even if with a different timing between North and South America. Important political events in many countries have brought about changes in government and policies.

After a brief reference to literature, including some definitions and classifications of volunteers (Sect. 2), some recent data on North and South America will be commented (Sects. 3 and 4).

The better criterion to classify the various areas examined is very simple: to distinguish between the ones where volunteerism has increased, and the ones in which it has not, diminishing or remaining constant. This hardly coincides with any geographical or cultural divide. In the first group we find, beside the world and Europe as a whole, Mexico, France and Italy, in the second Spain, UK, USA, Canada, Argentina, Brazil. It can be argued that the different intensity and length of the crisis is a prime determinant of these diverging trends: the last three countries were not particularly hit by the economic downturn in the period 2008–2014, and USA could recover quickly from it. The reason why in UK and Spain volunteers have diminished can be that these countries were not only stricken by the crisis, but also by a particularly severe cut in public expenditure towards non-profit and volunteer associations. In order to appreciate the impact of changing economic conditions, a particular interesting case is the one of Argentina, where volunteers increased enormously during the crisis (2001–2002), and diminished again during the "resurrection".

The qualitative features of the stock of volunteers, as already noted, has changed in the sense that all age groups and both genders are more or less equally represented, in every country. Another interesting constant across the various countries is that the majority of volunteers are employed.

© Springer Nature Switzerland AG 2019
J. Gil-Lafuente et al. (Eds.): AEDEM 2017, SSDC 180, pp. 58–83, 2019.
https://doi.org/10.1007/978-3-030-00677-8_6

This image of the volunteers is quite different from what we have been used to consider so far, that is, an aggregate where most people have a lower attachment to the labour market, and therefore a low value of leisure.

1 Introduction

Among the realities that emerge as a reaction of societies and people to times of crises, volunteerism has a paramount importance. In a preceding work on European data it has been found that during the crisis volunteers have increased in the whole Europe, with some exceptions (Schenkel et al. 2014) The necessity to reduce labour costs in public services and to increase the appreciation of the public to the mission and the social value of the organization that provides the service (whether public or private) seems to be the most immediate determinant of the expansion of volunteering. In addition to the debt crisis, another factor has increased the role of volunteers, both from the point of view of supply and demand: rising unemployment.

Many observers point out that the question of volunteerism has grown considerably before and especially during the crisis as a result of the retreat of the state and public administration as a producer of public goods and services (Vaughan-Whitehead 2013).

A set of question arises: why the crisis has increased the supply of and the demand for voluntary work in some countries, and has reduced them in some others? How Europe and the other Western countries trends differ?

In the following sections some theoretical considerations will be reminded in a brief review of the literature, in which some classifications of volunteers will be presented (Sect. 2). Next the most recent empirical evidence on volunteers' supply and demand will be commented and compared (Sects. 3–7). The data on the number and other characteristics of the volunteers have been collected in the years following 2008, i.e. after the beginning of the Big Recession, when the expenditure on public services has been reduced, unemployment have increased and employment has decreased. Moreover, important political events have happened in many countries, and changes in government and policies have followed.

2 Determinants of the Supply of and the Demand for Voluntary Work

What makes a human activity volunteer work? According to Tilly and Tilly (1994) it consists of voluntary "unpaid work provided to parties to whom the worker owes no contractual, familial, or friendship obligations". The same definition was adopted by the International Conference of Labour Statisticians (Bollé 2009), and was recently endorsed by the ILO (2011). More recently a larger definition was adopted by ILO itself, including also work done in benefit of relatives and friends, living in others households[1]. As pointed out by Wilson and Musick (1997), voluntary work, unlike that

---

[1] In this paper however "informal" voluntarism is excluded from the analysis.

in the labour market and in the informal sector, is "uncommodified" and "freely undertaken". For this reason motivations play a crucial role in determining what is or what is not voluntary work (Bekkers and Bowman 2008). The benefits to the volunteers are both of psychic type (e.g. greater happiness) and utilitarian (e.g. best physical conditions and a higher level of human capital) (Mellor et al. 2009; Uslaner 2008). In the second category the opportunity to improve one's skills is included. As far as sport is concerned, an important psychic motivation is the "love for sport" (Welt Peachey et al. 2014). From an economic point of view, the consumption and investment motives have been distinguished. The degree of satisfaction that voluntary activity provides, both when it is undertaken as consumption or as investment, is the main determinant of the retention of volunteers, that is the decision to continue with the same organization.

Volunteers in social services have been divided into four groups, according to their motivations and the determinants of the satisfaction that derives from their activities (Marino et al. 2009):

Group A: "they want to do something good, feeling useful, but cannot expect too much."

They are mainly persons out of the labour force, especially retirees, singles, but also separated and widowed, and have the lowest level of education. This group show the less ideological attitude.

Group B "they want to pursue their ideals in a challenging and rewarding work environment"

In the volunteers belonging to this group decision and commitment are outstanding. Consequently they show a lower propensity to consider volunteering as leisure and, in the choice of the organization in which to operate, they give a lower importance to the organization's ability to meet their expectations.

Group C "volunteers with interest"

Volunteers in this group have the goal of finding a job. Like everyone else, they are inspired by altruistic attitudes, but more than the others seem to have also individual extrinsic motivations.

Group D "volunteers for choice, conscious and generous"

This group is dedicated to volunteering with more continuity. They tend to show greater empathy in the relationship with recipients. This group is the archetype of the volunteer, a sort of ideal profile for which to strive.

3 Empirical Evidence

3.1 General Considerations

Before examining the evolution of the stock of volunteers, some caveats must be remembered. First of all, the number of volunteers strictly speaking corresponds neither to the supply nor to the demand of them, resulting from the encounter of the two "functions". It is not a priori granted that the demand for voluntary work is unlimited, and always exceeds supply. In fact, the volunteer work is unpaid (although not always completely) as far as the remuneration of the individual is concerned, but its use involves training and organizational costs, etc. However the data discussed in this and the following sections seem to be related to the mainly to the supply, for two sets of considerations. The first is that voluntary work supply is not unlimited. The other is that not only the output of voluntary work (entertainment, besides education, health, social services, essential infrastructure) is rationed on the supply side, but also the demand for such services is able to generate its own supply, stimulating the creation of more or less formal organizations dedicated meet these needs when they are notcompletely satisfied by the State or the market.

The second caveat is that the various sources – Census, Multipurpose/Use of Time Surveys and Opinion Polls-differ as far as their nature is concerned. However the results are often similar.

3.2   World

As an introduction to the more detailed data that will be presented afterwards, some charts are reproduced from two worldwide sources[2].

The first is the World Giving Index, in the period 2011–2015. The overall index (which contains, on the top of the rate of volunteering, also the one of giving money and helping a stranger), point to a high level and an increase in all the continents, without relevant distinctions (except for the outstanding rate of Oceania) (Figs. 1 and 2).

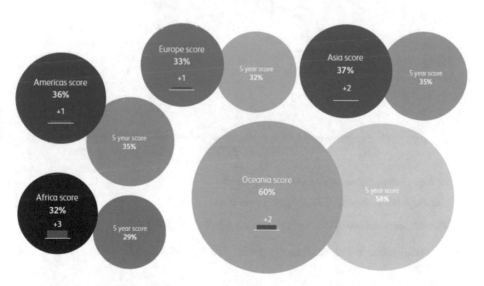

Fig. 1. World giving index score, continents, 2011–2015. Source: CAF (2016)

---

[2] In April 2015 the World Values Survey has made available times series data on volunteers and their motivations, across various European and non-European countries, which are of great interest for comparative purposes, which are not examined in detail here.

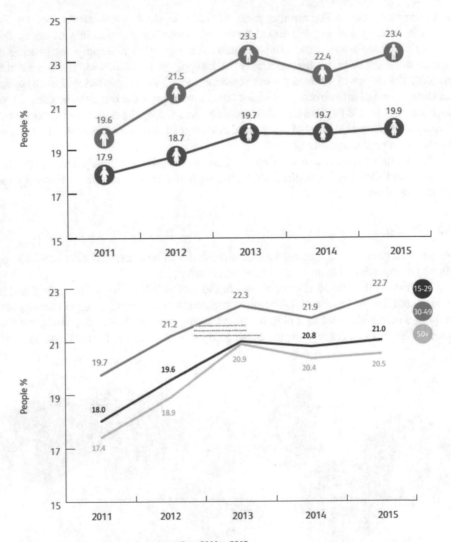

Data represents one-year scores for each year from 2011 to 2015.

Each one-year score is derived from the average of all the countries.

Data relate to participation in volunteering time during one month prior to interview.

Fig. 2. Participation in volunteering, by gender and age, World, 2011–2015. Source: CAF (2016)

The same source also indicates an increase in the world participation rate to volunteering, except a drop for female volunteers in 2014. Until now this drop has not found an explanation (Fig. 3).

The young people are the ones who volunteer more, and the old the ones who volunteer less.

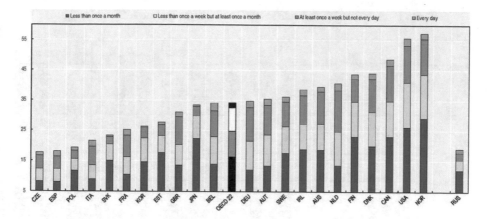

Fig. 3. Participation rates in formal volunteering, OECD countries, 2012

A table for OECD countries is included here, presenting data from the Use of Time Survey made by the same organization in 2012. It can be seen that the rates vary a lot among industrialized countries, with the former communist and Southern countries at the bottom, and the Scandinavian and Anglo-Saxon ones (except Great Britain) at the top.

4 Europe

4.1 Eurobarometer and Other European Data

Few months ago Lester Salamon in a seminar in Warsaw said "There are plentiful data on Third Sector in European Statistics, but Volunteering out of bounds or valued at zero" (Salamon 2016). However he was able to present the following table. These data can be compared to the others, remembering that also the informal volunteers are included. Broadly the same ranking appears as that resulting from the OECD data, with Scandinavian countries at the top, followed by Northern, Southern and Eastern countries, but with a great variability inside each group (Table 1).

For a comparison within the European countries, another option is still the 2011 Eurobarometer survey, carried on during the EU "Year of volunteering". More recently the survey has been dedicated to youth activities in 2014.

It is to be remembered that, even if the Eurobarometer data result from opinion polls carried out at the European level, many results are convergent with other types of survey: for instance, among the Italian respondents the percentage of volunteers is 26%, just above the European average, not much higher than the one resulting from the Italian Multipurpose Survey (2013) and Census (2011) (Istat 2013, and 2014a and b). The positive trend of participation in cultural associations (common to many countries) is another result shared by the Eurobarometer and the national sources. However, comparing the percentage of volunteers among the respondents in 2011 (24%) with the one in 2006 (34%) (European Commission 2007), a strong negative trend emerges, which is not confirmed by the national sources.

Table 1. Estimated Work-Force in the Third Sector, European Countries, 2014.

Country	NPIs					Coops & mutuals* (FTE)	Direct volunteers (FTE)	Total TS (FTE)	Total 2014 employment	NPIs as %of total employment			Other as % of total employment		
	Paid (FTE)	of which coops	Volunteers (FTE)	of which coops	Total (FTE)					Paid	Volunteers	Total	Coops*	Direct vol	TS
Northern Europe															
Austria	160.545	5.580	255.961	8.132	394.506	31.000	200.141	625.646	4.112.700	3,9%	5,7%	9,6%	0,6%	4,9%	15,2%
Belgium	451.951	1.219	129.639	350	581.590	6.774	185.867	774.230	4.543.500	9,9%	2,9%	12,8%	0,1%	4,1%	17,0%
France	1.535.368	34.632	680.000	17.150	2.215.368	214.622	139.572	2.569.562	25.802.200	6,0%	2,6%	8,6%	0,8%	0,5%	10,0%
Germany	2.397.618	74.723	1.307.580	40.752	3.705.199	415.129	1.479.321	5.599.649	39.871.300	6,0%	3,3%	9,3%	1,0%	3,7%	14,0%
Ireland	181.685	3.900	48.794	1.046	230.679	21.664	76.636	328.979	1.913.900	9,5%	2,5%	12,1%	1,1%	4,0%	17,2%
Luxembourg	22.657	174	18.070	139	40.727	967	9.664	51.357	245.600	9,2%	7,4%	16,6%	0,4%	3,9%	20,9%
Netherlands	854.045	16.565	488.632	9.433	1.346.677	92.027	281.756	1.720.459	8.236.100	10,4%	5,9%	16,4%	1,1%	3,4%	20,9%
United Kingdom	1.671.866	21.240	1.422.360	18.070	3.094.226	118.000	1.062.630	4.274.856	30.641.800	5,5%	4,6%	10,1%	0,4%	3,5%	14,0%
Southern Europe															
Cyprus	23.341	456	18.615	364	41.956	2.534	15.238	59.727	362.700	6,4%	5,1%	11,6%	0,7%	4,2%	16,5%
Greece	244.370	1.348	194.891	1.075	439.261	7.492	180.156	626.909	3.536.200	6,9%	5,5%	12,4%	0,2%	5,1%	17,7%
Italy	938.135	101.554	597.390	64.668	1.535.525	564.191	812.383	2.912.099	22.278.900	4,2%	2,7%	6,9%	2,5%	3,6%	13,1%
Malta	10.527	23	8.396	18	18.923	125	7.327	26.375	181.400	5,8%	4,6%	10,4%	0,1%	4,0%	14,5%
Portugal	175.092	4.625	55.680	1.471	230.772	25.696	166.405	422.873	4.499.500	3,9%	1,2%	5,1%	0,6%	3,7%	9,4%
Spain	722.223	58.176	348.830	28.099	1.071.053	323.199	914.613	2.308.864	17.344.200	4,2%	2,0%	6,2%	1,9%	5,3%	13,3%
Scandinavia															
Denmark	137.358	6.368	114.187	5.294	251.545	35.379	123.637	410.560	2.714.100	5,1%	4,2%	9,3%	1,3%	4,6%	15,1%
Finland	73.018	8.469	85.165	9.678	158.182	47.050	145.159	350.391	2.447.200	3,0%	3,5%	6,5%	1,9%	5,9%	14,3%
Norway	84.054		143.637		227.691		85.647	313.339	2.626.600	3,2%	5,5%	8,7%	0,0%	3,3%	11,9%
Sweden	194.128	15.913	355.741	29.162	549.869	88.408	155.830	794.107	4.772.100	4,1%	7,5%	11,5%	1,9%	3,3%	16,6%
Central and Eastern Europe															
Bulgaria	18.960	3.717	7.909	1.550	26.869	20.650	183.595	231.114	2.981.400	0,6%	0,3%	0,9%	0,7%	6,2%	7,8%
Croatia	70.512		29.412	-	99.924	-	107.770	207.694	1.565.700	4,5%	1,9%	6,4%	0,0%	6,9%	13,3%
Czech Republic	101.901	5.236	26.413	1.357	128.314	29.089	269.964	427.367	4.974.300	2,0%	0,5%	2,6%	0,6%	5,4%	8,6%
Estonia	20.652	887	8.614	370	29.266	4.925	38.951	73.122	624.800	3,3%	1,4%	4,7%	0,8%	6,2%	11,7%
Hungary	89.620	7.711	16.992	1.462	106.612	42.841	255.532	404.985	4.100.800	2,2%	0,4%	2,6%	1,0%	6,2%	9,9%
Latvia	34.170	40	14.253	17	48.422	220	50.334	98.977	884.600	3,9%	1,6%	5,5%	0,0%	5,7%	11,2%
Lithuania	7.415	807	3.093	337	10.508	4.486	74.919	89.913	1.319.000	0,6%	0,2%	0,8%	0,3%	5,7%	6,8%
Poland	212.900	22.842	172.700	18.529	385.600	126.900	1.243.900	1.756.400	16.033.200	1,3%	1,1%	2,4%	0,8%	7,8%	11,0%
Romania	28.107	3.094	36.576	4.026	64.683	17.167	515.288	597.157	8.613.700	0,3%	0,4%	0,8%	0,2%	6,0%	6,9%
Slovakia	19.943	2.348	7.637	899	27.580	13.045	146.491	187.116	2.363.100	0,8%	0,3%	1,2%	0,6%	6,2%	7,9%
Slovenia	42.972	309	17.924	129	60.896	1.714	33.813	96.424	916.800	4,7%	2,0%	6,6%	0,2%	3,7%	10,5%
Total EU+Norway	10.529.332	405.955	6.593.090	363.734	17.122.423	2.255.308	8.963.520	28.340.351	230.507.400	4,8%	3,0%	7,8%	1,0%	4,1%	12,9%

* Excluding coops and mutuals that are NPIs

Source: Salamon and Sokolowski (2016)

The dissimilarity between the areas of volunteer participation and the ones in which the volunteers themselves believe volunteerism is important, is perhaps the most interesting result of the 2011 survey (already commented in Schenkel et al. 2014). Solidarity, Welfare, health and the environment are in the first place in the ranking of areas which are thought to be the most important, but volunteers participate actively mainly in sport and cultural associations, and in NGO for humanitarian aid and neighbourhood associations. This points at a discrepancy between the demand for volunteers' services and the supply of volunteers, which can be read either as the result of the technical characteristics of the organizations involved, or a forerunner of a future trend of the demand for services, which has not yet found a sufficient supply to meet it.

The survey on Young People of 2014 (European Commission 2015) is the most recent official data source at the European level. Although young people do not constitute the major component of the volunteers in European countries, in a long-term perspective their preferences play a predominant role.

Again sport is one of the most popular area in youth engagement in voluntary activities, on average, and in almost all countries, above all in England, which hosted the Olympic Games in 2012 (Fig. 4).

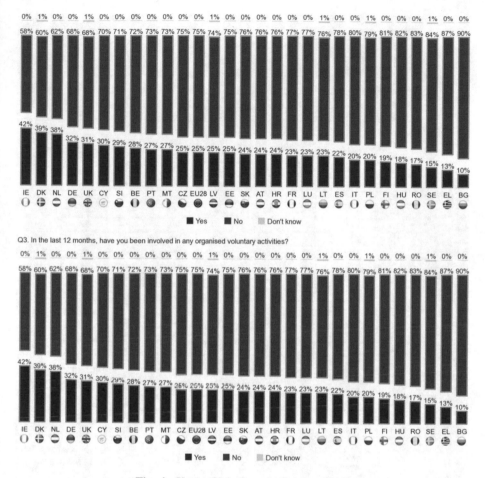

Fig. 4. Young Volunteers in Europe, 2015.

4.2 National Surveys

Some data are added concerning the European countries where updated data on volunteers have been published, and the comparisons with previous years are possible: Italy, Spain, England and France (Fig. 5).

4.2.1 Italy: The Census of Industry, Services and Non-profit Institutions

An overall increasing trend of voluntary labour supply is shown by the Italian Census data (Haddock 2014)[3].

---

[3] It is interesting to note that, according to another Italian source, i.e. the Time use surveys, the increase in the number of the volunteers have taken place since 2010, and so it is natural to put it in relation with the current crisis.

		Charity, humanitarian and development aid	Education, training or sport	Culture or art	Human rights	Religion	Animal welfare	Climate change or the environment	Politics	Other
	EU28	44%	40%	15%	13%	12%	9%	9%	8%	12%
	BE	43%	46%	23%	22%	7%	15%	10%	5%	17%
	BG	55%	25%	23%	5%	26%	10%	15%	8%	9%
	CZ	38%	36%	24%	13%	6%	17%	11%	7%	13%
	DK	33%	48%	14%	14%	8%	4%	6%	14%	16%
	DE	34%	44%	16%	8%	21%	3%	8%	9%	10%
	EE	29%	23%	18%	4%	1%	5%	21%	3%	15%
	IE	49%	41%	14%	20%	7%	9%	5%	7%	13%
	EL	48%	40%	14%	19%	1%	10%	22%	7%	0%
	ES	41%	46%	16%	19%	6%	15%	15%	14%	8%
	FR	43%	49%	18%	9%	8%	14%	10%	2%	12%
	HR	70%	24%	7%	15%	4%	13%	5%	9%	2%
	IT	49%	19%	7%	17%	11%	5%	7%	8%	24%
	CY	71%	18%	13%	11%	7%	15%	14%	5%	4%
	LV	45%	38%	34%	6%	3%	21%	28%	7%	10%
	LT	30%	26%	21%	9%	4%	11%	18%	4%	6%
	LU	42%	41%	22%	20%	2%	19%	22%	18%	13%
	HU	32%	29%	13%	8%	8%	16%	14%	8%	21%
	MT	47%	27%	16%	9%	18%	6%	7%	2%	18%
	NL	34%	45%	15%	6%	15%	7%	7%	5%	15%
	AT	47%	32%	18%	12%	8%	7%	6%	10%	13%
	PL	70%	23%	14%	14%	5%	14%	3%	2%	5%
	PT	68%	33%	20%	18%	7%	14%	5%	5%	3%
	RO	55%	23%	11%	3%	13%	9%	6%	7%	2%
	SI	45%	37%	29%	10%	9%	10%	9%	3%	4%
	SK	34%	27%	29%	5%	8%	10%	17%	2%	11%
	FI	37%	31%	19%	16%	11%	11%	6%	6%	20%
	SE	34%	32%	17%	28%	6%	12%	10%	25%	14%
	UK	47%	50%	12%	16%	13%	8%	8%	12%	14%

Highest percentage per country	Lowest percentage per country
Highest percentage per item	Lowest percentage per item

Fig. 5. Young Volunteers in Europe: Fields of activity (2014). Source: European Commission (2015)

Also the demand of voluntary labour seems to have grown in numerical and qualitative importance: the institutions that use volunteers show an increase of 10.6% over 2001, whereas the number of volunteers has increased by 43.5%. This dynamic corresponds both to the creation of new entities (associations, cooperatives, etc.) and the growth of the already existing ones[4]. Voluntary labour is a key input for the non-profit enterprises, given that over the 80% of non-profit institutions has made recourse to it, and represents the largest share (83.3%) of the human resources in the non-profit

---

[4] The Census data on the number of organizations indicate a 28% increase between 2001 and 2011. The results of the new Census (2017) will be shortly published.

sector. Further investigation at the micro level is needed to determine if this growth is due to the change in the nature of the organizations in the survey, or only in the nature of the labour demanded, while the organizations maintain substantially their original characteristics (legal form, type of business, size etc.). In favour of the first hypothesis, two facts seem to indicate a structural change in the industry: the growth of for-profit enterprises providing personal services, and the trend of the non-profit ones towards the concentration in big nationwide organizations.

The analysis of the economic activity shows that the non-profit enterprises are the majority in the fields of Social Services (with 361 non-profit institutions every 100 private companies) and Cultural, Sports, Entertainment and Fun Activities (with 239 institutions not profit every 100 private companies).

The following table shows the average number of volunteers for the organization according to age, gender, level of education. It appears that "men employed in the prime of life" make up the largest group, followed by women of the same age and condition. The prevailing level of education is high school graduation.

The use of foreign citizens and young people engaged in the civil service is increasingly common, given that on average the volunteers in these categories are almost 2 per organization.

In sum, the Census data contradict the widespread representations of volunteering as populated mainly by young and old people.

The detail of the Census data confirms the strong variability of regional participation. E.g.: if 80% of organizations employ volunteers on a national basis, this percentage varies from 67.7 (Lazio and Campania) to 92.8 (Trentino AA). Normalizing for the resident population of each region, the advantage of Trentino AA is more evident, across every type of human resource, except for the religious. The southern regions show a lower number of volunteers per resident population. Generally a strong volunteer sector is found in those regions where a long experience of non-profit organizations is established. Therefore the evolution and development of the non-profit organizations on a regional basis is characterized by an increasing role of the voluntary sector. The volunteer would be the expression of a mature non-profit sector that has managed to create a system, and not quite a task in which to park residual potential cooperators or social entrepreneurs. Volunteering is no longer the "apprenticeship" that has to carve out a role in the non-profit organization, but is to all intents and purposes a form of differently remunerated work, with the same dignity as the "traditional" paid work.

As for the sectors in which non-profit institutions and volunteers act, the Culture, sports and recreation are in the first place, with over 195 thousand institutions, representing 65% of the national total. The Social Welfare sector (which also includes civil protection activities) follows, with 25 thousand institutions (equal to 8.3% of the total). The field of labour relations and representation of interests, with 16 thousand institutions, constitutes the 5.4% of the total, close to that recorded by Education and Research, representing the 5.2% (15 thousand institutions). The other sectors include: Religion (2.3%), Philanthropy and promotion of volunteering (1.6%), cooperation and international solidarity (1.2%) and other activities (0.5%) (Figs. 6 and 7).

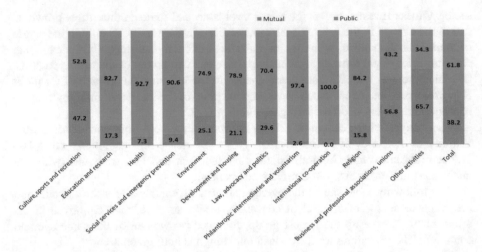

Fig. 6. Public benefit nonprofit institutions by sector of activity, Italy, (Percentage values), 2011. Source: Istat (2014)

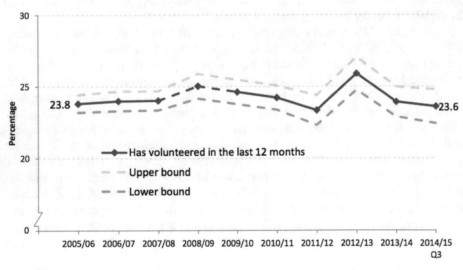

Notes
(1)   Confidence intervals range between +/-0.6 and +/-1.2 from 2005/06 onwards.
(2)   No data are available for 2009/10
(3)   The upper and lower bounds show the 95% confidence interval.

Fig. 7. Percentage of adults who had volunteered in the last 12 months, 2005/06 to 2014/15 quarter 3. Source: Department of Culture Media and Sport, 2015

4.2.2   Great Britain

A different picture emerges from the results about England, since until 2011 volunteerism has been shrinking, even if in 2012 Olympic Games gave a new incentive to volunteers participation, whose lasting effect was continued also in 2013. However in 2013 the numbers of volunteers is lower than in 2012 (but higher than in 2011), even if many interviewees have said that they were still volunteering in different forms, following their first experience in 2012 (Department of Culture Media and Sport 2015). It is possible to argue that a different composition of volunteerism by activities followed, due to a change in volunteers demand, coupled with a change in policy by the government (the "Big Society" project).

The same results have been found by the Citizenship Survey, which was interrupted in 2011, and then renewed under another name: Community Life Survey (Fig. 8).

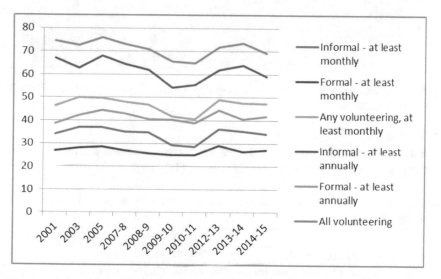

Fig. 8. Participation rates in Volunteering, England and Wales, 2001–2015. Source: Mohan (2016)

Other authors (Clark and Heath 2014; Lim and Lawrence 2014) agree with a overall decline of voluntarism in England.

It is also interesting to note that the decline involves above all the older age classes (Fig. 9).

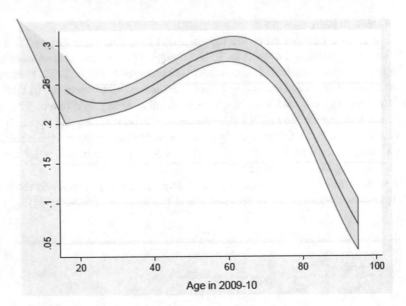

Age in 2009-10

Fig. 9. Predicted probability of formal volunteering by age, England and Wales, 2009–2010. Source: Mohan (2016)

4.2.3   France
In France a picture of increasing volunteers participation seems to emerge. However the hours of work per volunteers are decreasing (Table 2).

Table 2. Regularity and intensity of the voluntary commitment between 2010 and 2013 (percentage values)

	2010	2013	Δ2010–2013
1- Very occasionally	4,6%	6,7%	2,1%
2- At least a few hours each month	5,6%	7,5%	1,9%
3 - At least a few hours each week	8,4%	6,2%	−2,2%
4 - At least one day each week	4,1%	4,3%	0,2%
Regular weekly commitment (3 + 4)	12,5%	10,5%	−2,0%
All volunteers together	22,7%	24,7%	2,0%

Source: R&S (2014)

The same result are confirmed by a new Survey by a private non-profit organization, Ifop. The overall rate of participation has increased from 36% to 40% in 3 years from 2010 to 2013, and has not varied much in 2016, even if there was a decrease in the female rate[5] (Fig. 10).

---

[5] An extensive Census survey has been conducted in 2015, but only partial results have been published.

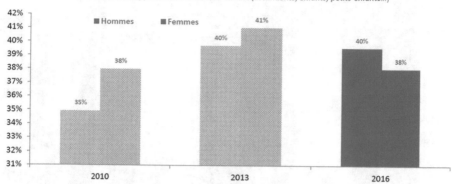

Fig. 10. Participation rate in volunteering by gender, France: 2010–2016

A further interesting result, which confirms the one of the Italian Census, is that there was an increase in the rate of participation of the population in the lower and central age classes (Fig. 11).

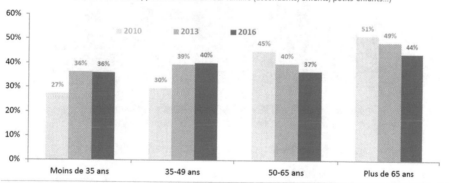

Fig. 11. Participation rate in volunteering by age, France: 2010–2016

4.2.4 Spain

In Spain between 2003 and 2011 active participation in volunteers associations seem to decrease together with the decrease in working activities (Figs. 12 and 13).

The same trend is confirmed by the results of some surveys, made by some independent consortium of Non-profit enterprises (PVE 2015).

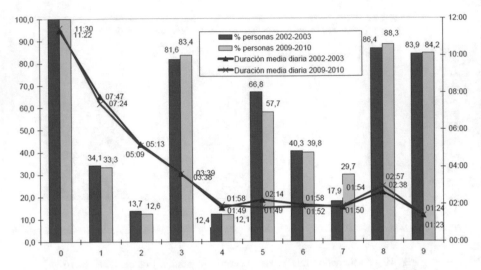

Fig. 12. Spain. Percentage of people performing the activity in the course of the day and average daily activity devoted by such persons, 2002–2003 and 2009–2010 ("Duracion media diaria" states for: Daily average length). Source: Instituto National de Estadistica, Encuesta del Empleo del Tiempo

Tabla 1. Resumen de datos sobre volumen del voluntariado en España. 2011-2015. Población mayor de 18 años.

	Barómetro CIS. 2011	Barómetro CIS. 2013	PVE 2014	PVE 2015
Población mayor de 18 años	2,83 %	2,68 %	9,8%	7,9 %
Estimación	1,09 millones	1,03 millones	3,74 millones	3,1 millones

Fuentes: Barómetros CIS: Estudio 2864, 2011 y Estudio 3005, 2013; La población española y su implicación con las ONG. PVE. 2014 y 2015 INE: Censos de población.

Fig. 13. Spain: Participation rate in volunteering 2011–2015

4.2.5   Americas

No overall data exist for the whole New World, similar to the European ones, even if the World Walue Survey produces maps and comparative Time Series of active participation in a variety of voluntary organizations for many North and South American countries. Instead some national results are available, which will be reported in the following sections.

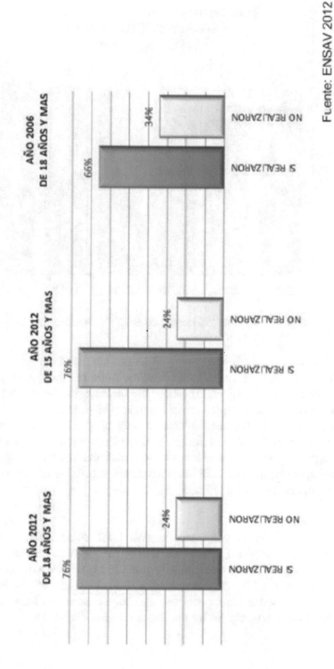

Gráfico 2: Acciones voluntarias en México

Fuente: ENSAV 2012

Fig. 14. Mexico: Number of volunteers "voluntary actions", 2006 and 2012. Source: Cemefi (ENSAV) (2012)

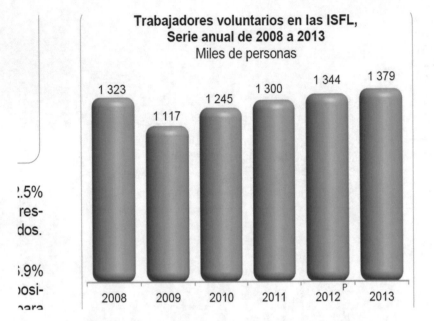

Fig. 15. Mexico: Volunteers, 2008–2013. Source: Inegi (2015)

4.2.6 Mexico

Two different sources are available for Mexico. Both, the independent (Cemefi 2012) and the official one (Inegi 2015), show an increase in the overall participation[6]. However the official data, which cover a longer period, show also a decline since the pre-crisis year, which has been overcome only in 2012. It has to be noticed that Mexico, together with US and Brazil, is one of the countries where the weight of religious organizations is higher. It has to be further investigated if the recent recovery of the participation rate has to be ascribed to the expansion of religious movements, like the Evangelic one (Figs. 14 and 15).

4.2.7 Canada

The estimated number of volunteers increased 7% since 2004, from 11.8 million Canadians to 12.7 million. However the rate of participation declined.

Canadians aged 35 to 44 were the only group to experience a decrease in the overall volunteer rate since 2010, from a rate of 54% in 2010 to 48% in 2013. This trend contrasts with the one in other countries, like Italy, France and Mexico, which, unlike Canada, have been severely hit by the Big Recession (Table 3).

---

[6] The last ENSAV survey (2016) suggest an increasing number of volunteers, i.e. 83% (Butcher Garcia-Colin 2016).

Table 3. Canada: Volunteer rate and hours, and by age, 2004–2013.

Table 1
Volunteer rate and volunteer hours, population aged 15 and over

	2013	2010	2007	2004
Volunteer rate				
Total population (thousands)	29,188	28,206†	27,000†	26,021†
Number of volunteers (thousands)	12,716	13,249†	12,444	11,773†
Volunteer rate (percentage)	44	47†	46†	45†
Volunteer hours				
Total annual volunteer hours (millions)	1,957	2,063	2,062	1,978
Average annual volunteer hours (hours)	154	156	166	168†

† significantly different from 2013 at p < 0.05
Sources: Statistics Canada, General Social Survey on Giving, Volunteering and Participating, 2013, and the Canada Survey of Giving, Volunteering and Participating, 2004, 2007 and 2010.

The total number of volunteers was lower in 2013 than in 2010 (12.7 and 13.2 million, respectively). This translates into a 4.0% decline in the total number of Canadian volunteers, despite the fact that the population aged 15 years and older increased by about one million during the same period (+3.5 %).

4

Chart 2
Distribution of volunteers, by age group

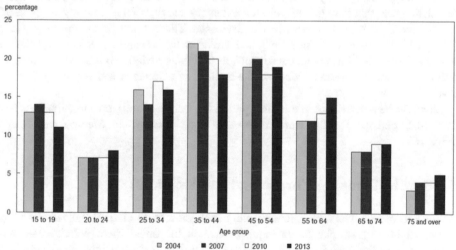

Sources: Statistics Canada, General Social Survey on Giving, Volunteering and Participating, 2013, and the Canada Survey of Giving, Volunteering and Participating, 2004, 2007 and 2010.

Source: Turcotte 2016.

4.2.8   United States

According to the calculations made by a private consortium of charities on official data, the rate of participation in the United States has been affected by a sudden drop in 2005, which has not recovered, until in 2013 a light but constant further decline followed. This negative trend is similar to the one experienced by other Anglo-Saxon countries (Canada, Great Britain), except that the decline is common across all genders, age classes, employment status, etc. The Third Sector organizations instead seem to have been robust against the decline of volunteerism (Salamon 2002) (Table 4).

4.2.9   Brazil

In Brazil the best available information about volunteers come from opinion polls, made by various private companies. The results are considerably constant across time. The participation rate fluctuates from 2002 until 2014 between 12 and 11% (Source: Datafolha 2014).

Another private source (Ibope) confirms the same participation rates. Another interesting result is the type of organizations where the volunteers carry on their activity. The weight of religious associations is largely predominant (Fig. 16).

4.2.10   Argentina

Also in Argentina the best available informations about volunteering come from a private (for profit) Agency: Gallup/TNS[7]. According to Gallup, in 2002 the 32% of the population told that they had exercised a voluntary activity in the current year. In 2004 – when the GNP started to rise, this rate decreased. The last poll (2014) indicates a rate (13%) which is the lowest since 1997, and similar to the Brazilian one. The explanation is that in 2002–2003 a "social cataclysm", as defined by Mario Roitter, was going on. With the return to a normal situation, the number of volunteers was reduced (Pizzarro 2015).

It has to be noticed that in the same period the weight of religious organizations has increased, perhaps for the same reasons of their increase in Mexico and Brazil (Fig. 17).

5   How to Compare Europe and the Americas?

At the end of this overview, probably the better criterion to classify the various areas examined is very simple: to distinguish between the ones where volunteerism has increased, and the ones in which it has not, diminishing or remaining constant. This hardly coincides with any geographical or cultural divide. In the first group we find, beside the world and Europe as a whole, Mexico, France and Italy, in the second Spain, UK, USA, Canada, Argentina, Brasil. It can be argued that the different intensity and length of the crisis is a prime determinant of these diverging trends: the last three countries were not particularly hit by the economic downturn in the period 2008–2014,

---

[7] A recent survey was made by the Government in the Buenos Aires area (Bocchicchio and Roiter 2013).

Table 4. USA: Number, hours, age and other characteristics of Volunteers, 2005–11 and 2011–15.

Table 6. Number, Hours, and Dollar Value of Volunteers, 2005–11

	2005	2006	2007	2008	2009	2010	2011
Per year							
Percentage of population volunteering	28.8	26.7	26.2	26.4	26.8	26.3	26.8
Number of volunteers	65.4 million	61.2 million	60.8 million	61.8 million	63.4 million	62.8 million	64.3 million
Total hours volunteered	13.5 billion	12.9 billion	15.5 billion	14.4 billion	14.9 billion	14.9 billion	15.2 billion
Average hours per volunteer	204	207	251	229	232	234	233
Median hours per volunteer	50	52	52	52	50	52	51
Per average day							
Percentage of population volunteering	7.1	6.5	7.0	6.8	7.1	6.8	6.0
Number of volunteers	16.5 million	15.2 million	16.6 million	16.2 million	17.1 million	16.6 million	14.6 million
Hours per day per volunteer	2.25	2.31	2.56	2.43	2.39	2.46	2.84
Value of volunteers							
Population age 15 and over	230.4 million	233.1 million	236.1 million	238.7 million	239.9 million	241.9 million	243.7 million
Full-time equivalent employment	7.9 million	7.6 million	9.1 million	8.4 million	8.8 million	8.8 million	8.9 million
Assigned hourly wages for volunteers	$16.13	$16.76	$17.43	$18.08	$18.63	$19.07	$19.54
Assigned value of volunteer time	$217.9 billion	$215.6 billion	$270.2 billion	$270.2 billion	$277.7 billion	$283.8 billion	$296.2 billion

Sources: Authors' calculations based on per year figures from U.S. Department of Labor, Bureau of Labor Statistics, Current Population Survey, Volunteer Supplement (2005-2011); per average day figures from U.S. Department of Labor, Bureau of Labor Statistics, American Time Use Survey (2005-2011); hourly wages from U.S. Department of Labor, Bureau of Labor Statistics, Current Employment Statistics (2011).
— = data not available

TABLE A

Number, Hours, and Dollar Value of Volunteers, 2008–14

	2008	2009	2010	2011	2012	2013	2014
Per year							
Percentage of population volunteering	26.4	26.8	26.3	26.8	26.5	25.4	25.3
Number of volunteers	61.8 million	63.4 million	62.8 million	64.3 million	64.5 million	62.6 million	62.8 million
Hours volunteered	8.0 billion	8.1 billion	8.1 billion	8.5 billion	8.5 billion	8.3 billion	8.7 billion
Average hours per volunteer	130	128	129	132	132	133	139
Median hours per volunteer	52	52	52	51	50	50	50
Per average day							
Percentage of population volunteering	6.8	7.1	6.8	6.0	5.8	6.1	6.4
Number of volunteers	16.2 million	17.1 million	16.6 million	14.6 million	14.3 million	15.1 million	16.0 million
Hours per day per volunteering	2.43	2.39	2.46	2.84	2.48	2.57	2.41
Value of volunteers							
Population age 16 and over	234.4 million	236.3 million	238.3 million	240.0 million	243.8 million	246.2 million	248.4 million
Full-time equivalent employees	4.7 million	4.8 million	4.8 million	5.0 million	5.0 million	4.9 million	5.1 million
Assigned hourly wages for volunteers	$18.08	$18.63	$19.07	$19.47	$19.75	$20.16	$20.59
Assigned value of volunteer time	$144.7 billion	$150.7 billion	$154.1 billion	$164.8 billion	$168.3 billion	$167.2 billion	$179.2 billion

Sources: Authors' calculations based on data from US Department of Labor, Bureau of Labor Statistics, Current Population Survey, Volunteer Supplement (2008–14); US Department of Labor, Bureau of Labor Statistics, American Time Use Survey (2008–14), and US Department of Labor, Bureau of Labor Statistics, Current Employment Statistics (2014).

(continued)

Table 4. (*continued*)

Table A. Volunteers by selected characteristics, September 2011 through September 2015

(Numbers in thousands)

Characteristics	September 2011		September 2012		September 2013		September 2014		September 2015	
	Number	Percent of population	Number	Percent of population	Number	Percent of population	Number	Percent of population	Number	Percent of population
Sex										
Total, both sexes	64,252	26.8	64,513	26.5	62,615	25.4	62,757	25.3	62,623	24.9
Men	27,354	23.5	27,238	23.2	26,404	22.2	26,375	22.0	26,498	21.8
Women	36,898	29.9	37,274	29.5	36,211	28.4	36,381	28.3	36,126	27.8
Age										
Total, 16 years and over	64,252	26.8	64,513	26.5	62,615	25.4	62,757	25.3	62,623	24.9
16 to 24 years	8,578	22.5	8,776	22.6	8,466	21.8	8,469	21.9	8,415	21.8
25 to 34 years	9,691	23.3	9,513	23.2	9,118	21.9	9,291	22.0	9,548	22.3
35 to 44 years	12,566	31.8	12,527	31.6	12,098	30.6	11,783	29.8	11,490	28.9
45 to 54 years	13,420	30.6	12,777	29.3	12,184	28.2	12,204	28.5	11,933	28.0
55 to 64 years	10,449	28.1	10,619	27.6	10,191	26.0	10,331	25.9	10,213	25.1
65 years and over	9,547	24.0	10,301	24.4	10,558	24.1	10,679	23.6	11,024	23.5
Race and Hispanic or Latino ethnicity										
White	54,432	28.2	53,778	27.8	52,685	27.1	52,201	26.7	51,986	26.4
Black or African American	5,934	20.3	6,316	21.1	5,637	18.5	6,094	19.7	6,086	19.3
Asian	2,304	20.0	2,524	19.6	2,525	19.0	2,513	18.2	2,596	17.9
Hispanic or Latino ethnicity	5,151	14.9	5,635	15.2	5,838	15.5	5,982	15.5	6,165	15.5
Educational attainment [1]										
Less than a high school diploma	2,461	9.8	2,177	8.8	2,204	9.0	2,100	8.8	1,900	8.1
High school graduates, no college [2]	11,049	18.2	10,527	17.3	10,138	16.7	10,075	16.4	9,576	15.6
Some college or associate degree	15,946	29.5	15,832	28.7	15,562	27.7	15,494	27.3	15,102	26.5
Bachelor's degree and higher [3]	26,218	42.4	27,202	42.2	26,244	39.8	26,619	39.4	27,629	38.8
Employment status										
Civilian labor force	45,249	29.1	44,974	28.7	43,162	27.5	42,780	27.3	42,563	27.0
Employed	41,881	29.6	42,083	29.1	40,401	27.7	40,497	27.5	40,701	27.2
Full time [4]	32,517	28.7	32,568	28.1	31,524	26.8	31,557	26.5	32,085	26.3
Part time [5]	9,363	33.3	9,515	33.4	8,877	31.7	8,940	31.7	8,616	31.1
Unemployed	3,368	23.8	2,891	23.8	2,761	24.1	2,283	24.0	1,861	23.3
Not in the labor force	19,003	22.5	19,539	22.4	19,452	21.9	19,977	21.8	20,060	21.4

[1] Data refer to persons 25 years and over.
[2] Includes persons with a high school diploma or equivalent.
[3] Includes persons with bachelor's, master's, professional, and doctoral degrees.
[4] Usually work 35 hours or more a week at all jobs.
[5] Usually work less than 35 hours a week at all jobs.
NOTE: Estimates for the above race groups (White, Black or African American, and Asian) do not sum to totals because data are not presented for all races. Persons whose ethnicity is identified as Hispanic or Latino may be of any race. Updated population controls are introduced annually with the release of January data. Data on volunteers relate to persons who performed unpaid volunteer activities for an organization at any point in the year ending in September. See the Technical Note for futher information.

Source: McKever (2015)

and USA could recover quickly from it. The reason why in UK and Spain volunteers have diminished can be that these countries were not only stricken by the crisis, but also by a particularly severe cut in public expenditure towards non-profit and volunteer associations. In order to appreciate the impact of changing economic conditions, a particular interesting case is the one of Argentina. In this country volunteers increased enormously during the crisis (2001–2002), and diminished again during the "resurrection". Another interesting case could be, when the data will be available, the one of Brasil, a country which is at the moment affected by a severe recession and a political change.

The qualitative features of the stock of volunteers, as already noted, has changed in the sense that all age groups and both genders are more or less equally represented, in every country. Another interesting constant across the various countries is that the majority of volunteers are employed.

This image of the volunteers is quite different from what we have been used to consider so far, that is, an aggregate where most people have a lower attachment to the

Fig. 16. Brazil: Fields of voluntary activity, 2015. Source: Ibope Inteligencia, 2012

Fig. 17. Argentina: Number of volunteers, 1997–2012, by age and other characteristics (2012). Source: TNS Argentina (2013)

labour market, and therefore a low value of leisure. With reference to the categories listed in Sect. 2, the volunteers should be mainly of type B *("they want to pursue their own ideals in a challenging and rewarding work environment")*, and, hopefully, of type D *("volunteers by choice, conscious and generous")*.

A different composition of the fields where volunteers exercise their activities emerges mainly for the outstanding presence of religious organizations in three countries: USA, Mexico and Brasil[8]. In the other countries, instead sport and art/culture are the most popular fields of volunteer activity, together with social services. The same is true also for many others countries whose statistics have not been examined in detail, since they do not permit comparisons in time, as Austria, Poland, Croatia, Switzerland, Norway, Sweden, Portugal.

Connected with the increasing popularity of volunteerism in Sport, Culture and Entertainment in the last years is the diffusion in many countries of "episodic volunteerism" (Güntert et al. 2015). Sport and artistic/cultural events, as Olimpic Games, Expo and so on, give a lot of possibility to volunteer for a limited engagement in time. In the Country Reports issued by the Center for Civil Society in 2016, among the most quoted barriers to the expansion of the third sector in Europe is not only the need to find more volunteers on a permanent basis (Austria, Poland, Croatia, France, Holland, UK), but also flexible ways of employing them (Germany, Holland) (Center for Civil Society 2016).

6 Conclusions and Next Steps of the Research

In the previous paper on the transformations of voluntarism in Europe (Schenkel et al. 2014) three elements were highlighted: the extension to functions so far carried out only by the state, and the need for structured organizations able to enhance the quality of the contribution of volunteer work; the increase of the operational capabilities of these organizations, which implies an increase in the demand for volunteers; the increasing professional skills (required and offered) in the supply of goods and services by Non Profit enterprises, which have to face competition both from other Non Profit and For Profit companies.

Beside that, another, complementary side was identified: the "playful" one. A community volunteerism is growing, intended not only to respond to immediate need and difficult situations exacerbated by the crisis, but also to produce a relational good used for internal consumption: fun. The Salamon et al. (2013)' distinction between "service" and "expressive" type of activity of volunteers goes in the same direction.

In this survey, where the geographical scope was extended, again some differences in countries trends in volunteers participation appear, but the overall picture remains the same. What can be added is that in the North American countries, where the role of the State in providing Services is traditionally limited, it is likely that volunteers have already reached an upper threshold. The reverse is true for Latin American countries,

---

[8] According to the World Value Survey these are also the countries where a particular high percentage of persons declare to be religious.

where the feeling that volunteers are "few" seems common (Piacentini 2016; Itaù Social 2015; La Naciòn 2015).

These hypothesis wait for some and more comparable data, hoping that the improved economic conditions in the Western world can offer a new environment to test them.

References

Bekkers, R., Bowman, W.: The relationship between confidence in charitable organizations and volunteering revisited. Nonprofit Volunt. Sect. Q. **38**, 884–889 (2008)

Bocchicchio, F., Roiter, M.: Trabajo voluntario_en la Ciudad_de Buenos Aires. Año 2010. Ciudad de Buenos Aires (2013)

Butcher Garcia-Colin, J., Verduzco Igartua, G.: Accion voluntaria y Voluntariado en Mexico. Fundacion Telefonica Mexico (2016)

Bollè, P.: Labour statistics: the boundaries and diversity of work. Int. Labour Rev. **148**, 183–193 (2009)

Borzaga, C., Gui, B., Schenkel, M.: Disoccupazione e bisogni insoddisfatti: il ruolo delle organizzazioni non-profit. Quaderni di Economia del lavoro **50**(1995), 100–129 (1995)

CAF (Charity Aid Foundation). CAF World Giving Index 2016. The world's leading study of generosity, October 2016

Casey, J.: Comparing nonprofit sectors around the world. What do we know and how do we know it? J. Nonprofit Educ. Leadersh. **6**(3), 187–223 (2016). http://dx.doi.org/10.18666/JNEL-2016-V6-I3-7583

Center for Civil Society: External and Internal barriers to Third Sector Development. Third Sector Impact project, various numbers (2016a)

Center for Civil Society: Third Sector Impact project Policy Brief, various numbers (2016b)

Department for Culture Media & Sport: Taking Part. 2014/15 Quarter 3. Statistical Release, 15 March 2015

Datafolha: Opinião_do brasileirosobreVoluntariado, S. Paulo, Fundaćao Itaù Social (2014)

European Commission: European Youth, Flash Eurobarometer 408 (2015)

European Commission: European Social Reality. Report, Special Eurobarometer (2007)

European Parliament: Special Eurobarometer of the European Parliament 75.2—Voluntary Work, Directorate-General for Communication, Directorate for relations with citizens. 'Monitoring Public Opinion' Unit (2011)

Freeman, R.B.: Working for nothing: the supply of volunteer labor. J. Labor Econ. **15**, 140–166 (1997)

Frisanco R.: Rapporto sul Volontariato in Italia, Fondazione Italiana per il Volontariato (2011)

Fundaćao Itaù Social: Recopilaciòn de estudios y valutaciones sobre voluntariado, S. Paulo (2015)

Güntert S. T., Neufeind, M., T. Wehner.: Motives for event volunteering: extending the functional approach. Nonprofit Volunt. Sect. Q. **44**, 686–707 (2015)

Haddock, M.: Italy releases nonprofit census data, shows tremendous sector growth. Comparative Non Profit Sector Project, Johns Hopkins University Center for Civil Society Studies, 20 October 2014

Ibope Inteligencia: Pesquisa Voluntariado no Brasil – 2011, S. Paulo, Red Brasil Voluntario (2012)

INE: Encuesta de Empleo del Tiempo (2011). www.ine.es

Instituto Nacional de Estadística y Geografía (México) INEGI: Sistema de Cuentas Nacionales de México: Cuenta satélite de las institucionessin fines de lucro de México 2013: preliminar: año base 2008 (2015)

International Labour Organization: Manual on the Measurement of Volunteer Work. ILO, Geneva (2011)

Istat: La rilevazione sulle Istituzioni non profit. Un settore in crescita (2013). www.istat.it

Istat: Attività gratuite a beneficio di altri (2014a). www.istat.it

Istat: Nonprofit Institution Profile Based On 2011 Census Results. Roma, 16 April 2014 (2014b). www.istat.it

Leś, E., Nałęcz, S., Pieliński, B.: Third sector barriers in Poland, TSI National Report Series No. 7. Seventh Framework Programme (grant agreement 613034), European Union. Brussels: Third Sector Impact (2016)

Marino, D., Michelutti, M., Schenkel, M.: The attitudes, motivations and satisfaction of volunteers. Int. J. Appl. Econ. Econom. **17**, 368–403 (2009)

McKeever, B.S.: Public Charities, Giving, and Volunteering. Urban Institute, Centeron Non profits and Philanthropy (2015)

Mellor, D., Hayashi, Y., Stokes, M., Firth, L., Lake, L., Staples, M., Chambers, S., Cummins, R.: Volunteering and its relationship with neighborhood and personal well-being. Nonprofit Volunt. Sect. Q. **38**, 144–159 (2009)

Menchik, P., Weisbrod, B.A.: Volunteer labor supply. J. Public Econ. **32**, 159–183 (1987)

Mohan, J.: Developmental trends in the British third sector: evidence on voluntary action by individuals. TSI Working Paper Series No.14 Seventh Framework Programme (grant agreement 613034), European Union. Brussels: Third Sector Impact (2016)

Musick, M.A., Wilson, J.: Volunteers: A Social Profile. Indiana University Press, Bloomington (2007)

OECD: How's Life? 2015: Measuring Well-being. The value of giving: Volunteering and well-being (2015). https://doi.org/10.1787/how_life-2015-9-en

Piacentini, P.: Trabalho voluntario no Brazil, Pré Univesp, n. 61 (2016)

Pizzarro, I.: Solidaridad: historia de un valor que se afianza en el país, LA NACION, Martes 14 de julio de 2015 (2015)

Plataforma del Voluntariado de España (PVE): Hechos y cifras del Voluntarido en Espania, Observatorio del Voluntariado (2015)

Putnam, R.D.: Making Democracy Work. Civic Traditions in Modern Italy. Princeton University Press, Princeton (1993)

R&S.: La France Benèvole en 2016, 13ème edition-Juin 2016 (2016)

Roiter, M.: Beyond Images and perceptions: Conceptualizing and Measuring Volunteerism in Buenos Aires. In: Butcher, J., Einolf, C.J. (eds.) Perspectives on Volunteering: Voices from the South, pp. 171–194. Springer, Cham (2017)

Salamon, L.M.: Putting NPIs and the Third Sector on the Economic Map of the World, 47° Session of the United Nations Statistical Commission, 8 March 2016, New York (2016a)

Salamon, L.M.: Putting the Third Sector/Social Economy on the Economic Map of Europe: The Statistical Revolution Seminar on the Development of Third/Social Economy Statistics, 27 October 2016, Warsaw (2016b)

Salamon, L.M.: The Resilient Sector: The State of Nonprofit America. In: Salamon, L.M. (ed.) The State of Non Profit America, pp. xi–xviii. Brookings Institution Press in collaboration with the Aspen Institute, Washington, D.C. (2002)

Salamon, M.L., Sokolowski, W.: The Size and Scope of the European Third Sector, TSI Working Paper No. 12, Seventh Framework Programme (grant agreement 613034), European Union. Brussels: Third Sector Impact (2016)

Salamon, L.M., Sokolowski, S.W., Anheier, H.K.: Social Origins of Civil Society: An Overview. Johns Hopkins University (2000)

Salamon, L.M., Sokolowski, S.W., Haddok, M.A.: Measuring the economic value of volunteer work globally: concepts, estimates, and a roadmap to the future. Ann. Public Coop. Econ. **82**, 217–252 (2011)

Salamon, L., Sokolowski, S.W., Haddock, M.A., Tice H.S.: The Global Civil Society and Volunteering. Comparative Non Profit Sector Project, Johns Hopkins University Center for Civil Society Studies (2013)

Sajardo Moreno, A., Serra Yoldi, I.: Advances recientes en la investigacion economic sobre el voluntariado: Valoracion economic trabajo voluntario, costes de gestion of voluntariado y voluntariado corporate, CIRIEC - España, Revista de Economía Pública, Social y Cooperativa, Decembre 2008, iss. 63, pp. 191–225 (2008)

Schenkel, M., Ermano, P., Marino, D.: Recent trends in the supply and demand of volunteers. Am. J. Ind. Bus. Manag. **4**, 319–331 (2014). http://www.scirp.org/journal/ajibm

Schiff, J., Weisbrod, B.: Competition between for-profit and non-profit organizations. In: Ben-Ner, A., Gui, B. (eds.) The Non-profit Sector in the Mixed Economy. The University of Michigan Press, Ann Arbor (1993)

Tilly, C., Tilly, C.: Capitalist work and labor markets. In: Smelser, N., Swedberg, R. (eds.) Handbook of Economic Sociology, pp. 283–313. Princeton University Press, Princeton (1994)

TNS Argentina.: Disminuye_la_ proporción_de_ _argentinos_que_ realiza_ trabajo_ voluntario. Estudio _de _opinión _pública_ TNS _Argentina _sobre _Voluntariado (2013)

Turcotte, M.: Volunteering and charitable giving in Canada. Minister responsible for Statistics Canada, Catalogue no. 89-652-X2015001 (2016)

Uslaner, E.M.: The foundations of trust: macro and micro. Camb. J. Econ. **32**, 289 294 (2008)

Vaughan-Whitehead, D.: Public sector shocks in Europe: between structural reforms and quantitative adjustment. In: Vaughan-Whitehead, D. (ed.) Public Sector Shock, pp. 1–42. Edward Elgar Publishing, Cheltenham (2013)

Weisbrod, B.A.: Towards a theory of the voluntary nonprofit sector in a three-sector economy. In: Phelps, E. (ed.) Altruism, morality and economic theory, pp. 171–195. Russell Sage, New York (1975)

Welty Peachey, J., Lyras, A., Cohen, A., Bruening, J.E., Cunningham, G.B.: Exploring the motives and retention factors of sport-for-development volunteers. Nonprofit Volunt. Sect. Q. **43**, 1052–1069 (2014)

Wilson, J., Musick, M.: Who cares? Toward an integrated theory of volunteer work. Am. Sociol. Rev. **62**, 64–713 (1997)

World Value Survey: Wave 6 2010–2014 OFFICIAL AGGREGATE v.20150418. World Values Survey Association (2016). (www.worldvaluessurvey.org)

Effect of the Spanish Sovereign Risk Premium on the IBEX 35. Evolution 2000–2016

Jessica Paule Vianez[✉], Paola Plaza Casado,
Sandra Escamilla Solano, and Miguel Ángel Sánchez de Lara

Universidad Rey Juan Carlos, Madrid, Spain
{jessica.paule, paola.plaza, sandra.escamilla,
miguel.sanchezdelara}@urjc.es

Abstract. In a global world such as we live today, where any economic, social or environmental news affects not only the country where it originates, but it extends as a domino effect, it has a consequence that the information disclosure by the public institutions are must adequate to mitigate the impact that may be in their investors. The risk premium has become a fundamental element in analyzing the confidence that is deposited in the countries. Every day sees its fluctuation according to the news which are disclosed in both the economic and general press. Therefore, the objective of this work will be to analyze the relationship that the Spanish risk premium has in the IBEX 35. To do this, the period 2000 to 2016 will be taken as the basis of study. Notably as a main conclusion that the relationship between both variables exists, but not with the expected intensity.

1 Introduction

The signing of the Maastricht Treaty in 1992 and the adoption of the euro as currency in 2002 were events of great importance for Europe, although in any moment they consider that future crises could endanger economic cohesion and integration.

The phases determined in the Delors report showed how the euro established the risk of exchange rates would be eliminated among the Eurosystem countries, but even though there should be no discrimination in terms of access to financing by the states, distrust generated by the subprime mortgage in 2007 and its global contagion effect and the bankruptcy of Lehman Brothers, implied that Europe will begin in 2008 a sovereign debt crisis that had serious consequences for the countries of the Mediterranean and Ireland, causing the countries called PIIGS (Portugal, Ireland, Italy, Greece and Spain) suffered rescues both financial and banking. This situation led to the generation of distrust on the part of investors and began an upward stage of the sovereign risk premium of all countries, it was not until structural measures were taken and confidence was returned to the markets when it began its decline.

Therefore, this work has a double contribution, on the one hand, at the theoretical level because it brings light to a financial issue about which there are hardly any works that relate the risk premium and the IBEX 35 and, on the other hand, at the This is useful given that an approximation is made with a regression model to check the effect of the sovereign risk premium on the IBEX 35. For this, a sample of data is obtained

J. Gil-Lafuente et al. (Eds.): AEDEM 2017, SSDC 180, pp. 84–96, 2019.
https://doi.org/10.1007/978-3-030-00677-8_7

from the year 2000 until 2016. Also, given that if there are differences between the period, it was decided to subdivide the time horizon into two sections, from 2000–2007 and 2008–2016. The structure of the present work is organized in three parts, the first a literature review of the variables to be analyzed, the second, an analysis is made of the effect of the risk premium on the IBEX 35 and, finally, the conclusions are established.

2   Theoretical Framework

One of the main aspects that are analyzed when looking at the evolution of a country, in addition to the macroeconomic indicators, are those indicators that, although they do not have the macroeconomic character, are relevant when it comes to knowing the health of the economy of a country. Country: the risk premium and the selective one of quoted companies, in the Spanish case the IBEX 35. These two variables are characterized because they move according to the impulses of the investors and, in short, give confidence to the financial markets, countries and large companies (Arrow 1972; North 1981; Fukuyama 1995; Stiglitz 1999). This trust allows the market participants to participate in the financial markets (Guiso et al. 2007; Bormann 2013; Pérez 2015).

To understand the relevance of the risk premium in our lives today, we must go back to the Delors report (1989) where the phases to complete the integration of the European Union (EU) were presented. It established the evolution that would follow all countries that joined the Eurosystem:

1. First phase (1990–1993): characterized by the signing of the Maastricht Treaty (1992) and the liberalization of capital transactions, cooperation among central banks, establishment of improvements in economic convergence and the free use of ECU.
2. Second phase (1994–1998): there is the creation of the European Monetary Institute (IME), both the ERM II is established to establish the exchange policy between the Eurosystem countries and the rest of the EU countries, as well as the economic convergence criteria for those countries that adopt the euro as the single currency.
3. Third phase (1999-present): the irrevocable fixation of conversion rates is established, the euro is introduced as currency on 1 January 2000, the establishment of the ESCB that implements the monetary policy and the Stability Pact enters into force and Growth.

The cession of the monetary and economic policy of the countries had, among others, an implicit objective, the suppression of the risk of the exchange rates between the member countries. For this, the Eurosystem countries had to adapt their legislation to the regulations issued by the EU (Álvarez 2013). All this process involved the elimination of this risk, but the acquisition of greater importance of the country risk premium. For Knight (1921) it is the additional compensation offered to investors to maintain an asset with a quantifiable level of risk, and which is associated with the level of uncertainty about their possible default.

On the other hand, Álvarez (2013) performs an analysis of the risk premium by establishing three periods that revolve around it:

1. First period 1990–1997: stage coinciding with the first two phases enunciated with the Delors report, characterized by a reduction in the risk premium due to the establishment of measures aimed at meeting the criteria for adopting the euro as currency. Many of the countries had to make structural reforms by having extremely high deficits.
2. Second period 1998–2007: coincides with the introduction of the euro as currency and with an economic period of relative stability.
3. Third period 2008–2015: characterized by the sovereign debt crisis experienced in the EU, and more specifically with the countries that during that period suffered economic and financial rescues (Greece, Ireland, Portugal, Italy and Spain). The financial crisis that begins with subprime mortgages leads to a credit crisis and balance (Amor 2011, p. 47).

It should be noted that from the beginning of the EU to the beginning of the subprime crisis and later the sovereign debt crisis, risk premia between the different countries were practically nil, reflecting the expectation that the monetary union process would allow to advance in the real convergence of those countries with the worst macroeconomic foundations towards the standards presented by the German economy, the reference of the UME (Amor 2011, p. 43). The sovereign debt crisis experienced since 2008 has placed the risk premium at a value to be considered, not only when obtaining financing but also as an indicator of confidence, since it is considered a measure of country risk (Favero et al. 1997; Düllmann and Windfuhr 2000; Geyer et al. 2004; Fontana and Scheicher 2010). The risk premium has become more important in recent years since a high value of the risk premium generates uncertainty in the investors of the financial markets and, therefore, can make it difficult to obtain financing from those countries, by increasing the yield of the 10-year bond (Pérez 2015). This crisis has highlighted the level of convergence among the different Eurosystem countries, making it clear that the degree of solvency and economic cohesion is not the same throughout the integration process, making clear the great imbalances between the countries of the North and South of Europe (Amor 2011).

Graph 1 shows the average return of the Spanish bond, it is remarkable that in 2011–2012 the highest values were recorded, linked to the Spanish financial crisis. This is where the question can be asked: why can the risk premium be a determining factor in the financing of the countries? This is because there are determining factors that affect the risk premium and implicitly in determining the yield of 10-year bonds.

For García and Werner (2015, p. 7) there are several fundamental factors that affect the determination of the risk premium of European countries:

– Macroeconomic indicators
– Emotional economy
– Working market

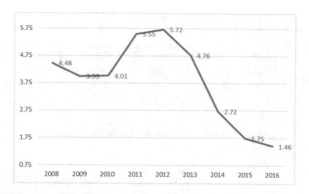

Graph 1. Average return of the Spanish 10-year bond. Source: Own elaboration based on data from the Bank of Spain

- Commerce
- Exchange rates
- Securities market indicators
- Prices

Pérez (2015, p. 86) considers 11 fundamental factors impact on the value offered of the risk premium:

1. Public finances
2. Volume of the debt
3. Inflation
4. Economic activity
5. Short-term interest rates
6. Interest rate differentials
7. Economic sentiment
8. Labor market
9. Asymmetric information
10. Speculation
11. Communication of monetary policy

Also, in Table 1 you can find the researches of other authors that relate the risk premium with different variables such as systematic, specific risks, portfolio management, etc.

As shown in Table 1, there are multiple factors that can affect the determination of the risk premium. When research has been sought on whether the risk premium affects the IBEX 35, a gap has been found in literature, and no research is found to analyse this relationship. Therefore, the contribution made with this research in the state of the art is more relevant.

Table 1. List of research on the risk premium and variables that have been used

Variables	Authors
Liquidity risk	Düllmann and Windfuhr (2000), Schulz and Woff (2009), Beber et al. (2009), Barrios et al. (2009), Martínez et al. (2016)
Credit Risk	Manganelli and Wolswijk (2007), Schuknech et al. (2009), Barbosa and Costa (2010); Barrios et al. (2009), Bernoth and Erdogan (2012), Martínez et al. (2016)
Public debt	Alesina (1992), Bernoth et al. (2004), De Andrés (2004), Mangenelli and Wollosijk (2007)
Déficit	Bernoth and Erdogan (2012)
Risk aversión	Codogno et al. (2003), Schulz and Woff (2009), Mody (2009), Attinasi et al. (2009), Barrios et al. (2009), Martínez et al. (2016)
Inflatión	Jacoby and Shiller (2007)
Unemployment	García and Werner (2015)
Informational asymmetry	Haugh et al. (2009)
Type of interest	De Andrés (2004), Ardagna et al. (2004), Schuknecht et al. (2009), Krugman (2012)
Institutional factors	Alonso and Trillo (2015)
Taxation	Haugh et al. (2009)

Source: Own elaboration

3   Empirical Research

3.1   Objectives, Sample and Data Collection

Given the relevance of both the Spanish risk premium and the IBEX 35 as indicators of the state of health of the economy, it is considered necessary to carry out an analysis on how the Spanish risk premium influences the IBEX 35. For this purpose, a study of descriptive statistics and a simple regression of the analysed variables to be able to contrast the proposed objective.

The regression model that arises is:

$$y = \alpha + \beta x + \varepsilon$$

Being:

- y: the daily returns of the IBEX 35
- x: the daily returns of the risk premium

The sample used for the present work is based on the official data published by the Madrid Stock Exchange to obtain the closing price of the IBEX 35 and in the case of the risk premium the values offered by the newspaper have been taken economic expansion. The period taken for the analysis includes from January 3, 2000 to December 31, 2016. To perform the regression of the model, it is decided to take the daily variations of the risk premium and the IBEX 35 for the entire period analyzed, as well as making a subdivision between the pre-crisis period (2000–2007) and the crisis period (2008–2016). The data processing is done through the statistical program SPSS v22.

3.2 Results

To understand the evolution of the risk premium and the IBEX 35, it is considered necessary to carry out a study of macroeconomic variables. Next, Table 2 expresses the GDP, Inflation and Deficit values of Spain for the period analyzed.

Table 2. Macroeconomic indicators Spain[a]

Years	PIB	Inflation	Deficit
2016	3.20	1.60	−4.54
2015	3.20	0.00	−5.13
2014	1.40	−1.00	−5.99
2013	−1.70	0.30	−7.00
2012	−2.90	2.90	−10.47
2011	−1.00	2.40	−9.61
2010	0.01	3.00	−9.38
2009	−3.60	0.80	−10.96
2008	1.10	1.40	−4.42
2007	3.80	4.20	1.92
2006	4.20	2.70	2.20
2005	3.70	3.70	1.21
2004	3.20	3.20	−0.04
2003	3.20	2.60	−0.37
2002	2.90	4.10	−0.41
2001	4.00	2.70	−0.55
2000	5.30	3.60	−1.02

[a]Data expressed in %
Source: Own elaboration based on data from the Bank of Spain, INE and expansion

The results of the evolution of both the Spanish risk premium and the IBEX 35 are shown in graph 2. Their analysis evidences the existence of a period of high variability of the two variables, which in its beginnings in the offered results is observed that there is practically no relationship between the variables, but it will be from 2008 when the bankruptcy of Lehman Brothers explodes, that the Spanish risk premium and the IBEX 35 will start moving inversely. Therefore, it is considered necessary to analyse the period between 2008–2016, with a temporary subdivision of 2000–2007 and 2008–2016.

The beginning of the 21st century is characterized by a marked period where the establishment of the monetary economic union, whose beginning in 1999 has its

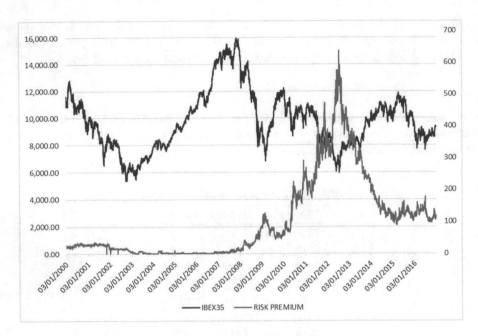

Graph 2. Evolution of the Spanish risk premium and the IBEX 35 period 2000–2016. Source: Own elaboration

culmination with the entry into force of the single currency on January 1, 2002. This period was for the countries of the Eurosystem a consolidation of its debt, as well as the strengthening of the economy and positioning at the international level, resulting in a decline in sovereign debt yields and, consequently, in spreads of the risk premium. The single currency was the elimination of the risk of exchange rates. All these factors implied that the Spanish risk premium would be placed at values close to 0 when Spain has the highest credit rating (Triple A), financed under the same conditions as Germany, it will not reach values higher than 40 points until the crisis begins American bankruptcy and Lehman Brothers.

With the outbreak of the subprime mortgage crisis in 2007, there was a growing distrust among the countries. In addition, the banking difficulties involved adding obstacles in the financing of the different economic agents, from the families that had acquired mortgages with Euribor values around 5%, the SMEs that stopped having access to business financing, to the countries themselves, which to acquire sovereign debt had to pay high levels of interest. On September 15, 2008, it marked a turning point between the Spanish risk premium and its relationship with the IBEX 35 (Graph 3).

If in the first years of the 21st century they were characterized by the economic boom and the expansion of the European countries, in the Spanish case that the year 2008 supposed the beginning of the Spanish crisis derived from the globalization of the economies, the preceding crisis of the USA and the Spanish property bubble. The Spanish economy began to weaken and with this the escalation of the Spanish risk

Graph 3. Evolution of the Spanish risk premium and the IBEX 35 period 2008–2016. Source: Own elaboration

premium and the decline of the Spanish selective began. In 2009 due to the economic deterioration that Spain was suffering and the distrust that was given to the markets because of the problems of the banking sector, Standard & Poor's decided to take away the triple A, reached maximum values of 127 points in terms of premium and 6,817.40 in terms of IBEX 35. Despite maintaining some stabilization, in the first quarter of 2010 the Moody's and Fitch agencies made the same decision as Standard & Poor's, surpassing, as can be seen in graph 2, the psychological barrier of 100 points, a value that promptly low in October 2016.

In 2011 the Spanish economy shows clear signs of exhaustion and, together with the change of government produced by the elections held on November 20, 2011 and the announcement of the plan of urgent economic measures to meet the objectives set from Brussels in macroeconomic terms (GDP, Deficit, inflation, etc.) meant taking drastic and austere measures. This entire process culminates with the bank restructuring process derived from the financial rescue suffered by Spain when, on July 24, 2012, the Spanish risk premium reached the value of 637 and the IBEX 35 5,956.30. Graph 4 shows the situation that Spain went through during 2012.

Then, in the Table 3 the analysis of descriptive by years for the period analysed can be observed.

This analysis allows us to observe how the risk premium until 2007 did not exceed values of more than 40 points, and in the period of the sovereign debt crisis the value of 637 was reached. For the IBEX 35 the value smallest can be observed in 2012 and the highest value in November 2007

As can be seen in Table 4, a descriptive study of the entire analyzed period is carried out, observing that the average of the risk premium is sensibly low due to the low values that are given up to 2007.

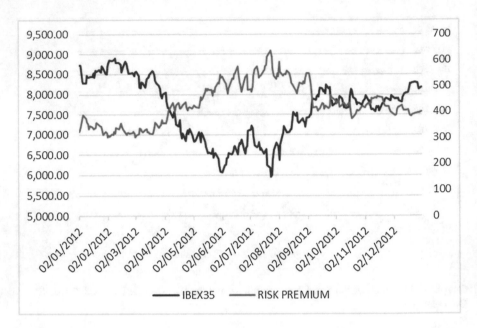

Graph 4. Evolution of the Spanish risk premium and the IBEX 35 year 2012. Source: Own elaboration

Table 3. Descriptive analysis by years

	RISK PREMIUM						IBEX 35				
Years	N	Average	Median	Standard deviation	Mín	Max	Average	Median	Standard deviation	Mín	Max
2016	257	124.86	126	16.41	97	179	8,720.00	8,716.00	335.78	7,645.50	9,412.80
2015	256	121.72	122	14.60	89	164	10,661.21	10,692.80	648.95	9,291.40	11,866.40
2014	255	149.63	143	27.57	106	216	10,455.78	10,432.90	352.94	9,669.70	11,187.80
2013	255	294.01	295	48.22	220	390	8,686.25	8,515.20	655.09	7,553.20	10,037.80
2012	256	431.68	421	78.60	301	637	7,619.50	7,781.65	750.85	5,956.30	8,902.10
2011	257	280.50	269	71.58	173	472	9,568.98	9,872.20	1,030.37	7,640.70	11,113.00
2010	256	151.70	168	58.85	58	291	10,421.52	10,431.50	694.59	8,669.80	12,222.50
2009	254	75.22	69	21.66	45	127	9,986.18	9,677.05	1,518.98	6,817.40	12,035.10
2008	253	37.55	32	18.53	7	88	11,874.96	11,999.80	1,818.64	7,905.40	15,002.50
2007	252	7.56	6	3.77	1	20	14,823.69	14,812.75	516.64	13,602.40	15,945.70
2006	254	1.96	1	1.55	1	8	12,166.80	11,866.40	1,061.78	10,665.60	14,387.60
2005	256	1.72	1	1.02	1	5	9,858.14	9,800.30	583.20	8,945.70	10,919.20
2004	251	2.75	2	2.38	1	11	8,163.62	8,096.40	321.04	7,578.30	9,100.70
2003	250	3.57	3	2.72	1	10	6,709.12	6,862.20	496.58	5,452.40	7,760.40
2002	250	15.12	16	3.45	1	21	7,086.44	6,875.95	949.67	5,364.50	8,554.70
2001	250	28.94	29	3.68	1	37	8,821.82	8,840.70	856.83	6,498.40	10,132.00
2000	250	27.20	28	3.68	16	36	10,971.77	10,949.60	838.44	8,864.30	12,816.80

Source: Own elaboration

Table 4. Analysis of descriptive period 2000–2016 (n = 4312)

	Rango	Minimum	Maximum	Average	Standard deviation	Variance	Asimetry		Curtosis	
	Statistical	Statistical	Statistical	Statistical	Statistical	Statistical	Statistical	Standard error	Statistical	Standard error
Spanish risk premium	636.0	1.0	637.0	104.049	127.2943	16,203.842	1.492	.037	1.633	.075
IBEX 35	10,581.2	5,364.5	15,945.7	9,801.999	2,151.3294	4,628,218.15	.591	.037	.137	.075

Source: Own elaboration

Table 5 shows the result of the Pearson correlation analysis between the Spanish risk premium and the IBEX 35, which shows how there is a negative or inverse correlation between the variables of average intensity.

Table 5. Pearson correlation period 2000–2016

		Spanish risk premium	IBEX 35
Pearson correlation	Spanish risk premium	1	−.287**
Pearson correlation	IBEX 35	−.287**	1

**The correlation is significant at the 0.01 level (2 tails).
Source: Own elaboration

Given that the variation of the risk premium was non-existent during the period 2000–2016 when performing the regression model for that period, it can be seen that there is no demonstrable significant relationship between the profitability of the IBEX 35 and the risk premium.

Table 6 shows the result of the Pearson correlation analysis between the Spanish risk premium and the IBEX 35 during the 2008–2016 period, which shows how there is a negative or inverse correlation between the variables of average intensity, greater than that recorded for the whole period.

Table 6. Pearson correlation period 2008–2016 in variations

		Var. IBEX 35	Var. risk premium
Pearson correlation	Var. IBEX 35	1.000	−.344
	Var. risk premium	−.344	1.000

Source: Own elaboration

In this way, as can be seen in Table 7, the IBEX 35 is explained by 11.8% of the Spanish risk premium.

Table 7. Summary regression model 2008–2016

Model	R	R squared	R squared tight	Standard error of the estimate
1	.344[a]	.118	.118	.0158854

[a]Predictors: (Constant), PREMIUM VARIATION
[b]Dependent variable: IBEX35 VARIATION
Source: Own elaboration

Table 8. Coefficients[a] of the linear regression for the period 2008–2016

Model		Non-standardized coefficients		Standardized coefficients	T	Sig.
		B	Standard error	Beta		
1	(Constant)	.000	.000		.579	.562
	PREMIUM VARIATION	−.092	.005	−.344	−17.534	.000

[a]Dependant variable: VARIATION IBEX35
Source: Own elaboration

In this way, once the regression of the period 2008–2016 is done, the values of the coefficients are obtained (Table 8), leaving the equation as follows:

$$IBEX35 = 0 + (-0.092) * RISK\ PREMIUM$$

4   Conclusions

Something that can be observed clearly throughout the work is that in a pre-crisis situation (period 2000–2007) there was no discrimination in sovereign risks in the EMU (Amor 2011). Due to the financial crisis initiated in 2007 by the subprime mortgages and the global economic instability that led to the bankruptcy of Lehman Brothers in 2008, the main consequence was evidence of a Europe at two levels (González and González 2013).

In the case of Spain, in addition, it was affected by a series of factors such as the housing bubble, an excess deficit, as well as a weakened financial system that had to be restructured and in some cases, rescued. All this together with a situation of global recession caused sovereign risk premium levels to reach excessively high values derived from a situation that generated distrust.

Regarding the literature found on the subject under study, it can be concluded that there are numerous studies on the risk premium and factors that can determine the risk premium, but hardly any on the effect that the risk premium exerts. about the IBEX 35.

From the empirical research it can be concluded (1) that as macroeconomic data improve over time, there is a decrease in the risk premium and an increase in the IBEX 35, due in large part to the confidence placed in the markets; (2) it is assumed that the risk premium exerts an important influence on the IBEX 35 but unlike what was initially believed, although there is an inverse relationship between the risk premium and the IBEX 35, it is not as intense as it was thought.

Given the relevance of the sovereign risk premium on investor confidence today, it is established as a future line of research to see if there are other variables that have an important effect on the IBEX 35, as well as to use other study techniques that allow us to better analyse the effect of the risk premium on the IBEX 35, such as the use of the volatility of the IBEX 35 or perform an ANOVA study.

References

Alesina, A., De Broeck, M., Prati, A., Tabellini, G.: Default risk on government debt in OECD countries. Econ. Policy **15**(15), 427–463 (1992)

Alonso, N., Trillo, D.: Riesgo soberano en la eurozona: ¿una cuestión técnica? Papeles de Europa **28**(1), 1–26 (2015)

Álvarez, M.: La prima de riesgo. Análisis de su evolución. Universidad de A Coruña, A Coruña, Spain (2013)

Amor, J.M.: La vulnerabilidad en los mercados de deuda soberana en la Unión Monetaria Europea. Boletín Económico ICE **863**, 43–51 (2011)

Ardagna, S., Caselli, F., Lane, T.: Fiscal discipline and the cost of public debt service: Some estimates for OECD countries. European Central Bank (ECB), Working Paper No. 411, Frankfurt, Germany (2004)

Arrow, K.: Gifts and exchanges. Philos. Public Aff. **1**(4), 343–362 (1972)

Attinasi, M., Chcherita, C., Nickel, C.: What explains the surge in euro area sovereign spreads during the financial crisis of 2007–09? European Central Bank Working Paper No. 1131, Frankfurt, Germany (2009)

Barbosa, L., Costa, S.: Determinants of sovereign bond yield spreads in the euro area in the context of the economic and financial crisis. Economic Bulletin Autumn 2010. Bank of Portugal, Lisbon, Portugal (2010)

Barrios, S., Iversen, P., Lewandowska, M., Setzer, R.: Determinants of intraeuro area government bond spreads during the financial crisis. European Commission's Directorate General for Economic and Financial Affairs, Economic Paper No. 388, Brussels, Belgium (2009)

Beber, A., Brandt, M.W., Kavajecz, K.A.: Flight-to-quality or flight-toliquidity? Evidence from the euro-area bond market. Rev. Financ. Stud. **3**(22), 925 957 (2009)

Bernoth, K., von Hagen J., Schuknecht, L.: Sovereign risk premiums in the European government bond market. European Central Bank, Working Paper No. 369, Frankfurt, Germany (2004)

Bernoth, K., Erdogan, B.: Sovereign bond yield spreads: a time varying coefficient approach. J. Int. Money Financ. **31**, 639–656 (2012)

Bormann, S.: Sentiment indices on financial markets: what do they measure? Economics Discussion Papers No. 2013/58, Kiel Institute for the Word Economy, Kiev, Ukraine (2013)

Codogno, L., Favero, C., Missale, A.: Yield spreads on EMU government bonds. Economic Policy, October, pp. 503–532 (2003)

De Andrés, J.: Un análisis de la curva de rendimientos en el mercado de deuda pública española a medio y largo plazo en el periodo 1993–2004. Revista Atlántica de Economía **3**(11), 1–30 (2004)

Dülman, K., Windfuhr, M.: Credit spreads between German and Italian sovereign bonds: do one-factor affine models work? Can. J. Adm. Sci. **17**(2), 166–181 (2000)

Favero, C.A., Giavazzi, F., Spaventa, L.: High yields: the spread on German interest rates. Econ. J. **107**(443), 956–985 (1997)

Fontana, A., Scheicher, M.: An analysis of euro area sovereign CDS and their relation with government bonds. European Central Bank (ECB), Working Paper No. 1271, Frankfurt, Germany (2010)

Fukuyama, F.: Trust: The Social Virtues and the Creation of Prosperity. The Free Press, Nueva York (1995)

Garcia, J.A., Werner, S.E.V.: Bond risk premia, macroeconomic factors and financial crisis in the euro area. European Central Bank (ECB), Working Paper No. 1938, Frankfurt, Germany (2015)

Geyer, A., Kossmeier, S., Pichler, S.: Measuring systematic risk in EMU government yield spreads. Rev. Financ. **8**(2), 171–197 (2004)

González, C., González, M.: Is there a contagion effect in the eurozone? Evidence from the sovereign risk premiums analysis. Cuadernos de CC.EE y EE **65**, 71–84 (2013)

Guiso, L., Sapienza, P., Zingales, L.: Trusting the stock market. European Corporate Governance Institute (ECGI). Working Papers Series in Finance, Working Paper No. 170, Brussels, Belgium (2007)

Haugh, D., Ollivaud, P., Turner, D.: What drives sovereign risk premiums? An analysis of recent evidence from the euro area. The Organisation for Economic Co-operation and Development (OECD), Economics Department, Working Paper No. 718, Paris, France (2009)

Jacoby, G., Shiller, I.: The determinants of TIPS yields spreads. J. Appl. Financ. **17**(2), 72–81 (2007)

Knight, F.H.: Risk, Uncertainty, and Profit. Harper, New York, USA (1921)

Krugman, P.: ¡Acabad ya con esta crisis!. Crítica, Barcelona, Spain (2012)

Manganelli, S., Wolswijk, G.: What drives spreads in the euro area government bond market? Econ. Policy **24**(58), 191–240 (2007)

Martínez, L.B., Teruel, M., Terceño, A.: Determinantes de Spreads Soberanos durante la reciente crisis financiera: el caso europeo. Cuadernos de Administración **29**(53), 77−100 (2016)

Mody, A.: From Bear Stearns to Anglo Irish: How Eurozone sovereign spreads related to financial sector vulnerability. International Monetary Fund (IMF), Working Paper No. 09/108, Washington, USA (2009)

North, D.: Structure and Change in Economic History. WW Norton, New York (1981)

Pérez, I.: El papel de la confianza en la evolución de la economía: evidencia empírica para el caso español. Universidad de A Coruña, A Coruña, Spain (2015)

Schuknecht, L., Von Hagen, J., Wolswijk, G.: Government risk premiums in the bond market: EMU and Canada. Eur. J. Polit. Econ. **25**, 371–384 (2009)

Schulz, A., Wolff, G.B.: The German SNG bond market: Structure, determinants of yield spreads and Berlin's forgone bailout. J. Econ. Stat. **229**(1), 61–83 (2009)

Stiglitz, J.: Whither reform? Ten years of transition. Annual Bank Conference on Development Economics. Washington, USA, 28–30 April 1999

The Revolution of Active Methodologies: Experiential Learning and Reflection in Higher Education

Verónica Baena-Graciá[✉], Miriam Jiménez-Bernal,
and Elisabet Marina-Sanz

Universidad Europea de Madrid, Madrid, Spain
veronica.baena@universidadeuropea.es

Abstract. Higher Education is immersed in a process of revolution with the introduction of active methodologies. Experiential learning is one of them, and it is based on a simple principle: learning by doing. However, any methodology consisting of the active participation of students requires a reflection upon the corresponding activities, their development and results in order for the students to be able to internalize it. The aim of this communication is to present the preliminary analysis of a survey on methodologies conducted among students of the Degree in Primary Education. The questions were focused on students' knowledge and definition of several active methodologies, as well as on their ability to identify activities related to them and carried out in different subjects. Results show that students seem to associate Experiential Learning with very specific activities and with their internships, so explicit discussion on this methodology turns out to be an essential part of the process.

1 Introduction

The European Space for Higher Education (ESHE) emphasizes in several documents those aspects related to the organization of Higher Education, as well as those referring to the acquisition and measurement of knowledge. However, it is noticeable that some aspects that are essential to personal and individual development and to the promotion of life in democratic societies are not specified (Huber 2008). Thus, the current educational framework and the success of constructivism as ap pedagogy have determined, on the one hand, a turn into a paradigm focused on significant and contextualized learning and, on the other hand, a practical application and demonstration of the acquired and developed skills and competencies. In this sense, current curriculum designs and syllabus face the theoretical-practical challenge to overcome the traditional dichotomy, that is, they deal with the difficulty of being able to open to professional

E. Marina-Sanz—This research was supported in part by funds received from the David A. Wilson Award for Excellence in Teaching and Learning, which was created by the Laureate International Universities network to support research focused on teaching and learning. For more information on the award or Laureate, please visit www.laureate.net.

© Springer Nature Switzerland AG 2019
J. Gil-Lafuente et al. (Eds.): AEDEM 2017, SSDC 180, pp. 97–104, 2019.
https://doi.org/10.1007/978-3-030-00677-8_8

knowledge by means of the implementation of actions integrating work and world of life, working subject and productive knowledge (Fernández 2009).

According to these arguments, we could state that active methodologies have been more and more relevant during the last decades, what can be easily understood taking into account the educational tendencies and social demands in our current society (characterized, among many other questions, by overexposure to information and continuous progress and advances, that require to an almost constant updating). In fact, the promotion of skills related to critical thinking and analysis, as well as to the application of theoretical knowledge to practical situations, occupies an outstanding position in this Society of Knowledge (Castells 2002), where the overload of available information makes it necessary to distinguish between what is relevant and what is not. Those skills have become more important than the mere acquisition of theoretical concepts.

In this social and educational scenario, the school is not anymore the only place where learning takes place, and neither can it intend to assume the educational function of society on its own (Aguirre and Vázquez 2004 in Martín n/d). In this sense, Experiential Learning as an active methodology offers an interesting and unique opportunity to connect theory and practice. Also, there is an added value: this methodology involves students in their learning processes, promoting their participation, increasing their motivation and optimizing their academic performance and results (Romero 2010). Thus, and according to the Experiential Learning Theory—which determines that experimentation is a decisive factor in individuals' learning, several works have arisen claiming for the implementation of experiential learning in the Higher Educational classroom (i.e., Mattera et al. 2014) and vindicating the potential of experience to foster knowledge, on the basis of significant learning—that is, the fact that individuals learn when they find a significant relation between their interaction and the context or environment (Dewey 1938, cited by Romero 2010).

However, experience is not enough if we are to ensure learning, since this is intrinsically linked to a process of personal reflection, where meaning is constructed upon the experience lived (Smith 2001). Reflection is, then, a key to guarantee the construction of meaning upon experience.

2 Objectives and Methodology

The research project that is being carried out by the authors of this paper focuses on Experiential Learning as a source for learning that provides students with the possibility of deducing theory from practice, of applying theoretical knowledge to practical situations and, in general terms, of getting involved in their learning process to increase their motivation and academic performance. This reflection upon the methodologies used and the activities being developed is important in every professional field, including of course those areas related to Economics, Marketing and Business, but it is even more relevant when it comes to future teachers.

The main aim of this paper is, then, to present some preliminary results on the analysis of surveys and focus groups carried out among students from the Degrees in Primary Education, International Relations and Business. Surveys were taken by eight students from the degree in Primary Education before and after the implementation of different active methodologies by means of various activities, while focus groups took place only after the end of the course and participants where sixteen students from the degrees in International Relations and Business. Both tools were intended to foster reflection and discussion on the application of diverse methodologies in the Higher Education classroom.

Since the methodology used by the research group is a mixed one, both qualitative and quantitative tools are employed in this study. The results at this stage of the experience, in particular, will be merely qualitative, based on the answers provided by students to a brief survey on methodologies and their implementation in the classroom through activities—both before and after the actions were developed, as well as on those provided by students in two focus groups that were carried out at the end of the first term.

3   Description of the Experience

During the first two terms of the 2016/17 academic year, several actions were taken in the degrees in Primary Education, International Relations and Business, all framed by the contents and skills to be developed in subjects as Didactics in Language and Literature and Communication Skills. Specifically, in the Communication Skills course, Business students practiced negotiation skills through role plays and were required to make a research on the needs of students in their first year regarding teamwork and oral communication, among others, while in the degree in Primary Education, the activities included materials creation and activities' design, thus promoting—in both cases—skills such as entrepreneurship and teamwork, highly valued in professional settings.

The students from the degree in Primary Education developed several activities corresponding to different methodologies, even though we will focus on Experiential Learning. They were asked to complete the survey before and after carrying out all of the activities, that is, answers to the surveys are more than a month distant from one another. After the second survey, aimed at checking whether students were able to relate their experiences to the methodologies, a debate was conducted in order to reflect upon all of them and their advantages, disadvantages and significance, focusing specially on Experiential Learning.

The students from the degrees in International Relations and Business participated in two different focus groups, one carried out with students who were native speakers of Spanish (international and local students) and another one carried out with non-native speakers of Spanish, where, upon many other questions, they discussed about their views on active methodologies. Focus groups were transcribed and the answers analyzed using Discourse Analysis techniques, which allow the research group to identify relevant topics and standpoints—mainly active versus passive positions adopted by students.

3.1   Survey and Data Collection

In the case the students from the degree in Primary Education, questions posed on the survey. Eight students answered them: six completed just the first survey and one of them only the last one, at least regarding Experiential learning—the methodology we are focusing on. Questions, which were also the trigger for the debate, included the following ones:

– Do you know the methodology called Experiential Learning?
– How would you define it?
– Has it been implemented in any subject or course you have taken so far?
– In which activities have you observed it and what kind of contents were worked through this methodology?

In the degree in International Relations and Business, two focus groups were carried out. Both of them focused on the same topics and eight students participated in each one of them. The main questions regarding the topics at hand included the following ones:

– Have you found any differences in methodologies, activities and teaching methods between University and High School or between your previous university and this one?
– Do these methodologies influence somehow your motivation to participate actively in the classes?

3.2   Experiential Learning: Activities Developed

Specifically, in order to implement Experiential Learning two projects and a teamwork activity were carried out in the Didactics in Language and Literature course. The teamwork activity objective was to create audiovisual resources addressed to a Primary classroom, so as to promote reading and Education in Values, and students were required to participate in the design, recording and production stages.

Regarding the projects, the first one consisted on writing a brief text about Madrid. Its goal was to provide a teacher with real texts, correctly written and adapted to the reading level of their addressees (students of Spanish as a Foreign Language, level B1, from Germany), so that she could use them in a session about the city with appropriate samples. The teacher offered feedback on the texts, assessing their usage options.

The second one had an NGO as external partner. In this case, the objective was to design an activity for the promotion of creative writing and another one based on interactive performances. They were addressed to children from an association in Madrid that fosters the integration of migrant families in the community and social life in the neighborhood.

The description offered to the NGO involved objectives for the students—being able to design activities addressed to Primary education children and to put them into practice in an appropriate way, and for children—having fun while learning about literary concepts and working on their literary creativity and cooperative learning. The assessment sheet delivered to the NGO included questions about the objectives, addressees, timing, development and, in general, any items assessed by the teacher (has the activity been entertaining? have the resources employed been sufficient? have the volunteering teachers been flexible?, have they adapted to the situation?, have the objectives for children been fulfilled—fun, creativity fostering and so on?). Thus, assessment is not only the teachers' or the students' responsibility.

The experiential activities implemented with students from the degree in International Business and International Relations included role plays to develop negotiation skills, where students played different roles and tried to achieve a common objective, a research about the current situation regarding refugees, which included exhaustive documentation and analysis and a proposal for possible actions to be taken, and a group activity in collaboration with students from another course in the same degree, where they carried out a needs analysis regarding students in their first year and communication skills—by means of a survey they designed—and then defined a proposal for workshops that could be implemented in High Schools by a hypothetical education related company.

The role play aimed at developing negotiation skills was not only based on the Experiential Learning methodology, even if it was just a simulation, but also on gamification, which is being more and more utilized in Higher Education. Thus, students were given descriptions of their characters, which were built upon races from The Lord of the Rings—elves, magicians, hobbits, dwarfs and humans, and the descriptions included data such as the number of languages they spoke, prejudices and teamwork skills. Two members of every race were debating and the rest of groups were supporting them, being allowed to provide them with specifications and suggestions by means of notes. They also decided their final objective, and the only instruction in that sense was there needed to be a final agreement.

The research on the situation of refugees and the proposal for actions to be taken was previous to the final project, so it contributed to the development of the necessary skills by students—documentation and ICT use, time management and planning, interpersonal skills and so on.

Finally, the project in collaboration with students from a different course allowed students from the degree in International Relations and Business to develop several skills and contributed to their integration in the University, since they were attending their first year. The contents of the course (Communication Skills) and the competences they need to develop—entrepreneurship, interpersonal skills, oral and written communication, time management, autonomy, and digital skills, among others—were included in the activity design.

3.3 Debate and Focus Groups

After carrying out the activities and the post- survey, the questions were reviewed to provide deeper explanations on the methodologies included in the survey and to reflect upon the previous ideas of students. Definitions were offered and the implications of every methodology were exposed, after which the activities developed were referred to and their relations to the surveys and methodologies, as well as to skills acquisition, were shown.

Notes taken during the debate indicate that students have notions about the concepts and methodologies involved, even though they are not always able to provide an appropriate definition to them. If we go a step further, that is, analyzing the activities and associating them with the methodologies, seems to be more difficult to them. However, when the relation is made explicit, they seem to understand and identify it clearly.

The focus groups carried out were recorded and then transcribed. The questions, as shown in the previous section, focused on how methodologies were applied and perceived by students. Some of the answers to the questions indicate that there is a tension between the habit of traditional methodologies and their benefits and the perceived advantages of the new ones.

The use of active methodologies, in contrast with the traditional ones—where lectures were the essential method to transmit contents and the teacher was the center of the teaching and learning process, is supposed to foster the development of skills and to contribute to the students active role in their own learning. However, the reflection upon the methodologies and how they are used is necessary, because students need to develop their analytic and critical thinking skills. Thus, debates, surveys and focus groups contribute to the acquisition and improvement of those competencies and to a deeper comprehension of how the learning process works, essential for any professional in a world where lifelong learning and continuous training are required.

It has asked the companies values 0–5 items of country brand image and proceeded to transform them into percentages (Table 1). The results show that Spain is perceived as a country with friendly people (73.7%), creative (68.6%) and skilled (64.2%). These data reinforce the fact that one of the best assets available to Spain is its human capital. By contrast, the worst rated aspects are those related to the image of an innovative country (46.4%) and technologically advanced (45.5%).

4   Results

Firstly, we will present the results of the survey, comparing the pre- and post- ones. The answers to the survey considered valid are provided in the following table.

As it can be observed, students are not able to relate, a priori, Experiential Learning with the activities developed in Didactics in Language and Literature or in other courses where it has also been implemented. Answers indicate that they associate it specifically with their internship periods and with some other subjects and courses that, traditionally, have been considered as more practical and experiment-related.

Concerning the focus groups, some of the most interesting answers indicated that new methodologies and assessment tools are perceived in a positive way:

'Yes, I think it is **positive**. What is **nonsensical** is attending a course, not studying for four months, studying everything in one week, letting it fly in the exam and forgetting it. What is the use of that?' (A1)

even though this is not necessarily different from the methods and opportunities offered in the international universities were some of the students came from:

'In Morocco it's the same. There are a lot of differences between here and Morocco, but if you are a **good student** […] here they help you and they offer you opportunities for a good career' (I5)

Table 1. Survey results

Student	Pre	Post
C.D.B.	Yes. It consists of learning while you are experiencing what the teachers are explaining about the topic Activities: watching the plants in the university campus and commenting upon them (Sciences)	Yes. It consists of acquiring knowledge by means of experiences, visualizing reality
C.B.	Learning based on experience, significant learning Internship	Learning applied to practice situations to achieve significant learning
S.G.	Learning through experience In the labs	-
J.M.	Learning based on practical work	-
S.C.	Learning through experiences In Sciences and Psychology	It is the one where students learn through experience, making experiments
L.H.	It fosters the ability of people to learn from their own experience	The student's knowledge, developing skills and reinforcing values
C.C.	Learning by experimenting In Science	It consists of acquiring a knowledge making the student live in the first person the experience of their learning
N.C.	-	Learning skills, knowledge and values from our own experience

5 Discussion, Conclusion and Implications

The activities described in this paper not only allow the acquisition of theoretical concepts by students on the basis of practical work, but also contributed to increasing their motivation and academic performance. In the case of future teachers, the reflection upon the methodologies used became even more relevant, since they need to identify and know how to implement them, even though it is a necessary tool for any student that is responsible for their learning process.

Also, it contributed to a closer relation with the professional field reality, allowing students to be protagonists of their own learning process and, thus, making it possible for the results to be coherent with the strategic objectives established for the European Space for Higher Education. In fact, the authors of this paper consider this to be one of their main contributions to the academic field. Also, this kind of activity is a great opportunity to establish a relation between employers and possible employees (current students).

Although these activities need some time to achieve an appropriate degree of consolidation, results obtained so far have been highly satisfactory in both directions: from the point of view of the institution and organisms (possible future employers included) and from the involved students' and teachers' perspective.

In this sense, it is important to emphasize the role of experience in the learning process. Specifically, it is essential for students to find themselves in a real situation of professional development, facing the demands or requests of third parties—other than the teacher—and adapting to the norms of a real institution or organism. This fact increases the personal involvement of students and their relation with the learning process becomes more significant, intense and committed. This way, the students turn into pro-active agents.

Finally, the collaborating entities (NGOs, companies and so on) benefit from the students' talent, and they are able to identify possible employees for their organizations, while for the teachers' involved in the courses where these methodologies are implemented, Experiential Learning allows them to adapt scientific and academic theories to real contexts, providing them with an opportunity to offer non simulated situations and to obtain feedback from external partners in order to improve the activities for future occasions.

References

Castells, M.: La dimensión cultural de Internet. Institut de Cultura: Debates Culturales (2002). http://www.uoc.edu/culturaxxi/esp/articles/castells0502/castells0502.html

Fernández, E.: Aprendizaje experiencial, investigación-acción and creación organizacional de saber: la formación concebida como una zona de innovación profesional. In: Cultura e Investigación en el Espacio Europeo de Educación (coord.). Henar Rodríguez Navarro, Javier J. Maquilón Sánchez and Eduardo Fernández Rodríguez). REIFOP: Revista electrónica interuniversitaria de formación del profesorado, vol. 12, no. 3, pp. 39–57 (2009). https://dialnet.unirioja.es/servlet/articulo?codigo=3086514

Huber, G.L.: Aprendizaje activo y metodologías activas. Revista de Educación (número extraordinario, pp. 59–81) (2008). www.revistaeducación.mec.es/re2008/re2008_04.pdf

Martín, B (s/f): Contextos de aprendizaje: formales, no formales e informales. Consejo Nacional de Investigaciones Científicas and Técnicas. http://www.ehu.eus/ikastorratza/12_alea/contextos.pdf

Mattera, M., Baena, V., Ureña, R., Moreno, M.F.: Creativity in technology-enhanced experiential learning: Videocast implementation in higher education. Int. J. Technol. Enhan. Learn. 6(1), 46–64 (2014)

Romero, M.: El aprendizaje experiencial y las nuevas demandas formativas. Revista de Antropología Experimental, Especial Educación 10(8), 89–102 (2010)

Smith, M.K., Kolb, D.A.: Experiential learning. In: The encyclopedia of informal education (2001). http://www.infed.org/b-explrn.htm

Value Investment Using Stock Index

Luis Javier Saz Peñas, Raúl Gómez Martínez,
and Camilo Prado Román[⊠]

Universidad Rey Juan Carlos & European Academy of Management
and Business Economics, Madrid, Spain
ljsp015@gmail.com, {raul.gomez.martinez,
camilo.prado.roman}@urjc.es

Abstract. Portfolio management has been separated into the value investment approach, where managers select for their portfolios those assets that are cheaper according to their stock ratios, and growth management, which is based on the potentiality of the company. The Russell index segregate in their composition the stocks that are classified in one category to another. This study analyzes the profitability of the Russell value and growth indexes from 1979 to the present day. The study shows that value investment presents a better relation between profitability and risk which should be justified by ratios level and by the market's consideration.

Keywords: Value investment · Russell index · Market efficiency
Stock market · Enterprise value

1 Introduction

Value premium has been well studied by the financial theory since it was formulated by Benjamin Graham (1965). According to this author, investing in stocks quoted below its fair value (margin of safety) would beat the market.

As far as Capital Asset Pricing Model (CAPM) (Sharpe 1964) became a milestone to establish the relation between the performance and risk, value investing has been a challenge for this theory due to the fact that according some research a selection of stocks chosen according value criteria (Basu 1977; Lakonishok et al. 1994; Fama and French 1992) would beat the market with a similar amount of risk.

This investment style has found an explanation in the Asset Pricing Theory (APT) formulated by Ross (1976). According to this model assets risk can not be measured by a single factor (beta according to CAPM) but by a set of factors that contribute to the total risk.

Risk factors can be based on different considerations such as statistical by the use of principal component analysis, a second explanation would be based on macroeconomical factor and finally a third one would use fundamental factors that can be assimilated to different management styles.

This third explanation was studied in detail by Fama and French (1992, 2014). In these papers these authors identified Value, Size, Quality and Investment as the main sources of risk. Other popular styles are Growth, Yield or Momentum.

© Springer Nature Switzerland AG 2019
J. Gil-Lafuente et al. (Eds.): AEDEM 2017, SSDC 180, pp. 105–114, 2019.
https://doi.org/10.1007/978-3-030-00677-8_9

According to MSCI only factor that present a sizeable return and reflect exposure to systematic source of risk for long periods of time can be considered as risk premia factors.

The reason behind the academic interest in factor investing and in smart beta particularly has been well explained by the rate of growth of this investment style. According to FTSE Russell the amount of this type of investment is bigger than 1 Tn USD and 3 out 4 managers have expressed interest about it.

The explanation of this shift from the traditional investment process to the smart beta one is the popularization of Exchange Traded funds or ETF's that provide to investors with exposure to previous mentioned factors by paying lower fees than traditional funds.

Some statistics offered by Dimson, Marsh and Staunton estimate that, by the end of 2016, the number of ETFs could be higher than 6000 and the Asset under management would be around 3.5 Trillions of USD.

This article studies this value premium under a new approach by using stock indexes, in this case Russell indexes, that can be easily replicated by the investor using ETF's. Previous research (Rosenberg et al. 1984; Fama and French 1992; Lakonishok et al. 1994) has shown than a portfolio with a low price to book value offers higher return than the one built with a higher ratio. This analysis has also been studied using other ratios such as Price to earnings ratio (Basu 1977), dividend yield or price to cash flow.

Two schools of thought explain this outperformance, the first lead by Fama and French suggests that companies with a low price to book value are riskier than its peer with a lower ratio. The second school explains that the market does not follow these stocks due to its previous underperformance (Lakonishok et al. 1994).

2   Hypothesis and Methodology

This article analyzes this value premium under a new approach by using stock indexes, such as Russell 1000 and Russell 2000 that can be replicated by investors with ETF's. According to FTSE Russell: "Russell 3000 is a market capitalization index maintained by the FTSE Russell that provides exposure to the entire U.S. stock market. The index tracks the performance of the 3,000 largest U.S.-traded stocks which represent about 98% of all U.S. incorporated equity securities. Russell US Indexes are the leading US equity benchmarks for institutional investors. This broad range of US indexes allow investors to track current and historical market performance by specific size, investment style and other market characteristics.

All Russell US Indexes are subsets of the Russell 3000® Index, which includes the well-known large cap Russell 1000® Index and small cap Russell 2000® Index. The Russell US Indexes are designed as the building blocks of a broad range of financial products, such as index tracking funds, derivatives and Exchange Traded Funds (ETFs), as well as being performance benchmarks".

This article is focused in the Russell indexes and especially in Russell 1000 and Russell 2000 subsets because these indexes are focused in the United States market and the results can be easily compared to Fama and French (1992) database.

Theorical basis of value investing has focused on the segmentation of the market between "cheap" stocks and "expensive" stocks based on a set of criteria such as price to book value, earnings to price or price to cash flow.

Value premium was calculated based on the difference between value portfolio and growth portfolio (high minus low or HML in Fama and Frenc terminology. One of the main weakness of this approach for an investor could be how to short the expensive portfolio of the value premium especially in small and illiquid stocks.

The purpose of this paper is to analyze if value premium persists if value and growth portfolios are replaced by value and growth indexes from Russell universe. The main advantage of this approach is that only two assets are needed as far as long position could be taken by buying an ETFs while short position could be replicate by an inverse ETF or by taking a short position in the direct ETF.

Therefore, hypothesis to be tested is:

H1: Does value premium persist if the value and growth portfolios are replaced by value and stock indexes?

To test this hypothesis, we have compared the returns obtained by six assets based on different indexes calculated by Russell Investments, that are subsets of Russell 3000:

(a) Stocks that comprise Russell 1000 (1000 biggest companies by market capitalization of Russell 3000) under three different criteria: Russell 1000 Value, for value stocks, Russell 1000 growth, for growth stocks and Russell 1000. This index comprises approximately 90% of the market cap of Russell 3000.

(b) Stocks that comprise Russell 2000 (2000 smallest companies by market capitalization of Russell 3000) under three different criteria: Russell 2000 Value, for value stocks, Russell 2000 growth, for growth stocks and Russell 2000. This index comprises approximately 10% of the market cap of Russell 3000.

Hypothesis H1 will be accepted if return of indexes Russell value is higher than the rest of indexes during the period of control. If this hypothesis is accepted we will test if its size is comparable to Fama and French results.

3   Data

FTSE criteria to consider companies as "vale" is price to book value, furthermore as growth criteria FTSE has chosen growth forecast for the next 2 years published by Institutional Brokers Estimate System (I/B/E/S) and the historical growth for the last 5 years. By combining these criteria an algorithm gives a mark to every stock depending on its bias to value and growth.

The source of the data is Bloomberg. Return have been measured since first data are available (January 1979). In order to analyze the risk, we have used standard deviation. Additionally ten years rolling periods. Therefore 450 monthly observations per index has been studied since 1979.

4   Results

The following analysis about value premium tries to solve the theoretical construction of the value and growth portfolios of the previous research based in a long position in a value portfolio (value portfolio) minus a short position in a growth portfolio, HML in Fama and French terminology.

In this article we will study this value features supporting our conclusions in the performance of a subset or Russell indexes, the Total return value Index and the Total Return Growth index. These indexes fit this kind of research because two categories, according to the size can be identified, Russell 1000 for the big companies and Russell 2000 for the smallest ones.at the same time a further split based on value ratios is categorized.

An additional advantage in this approach is that long and short portfolio can be replicated by buying Exchange Trade Funds, direct ETFs would be used for the long position while inverse ETFs would be the instrument for the short one, if no inverse ETF would be available a synthetic position by selling short the ETF would be considered.

Value premium existence can be noted in Figs. 1 and 2 both in Russell 1000 index (big companies) and Russell 2000 (small companies).

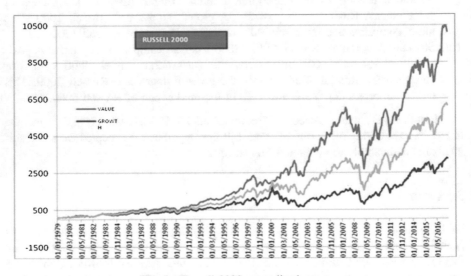

Fig. 1. Russell 2000 normalized return

Yearly return since January 1979 is detailed in the Table 1:

In order to test the first hypothesis, we have performed a risk analysis of these portfolios, we have used standard deviation as the method of comparison in line with Value at Risk analysis, and we will not use Beta of the portfolios due to the drawbacks of this method.

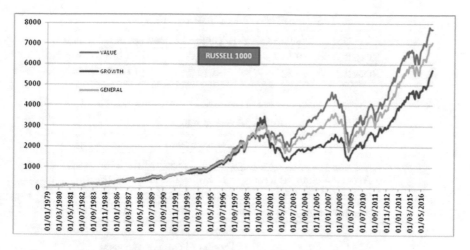

Fig. 2. Russell 1000 normalized return

Table 1. Russell indexes. Return analyzed

	Value	Growth	General
Russell 1000	11.99%	11.12%	11.74%
Russell 2000	12.79%	9.49%	11.32%

Table 2. Sharpe ratios

	Russell 1000		
	Value	Growth	General
Return	11.99%	11.12%	11.74%
Volatility	14.49%	16.96%	15.07%
Ratio Return/Volatility	0.83	0.66	0.78
	Russell 2000		
	Value	Growth	General
Return	12.792%	9.495%	11.323%
Volatility	17.26%	22.47%	19.35%
Ratio Return/Volatility	0.74	0.42	0.59

Table 3. Return on growth years

	Value - Growth	Value - General
RUSSELL 1000	1.74%	0.84%
RUSSELL 2000	3.73%	1.95%

Table 4. Return on negative growth years

	Value - Growth	Value - General	Rent value
Russell 1000	−9.65%	−4.84%	10.37%
Russell 2000	−6.62%	−3.14%	17.45%

Table 5. Average return

	Russell 1000		
	Value	Growth	General
Average Return 10 Years	11.36%	10.11%	10.94%
Worst Period 10 Years	−1.23%	−5.58%	−3.02%
Number of negative periods	4	27	21
	Russell 2000		
	Value	Growth	General
Average Return 10 Years	11.65%	7.59%	9.82%
Worst Period 10 Years	3.90%	−3.36%	1.22%
Number of negative periods	0	18	0

The figures obtained for the Sharpe ratios of every index relating performance and risk are summed up in the following table:

We have extended the concept of risk and we have analyzed how value investment has performed in negative years of the stock market and furthermore years with a decrease in the economy measured by the GDP.

Results are as follow for the GDP:

Second analysis is focused on negative years of the stock market and the results are:

Finally, we want to see how these strategies have performed analyzing its performance for long period of times in line with Benjamin Graham concept (1934) that defines "risk as the loss of value of an investment". To analyze this feature, we have studied long term performance in period of ten years rolling of a value portfolio vs a growth portfolio and an index. Results are as shown as follow:

To present this performance of value vs growth we have estimated the long-term ratio of these strategies in 10 year rolling periods for Russel 2000. The result is (Fig. 3):

Keeping in mind previous results, we could consider that there is a value premium, we estimate the size of this premium in the Table 6:

Once the existence of this premium has been stated we will compare its size with Fama and French results in order to conclude if an investor could obtain this value premium in an efficient way by taking a long and short position in these indexes.

Fig. 3. Value vs Growth

Table 6. Prime value

	Prime value	
	Value vs Growth	Value vs General
Russell 1000	0.87%	0.25%
Russell 2000	3.30%	1.47%

First at all it is necessary to establish some methodological differences between Russel and Fama and French. These differences can be summarized as:

- FTSE Russell uses different ratios such as Price to Book Value, historical growth and forecasted growth in order to obtain a composite value score (CVS) and according to this CVS a mark is assigned to every stock and the number or shares of this stock is split between both indexes. Therefor Fama and French criteria can be considered more restrictive than Russell
- Fama and French deal with size bias by equally weighting big and small companies:

$$HML = (BV + SV) - (BG + SG)$$

As can be seen, the classification of Fama and French is stricter. On one hand, strock can not be categorized as value or growth at the same time. On the other hand, it compares the extremes of the scale (30% more expensive vs 30% cheaper) while Russell would consider intermediate values.

To compare the results of both approaches, we need to generate Fama and French prime value as the arithmetic average of return (HML) since 1980 and we have compared these results with Russell 3000 total return value and growth since the same year (Fig. 4).

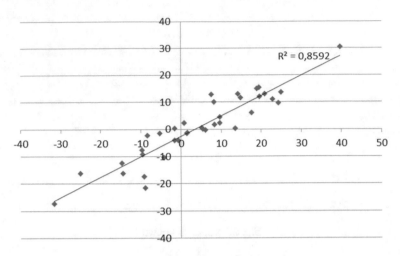

Fig. 4. Fama and French vs Russell 3000 prime value

Table 7. Prime value statistical significance

	Valor prima	t stats
Fama y French	4.56	1.86
Russell 3000	0.88	0.44

Our conclusions can be established as:

– Relation between both premium is very high as can be seen in this graph, so we can conclude that indexes capture value feature.

We can appreciate differences in the size of the value premium and in its statistical significance (Table 7).

The size of the value premium estimated by Fama and French is higher than the performance an investor could obtain by taking a long position in a value index and a short position in a growth index. From our point of view this difference can be explained by the methodologies used in the calculations. Investors should be aware of Fama and French stricter split between value and growth and the different exposure to size factor. In fact, the number of components in Russell 3000 value is 2090 stocks, while the first 30% of the index (Fama and French methodology) would be 1000 stocks.

As a conclusion value premium maximize its value if more extreme portfolios are constructed and size factor is considered.

5   Discussion, Conclusion and Implications

Firsts conclusions of this analysis show that an investment in value indexes such as Russell 1000 and Russell 2000 would have beaten the general index and the growth index, therefore value premium existence is established.

A second conclusion of this article, in line with previous research, is that value premium is more relevant in small companies than in big companies. This spread between Russell 2000 and Russell 1000 value premium is 2.43 points.

This value premium can be explained through two approaches.

- Value premium reflects a reward for assuming a higher risk by investing in lower quality companies being this premium a proxy for distress risk.
- Market extrapolate past rates of growth willing to pay higher multiples for the best performing companies in the past vs the underperformers.

As can be seen in Table 2, value strategies offer a higher return with a lower risk, measured by volatility in comparison with growth strategies or in the general index, therefore can be established than value strategies are more efficient in terms of risk to reward.

Table 3 shows that value investing can outperform growth in years in which stock market decline. Therefore, value investing should not be considered as speculative due to its defensive bias.

As Table 4 shows value investing underperform growth investing in years with a decline in GDP in the United States. A more in depth analysis shows than this underperformance does not imply a negative return of this style of investing but a relative underperformance due to the leading character of stocks markets vs the economy.

Main conclusions stated in Table 5 are:

- Value portfolio have outperformed a growth portfolio or a general market portfolio.
- If we define risk as permanent losses in long horizon periods (10 years) value investing shows small losses in big companies' indexes and no losses in small capitalization indexes.
- Number of negatives occurrences is smaller in value investing.
- Value premium existence is stated for the period with two episodes of extreme underperformance, as can be seen in the ratio between value and growth, these episodes are dotcom bubble during late 90s and global financial crisis since 2008.

It is relevant to note that this value premium is not a constant during the period of this study and a value investor could suffer period of underperformance.

To sum up main conclusion of this article are:

- It is shown than value premium exits and could be extracted by an investor through value and growth indexes.
- Size of this premium is higher for small companies.
- Value investing are not riskier (being risk measured as standard deviation), so value investing is more efficient in risk to reward consideration.
- Value stocks do not underperform in years of stock market declines.

- Size of this premiums is not constant in the time and investor could suffer period of underperfornance.
- Value premium obtained by investing in stock indexes is smaller than stated by Fama and French due to methodological differences in the construction of portfolios.

References

Basu, S.: Investment performance of common stocks in relation to their price-earnings ratios: a test of the efficient market hypothesis. J. Financ. **32**(3), 663–682 (1977)

Fama, E.F., French, K.R.: The cross-section of expected stock returns. J. Financ. **47**(2), 427–465 (1992)

Fama, E.F., French, K.R.: A five-factor asset pricing model. Fama-Miller Working Paper (2014)

Graham, B., Dodd, D.L.: Security Analysis: Principles and Technique. McGraw-Hill, New York (1934)

Graham, B.: The Intelligent Investor: A Book of Practical Counsel. Prabhat Prakashan, New Delhi (1965)

Lakonishok, J., Shleifer, A., Vishny, R.W.: Contrarian investment, extrapolation, and risk. J. Financ. **49**(5), 1541–1578 (1994)

Rosenberg, B., Reid, K., Lanstein, R.: Persuasive evidence of market inefficiency. J. Portf. Manag. **11**, 9–17 (1984)

Ross, S.A.: The arbitrage theory of capital asset pricing. J. Econ. Theor. **13**, 341–360 (1976)

Sharpe, W.F.: Capital asset prices: a theory of market equilibrium under conditions of risk. J. Financ. **19**(3), 425–442 (1964)

Board Resources and Firm Performance in SMEs

Marta Domínguez-CC[✉] and Carmen Barroso-Castro

Universidad de Sevilla, Seville, Spain
martad@us.es

Abstract. The objective of this paper is to analyse whether the knowledge, experience and capabilities of the board of directors aid the expansion of growth-oriented SMES. The sample consists of 200 board members of 32 firms listed on Spain's alternative investment market. Our results show that average board tenure, educational level and CEO experience all affect firm growth. This article highlights the particular importance of human capital for these firms and how it can improve their performance.

1 Introduction

The influence of the board of directors on firm performance has been widely analysed in the literature (Barroso et al. 2011), although there is little consensus on which of the directors' characteristics create an effective board (Johnson et al. 2013). This is an important question because company performance is largely dependent on the board's ability to improve its effectiveness (Bird et al. 2004). The literature has focused on the influence of the board in large firms, and it is absolutely essential to investigate other research contexts (Johnson et al. 2013). A particular characteristic of small and medium-sized enterprises (SMEs) is the significant number of shortcomings in their governing bodies. Moreover, these firms form the basis of the production network of many economic environments, and research into their future is therefore of particular relevance.

The aim of this work is to analyse the influence of the board among firms listed on the Spanish alternative investment market (MAB, which specialises in growth-oriented SMEs). As with the study of large firms quoted on the stock exchange (Barroso et al. 2011), we analyse which aspects of human and social capital found in the boards of these SMEs can help to explain their results. We define the human and social capital of the board as the set of knowledge and experience of the directors, as well as their collective ability to access the information and resources required to fulfil their roles in an effective manner (Kor and Sundaramurthy 2009).

Prior studies have focused on only part of the board, such as the directors who sit on particular committees, or external board members (Kor and Sundaramurthy 2009). However, it would make more sense to analyse the entire board, since the whole group takes the decisions that will facilitate the firm's growth.

This study makes important contributions to the literature: (i) it highlights the role of the board in the functioning of the firm, demonstrating how human and social capital

© Springer Nature Switzerland AG 2019
J. Gil-Lafuente et al. (Eds.): AEDEM 2017, SSDC 180, pp. 115–125, 2019.
https://doi.org/10.1007/978-3-030-00677-8_10

are key elements that allow this governing body to operate; (ii) it affects the relation between the board and the performance variables, specifically, firm growth; and (iii) it develops the field of study of boards of directors in a rarely researched context (SMEs).

2   Conceptual Framework

The literature has identified the board of directors as a valuable resource for a firm (Macus 2008), recognising its active role in a firm's performance (Pugliese et al. 2009). The normal functioning of the board requires the roles- control, service and resource provision -to be carried out simultaneously, moreover, it is extremely difficult to isolate each one in the course of day-to-day operations (Hillman and Dalziel 2003).

Given that boards meet only occasionally, their effectiveness depends, to a great extent, on participation, interaction and the exchange of information that takes place during board meetings (Forbes and Milliken 1999). Hillman and Dalziel (2003) combine the agency and resource dependency theories and propose that the board's human and social capital positively influences the fulfilment of all of its roles. It is not enough for board members simply to possess knowledge and capabilities; these elements must be integrated into the firm's internal processes (Macus 2008; Barroso-Castro et al. 2017) and the board must improve its motivation and ability to apply this knowledge to the decision-making process (Tian et al. 2011). It should be noted that the majority of these contributions are theoretical (Macus 2008; Hillman and Dalziel 2003, for example) or they relate to large companies. However, the literature also recognises the importance of the board for SMEs (Gabrielsson and Huse 2010; Yar Hamidi 2016; Calabrò et al. 2009), and for family-run SMEs in particular (Calabrò et al. 2016; Johannisson and Huse 2000).

The resource and capabilities theory (Barney 1991; Eisenhardt and Martin 2000) explains how board members' human and social capital can be integrated into the fabric of the firm and thereby achieve results that distinguish it from its competitors. From this perspective, the extent to which the board fulfils its role depends largely on the knowledge and capabilities of its members (Yar Hamidi 2016) and the combination (Kor and Sundaramurthy 2009) and assimilation of this knowledge into the firm's internal processes. This theory is particularly relevant to our study for a number of reasons: (i) it focuses on growing SMEs, where the governance structures are weakest (Mallin and Ow-Yong 1998) and the asymmetry of information is most marked (Palacín-Sánchez and Pérez-López 2016) making the board's role more important (Calabrò et al. 2009; Gordon et al. 2012) as it offsets the limited knowledge and capabilities of the management team (Hendry 2005) and provides access to resources and financing (Gabrielsson and Huse 2010); (ii) service is uppermost for this type of firm, which is the most prominent role in this theoretical perspective (Van den Heuvel et al. 2006); and (iii) the board's role of resource provider includes the requirement to act as advisor, to improve the firm's legitimacy, to provide access to basic resources such as capital, and to be the firm's link to the its environment and other stakeholders. These are all vital roles for growing SMEs (Calabrò et al. 2009).

Board tenure is linked to firm-specific knowledge. This knowledge can help to speed up decision making or ease the transmission of the firm's working patterns to new production or sales units. Tenure can therefore be seen as beneficial to the firm's performance (Miller 1991).

However, tenure can also have negative effects on performance. Directors operate and integrate their knowledge based on routines (Nelson and Winter 1982; Ocasio 1999; Zahra and Filatotchev 2004), which brings stability, but at the same time, might hinder and limit organisational actions (Nelson and Winter 1982). Board members' limited rationality may impede decision making, which might revolve around choices from a menu of routines learned over time (Winter 1995). Decisions regarding the firm's future must be based on the current environment however, without underestimating the environmental signals.

Board tenure also contributes to emphasise groupthink (Janis 1982). Communication within the group decreases, suppressing key information for the development of new strategies, and there tends to be a reaction against the introduction of strategic change (Golden and Zajac 2001), such as the firm's growth orientation. On the basis of these arguments, we propose:

Hypothesis 1: Board tenure has a negative effect on firm growth.

Functional diversity in the composition of the management teams makes it easier to attend simultaneously to the contradictory aspects associated with the exploration and exploitation of the firm's resources (Boeker 1997; Virany et al. 1992). The exploration of resources refers to the analysis of the environment in order to identify opportunities, while exploitation is more closely linked to the selection of strategies that have been successful for the firm in the past, in order to exploit the knowledge acquired during this period. For firms with an eye to growth, the exploration of resources, which creates internal variety, leading to new markets and products (March 1991; Rosenkopf and Nerkar 2001), is just as important as exploitation, which ensures operational efficiency (Beckman et al. 2004; March 1991).

Board diversity can also prevent possible problems in the functioning of interdependent teams that meet only sporadically (Forbes and Milliken 1999). Some studies stress the importance for SMEs to include directors with particular areas of specialisation (Calabrò et al. 2009; Gabrielsson and Winlund 2000). Functional diversity in the board's composition favours its complementarity, allowing board members' knowledge to be integrated, building on each director's specific capabilities. This should be encouraged in order to enable the generation of alternative ideas (Minichilli and Hansen 2007) and improve board effectiveness (Adobor 2004) and firm growth.

Based on these arguments, we propose:

Hypothesis 2: The functional diversity of the board has a positive effect on firm growth.

Human capital is defined as individuals' knowledge and capabilities that have been acquired through their education, training and experience (Becker 1993). While the level of education can be viewed as a demographic variable, it indicates the level of the board's experience or skills (Johnson et al. 2013). We therefore consider it to be part of the board's human capital because it determines the level of knowledge and skills that are available for effectively performing its roles. The educational level equips the board to seek alternative solutions to the organisation's problems (Wincent et al. 2009). The directors' knowledge and skills are the basis of their ability to carry out their service role effectively (Calabrò et al. 2009), which is a priority for SMEs (Van den Heuvel et al. 2006), while the lack of training can cause problems for the firm's functioning in the short term (Bennett and Robson 2004). Directors with higher levels of education are more likely to participate and to support complex decisions, because they can turn to their acquired knowledge to make these decisions. Furthermore, the educational level is also related to the openness to innovation (Wincent et al. 2009) that is characteristic of the environment of growth-oriented SMEs. These arguments allow us to propose the following hypothesis:

Hypothesis 3: The academic level of the board is positively related to firm growth.

Board members' prior experience as CEO can help the firm's growth. The CEO has the highest executive responsibility and is therefore used to taking decisions in a complex environment, has a global vision and understands the importance of organising resources in such a way that the senior management functions as a team (Sirmon et al. 2011). Directors with CEO experience acquire tacit knowledge relating to the solution of strategic problems, which is not easy to imitate. This intangible resource, acquired from their experience as CEO, can be uniquely integrated into the firm's resources, creating key competences (Barroso et al. 2011) and making it easier to take bold decisions, such as the firm's growth orientation.

CEOs tend to belong to business élites (Barroso et al. 2011), which can improve their access to fundamental resources (Jaw and Lin 2009), especially for SMEs (Gabrielsson and Huse 2010), increasing the range of possibilities for growth and reducing internal information bias. On the basis of these arguments, we propose:

Hypothesis 4: Board members' experience as CEO has a positive effect on firm growth.

International experience can lead to the acquisition of tacit knowledge, which is a source of competitive advantage (Barney 1991) in managing the firm's growth. Directors' international experience creates vital links to official institutions, foreign companies and other networks in foreign markets (Barroso et al. 2011). These international networks provide access to and the interpretation of information which reduces uncertainty and lowers the resistance to taking decisions concerning international growth (Zahra et al. 2007). Similarly, board members with international training have significant experience they can draw on in their proactive search for growth opportunities (Barroso et al. 2011). We therefore propose:

Hypothesis 5 The international experience of the board is positively related to firm growth.

3 Methodology

3.1 Sample

The sample consists of every firm listed on MAB in 2015, with the exception of one firm, whose quotation was suspended. The sample analysed therefore comprises the

200 board members of the 32 active firms in 2015. This type of market is largely oriented towards SMEs undergoing expansion. While Spain has only recently set out on this path, there are a number of markets that MAB is modelled on, such as London's AIM (active since 1995) and Alternex in Paris (2005), which have been in existence for longer.

Access to finance continues to be a crucial driver for competitiveness, innovation and growth, especially for SMEs. Most SMEs tend to turn in the first instance to financial intermediaries, but this method of financing may be insufficient and is a very expensive way of funding ambitious growth projects. In these circumstances, MAB may be a good alternative, as its regulation, costs and admission processes are geared towards SMEs. These are the companies that can contribute the most to the country's economic growth. In our analysis of these MAB companies, we are analysing the most innovative SMEs and those with the greatest potential. These are generally innovative companies belonging to sectors with strong links to new technologies: electronics and software; biotechnology; telecommunications and engineering, among others.

The composition and behaviour of Spanish boards reveal a number of character-istics that distinguish them from other European countries (Heidrick and Struggles 2009): (i) their unitary system of governance, which is similar to that of Italy, the United Kingdom and Portugal, consisting of a single board formed of executive and non-executive directors; (ii) their low level of independence, with around 30% of external or non-executive directors; and (iii) the high concentration of ownership among Spanish firms that is much higher than in other European countries. Tenure on Spanish boards is generally long and in recent years the number of meetings per year has begun to increase (11.4, compared to 9.6 in Europe), which reflects the growing involvement of boards in this country. The SMEs listed on MAB aim to comply with the codes of good governance expected of firms quoted on the primary exchange. In 2015, therefore, the level of board independence among MAB companies was higher than the average for Spanish firms, with 26% of internal directors in 2015. The average board tenure for these MAB-listed firms is four years.

The data were obtained from the MAB admission documents published on its website. After analysing important facts relating to the various changes that have taken place over the years, we were able to determine the composition of the board of each firm in 2015. We added to the information relating to the directors' knowledge and capabilities by accessing other sources, such as LexisNexis, professional social net-works (LinkedIn) and economic websites (Bloomberg). Data on firm growth between 2015 and 2016 were obtained from the financial analyses available from MAB.

3.2   Dependent Variable

Prior studies indicate that sales variance is an excellent performance indicator (Boeker and Goodstein 1993), and even that it is the indicator most often used to measure business success (Kor and Sundaramurthy 2009), because in dynamic environments it involves the firm's ability to maintain continuity and speed in both innovation and commercialisation (Eisenhardt and Martin 2000), enabling it to continue its operations and investment (Kor and Sundaramurthy 2009). We therefore measure firm growth as the rate of sales variance in the year 2016.

3.3   Independent Variables

Average board tenure is calculated on the basis of the tenure of each director. To take this measurement, we calculated the number of years that the director has served on the board as the difference between 2016 and the year of their appointment. We then calculated the average of all of the directors' years of tenure (Barroso et al. 2011; Golden and Zajac 2001; Kor and Sundaramurthy 2009).

With regard to general management experience, if a director has CEO experience, this is coded as 1, and 0 in the opposite case.

To calculate the functional diversity of the board we followed to Tuggle et al. (2010). Taking account of the current and previous experience of the board members in the fields of production or engineering, finance, commerce or marketing, human resources or other areas, we calculated the Blau heterogeneity index.

BlauIndex $= 1-\Sigma p_i^2$; where p is the proportion of directors with functional specialisation i.

The international experience of each board member includes both their work experience and training abroad. To achieve this we created a dichotomous variable where 1 indicates that the director either gained higher education abroad or has occupied or is currently occupying a post abroad (Barroso et al. 2011). Otherwise, it is coded as 0.

The information on each director for every variable was subsequently added at board level. First, the percentage of board members possessing each of the resources analysed was calculated (Barroso et al. 2011; Tian et al. 2011). This information led to a dichotomous variable to indicate the board's possession of each resource. If the percentage of board members with a particular resource is above 30%, it is coded as 1, and 0 in the opposite case.

With regard to educational level of the board was measured as the proportion of directors with a university degree (Wincent et al. 2009). The majority of boards had a high proportion of university graduates. It was not therefore possible to use the 30% figure to analyse the effect of this characteristic on firm growth, and instead we used the average value (0.8). If the percentage of directors with a university degree is above 80% it is coded as 1, otherwise it is coded as 0.

Prior literature indicates that the relation between corporate governance and performance may be affected by other variables. We therefore introduced firm size (Barroso et al., 2011), sectoral activity (Domínguez-CC and Cauzo 2015) and firm age (Barroso et al. 2011; Zahra et al. 2007) as control variables. Size was measured as the logarithm of sales in 2015. Firm age was measured as the number of years that it has existed in the market.

The sector is also an important control variable for measuring board effectiveness (Johnson et al. 2013). The firms in our sample belong to the following sectors: (1) Renewable energy; (2) Pharmaceutical and biotechnology products; (3) Electronics, software and telecommunications; (4) Other: engineering, services, commerce. We created three dummy variables for sectors 2, 3 and 4, using sector 1 as our reference.

4 Results

Table 1 sets out the descriptive analysis of the data and correlations. The variance inflation factor (VIF) of the independent variables was below 3, allowing us to rule out any problems of collinearity.

Table 1. Descriptors and correlations

	Mean	Stand. Dev.	Min.	Max.	1	2	3	4	5	6	7
1. SalesGrowth	0.126	0.753	−0.88	3.93	1						
2. Tenure	4.224	3.137	1	18	−0.1	1					
3. InternationalExp	0.507	0.338	0	1.29	0.16	0.116	1				
4. CEOExp	0.436	0.219	0.11	0.89	−0.1	−0.05	0.06	1			
5. FunctionalDiv	0.301	0.339	−0.36	0.89	0.04	0.315*	−0.1	−0.1	1		
6. AcademicLevel	0.798	0.251	0.17	1	0	−0.15	0.23	0.14	−0.3	1	
7. Size	14.95	2.449	8.4	18.23	−0.2	0.267	0.05	0.24	−0.1	0.19	1
8. Age	12.82	7.659	2	31	−0.2	0.425**	0.14	0.13	0.13	−0.2	0.356**

*Significance = 0.10 (bilateral)
**Significance = 0.05 (bilateral)

Table 2 shows the results of the regression analysis. The analysis confirms Hypothesis 1 ($\beta = -0.594$ model 5) that suggests that board members' average tenure is detrimental to firm growth. Hypothesis 3 indicates that the board's academic attainment has a positive effect on firm growth. The results obtained show a significant relation ($p < 0.05$) but one that is the inverse of our hypothesis. Hypothesis 4 proposes that the board's experience of general management is positive for firm growth and our analysis confirms this relation ($\beta = 0.924$). Our results do not support hypotheses 2 and 5.

Table 2. Regression analysis

Variables	Mod. 1		Mod. 2		Mod. 3		Mod. 4		Mod. 5	
	β	p-value	β	p-value	β	p-value	β	p-value	β	p-value
Constant	−0.227	0.78	−0.276	0.74	−0.269	0.742	−0.291	0.722	−0.192	0.802
FirmAge	0.170	0.447	0.132	0.578	0.092	0.695	0.095	0.687	0.014	0.952
Size	0.059	0.782	0.084	0.705	0.096	0.662	0.102	0.642	0.237	0.28
Sector 2	−0.252	0.178	−0.289	0.194	0.482	0.373	0.456	0.112	0.365	0.473
Sector 3	0.402	0.444	0.345	0.525	0.243	0.651	0.305	0.575	−0.006	0.991
Sector 4	0.707	0.251	0.693	0.246	0.623	0.266	0.605	0.283	0.469	0.375
Tenure	0.644	0.259	0.663	0.224	−0.418	0.101	−0.408	0.112	−0.594	0.026
CEOExp.			0.126	0.593	0.184	0.436	0.661	0.246	0.924	0.099
FunctionalDiv.					0.326	0.176	0.375	0.132	0.385	0.101
InternationalExp							−0.545	0.355	−0.595	0.283
AcademicLevel									−0.432	0.055
R	0.368		0.382		0.461		0.494		0.606	
R squared	136		0.146		0.213		0.244		0.368	
Adjusted R	−0.072		−0.103		−0.061		−0.066		0.067	
Standard Error	0.524		0.532		0.522		0.523		0.48	
Nº of observations	32		32		32		32		32	

5 Discussion and Conclusions

This study makes a number of contributions to the literature: (i) it is a more in-depth investigation into the relation between the human and social capital of the board and firm growth in a field that is infrequently studied –SMEs that list on alternative investment markets; (ii) it affects the active role of the board in decision-making and the firm's results; (iii) in line with previous studies (Macus 2008; Barroso et al. 2011; Pérez-Calero et al. 2016), it highlights the role of the resources and capabilities theory in explaining how the board's resources and capabilities can be integrated into the firm's internal processes to produce unique results for the firm.

The results obtained, in line with other works (Kor and Sundaramurthy 2009), show that the tendency of some boards to extend their average tenure has negative effects on sales growth. Long-serving directors in this role become much more inflexible and less proactive when faced with innovative proposals, as board members' human and social capital deteriorates over time (Lester et al. 2008). The relation between the board's academic achievement and firm performance are contrary to our expectations. These unexpected results might be explained by the high academic attainment of our sample that was recorded by the method used (0.8) and the particular characteristics of these firms. In future investigations, aspects linked to the average should be refined and a third explanatory variable included for this relation.

Our study point out that board members with experience as CEO are better able to advise and act as consultants on strategic matters and to take decisions in complex environments. Furthermore, directors who have enjoyed these positions may wish to maintain the prestige (Johnson et al. 2011), taking an active part in the board and contributing to the firm's growth.

Our investigation does not confirm that the board's functional diversity and international experience affect SME growth. We consider that our results are explained by the peculiarities of the functional composition of these boards and the directors' lower level of international experience (despite being firms that are likely to internationalise). These two aspects should be analysed more thoroughly in the future, as they will have a considerable effect on the evolution of these firms.

This study has some limitations. Although it includes every firm that was quoted on MAB in 2015, the number of observations is limited (with regard to the number of firms, rather than the number of directors). It would also be of interest to include a longitudinal rather than a cross-sectional study and to refine some of the measurements used for our variables. These limitations all point to future lines of research.

References

Adobor, H.: Selecting management talent for joint ventures: a suggested framework. Hum. Resour. Manag. Rev. **14**(2), 161–178 (2004)

Barney, J.: Firm resources and sustained competitive advantage. J. Manag. **17**, 99–120 (1991)

Barroso, C., Villegas, M.M., Pérez-Calero, L.: Board influence on a firm's internationalization. Corp. Governance: Int. Rev. **19**(4), 351–367 (2011)

Barroso-Castro, C., Villegas-Periñan, M.M., Dominguez, M.: Board members' contribution to strategy: the mediating role of board internal processes. Eur. Res. Manag. Bus Econ. **23**(2), 82–89 (2017)

Becker, G.S.: Human Capital. Columbia University Press, New York (1993)

Beckman, C.M., Haunschild, P.R., Phillips, D.J.: Friends or strangers? Firm-specific uncertainty, market uncertainty, and network partner selection. Organ. Sci. **15**(3), 259–275 (2004)

Bennett, R.J., Robson, P.J.A.: The role of boards of directors in small and medium-sized firms. J. Small Bus. Enterp. Dev. **11**(1), 95–113 (2004)

Bird, A., Buchanan, R., Rogers, P.: The seven habits of an effective board. Eur. Bus. J. **16**, 128–135 (2004)

Boeker, W.: Strategic change: the influence of managerial characteristics and organizational growth. Acad. Manag. J. **40**(1), 152–170 (1997)

Boeker, W., Goodstein, J.: Performance and successor choice: the moderating effects of governance and ownership. Acad. Manag. J. **36**(1), 172–186 (1993)

Calabrò, A., Brogi, M., Torchia, M.: What really matter in the internationalization of small and medium-sized family businesses? J. Small Bus. Manag. **54**(2), 679–696 (2016)

Calabrò, A., Mussolino, D., Huse, M.: The role of board of directors in the internationalisation process of small and medium sized family businesses. Int. J. Glob. Small Bus. **3**(4), 393–411 (2009)

Domínguez-CC, M., Cauzo-Bottala, L.: Consecuencias de un cambio de CEO: efectos a largo plazo sobre el equipo directivo y el rendimiento de la empresa. Span J. Finan. Account. **44**(1), 72–96 (2015)

Eisenhardt, K.M., Martin, J.A.: Dynamic capabilities: what are they? Strateg. Manag. J. **21**, 1105–1121 (2000)

Forbes, D.P., Milliken, F.J.: Cognition and corporate governance: understanding boards of directors as strategic decision-making groups. Acad. Manag. Rev. **24**, 489–505 (1999)

Gabrielsson, J., Winlund, H.: Boards of directors in small and medium-sized industrial firms: examining the effects of the board's working style on board task performance. Ent. Reg. Dev. **12**, 311–330 (2000)

Gabrielsson, J., Huse, M.: Governance theory: origins and implications for researching boards and governance in entrepreneurial firms. In: The Historical Foundations of Entrepreneurship Research, pp. 229–255. Edward Elgar, London (2010)

Golden, B.R., Zajac, E.J.: When will boards influence strategy? Inclination × power = strategic change. Strateg. Manag. J. **22**(12), 1087–1111 (2001)

Gordon, I.M., Hrazdil, K., Shapiro, D.: Corporate governance in publicly traded small firms: a study of Canadian venture exchange companies. Bus. Horiz. **55**(6), 583–591 (2012)

Heidrick & Struggles: Board in Turbulent Times: corporate Governance Report 2009. Heidrick & Struggles International, Inc., Paris (2009)

Hendry, J.: Beyond self-interest: agency theory and the board in a satisficing world. Br. J. Manag. **16**(s1) (2005)

Hillman, A.J., Dalziel, T.: Boards of directors and firm performance: integrating agency and resource dependence perspectives. Acad. Manag. Rev. **28**, 383–396 (2003)

Janis, I.L.: Groupthink: Psychological Studies of Policy Decisions and Fiascoes (1982)

Jaw, Y.L., Lin, W.T.: Corporate elite characteristics and firm's internationalization: CEO level and TMT- level roles. Int. J. Hum. Resour. Manag. **20**, 220–233 (2009)

Johannisson, B., Huse, M.: Recruiting outside board members in the small family business: an ideological challenge. Ent. Reg. Dev. **12**(4), 353–378 (2000)

Johnson, S.G., Schnatterly, K., Hill, A.D.: Board composition beyond independence: social capital, human capital, and demographics. J. Manag. **39**(1), 232–262 (2013)

Johnson, S., Schnatterly, K., Bolton, J.F., Tuggle, C.: Antecedents of new director social capital. J. Manag. Stud. **48**(8), 1782–1803 (2011)

Kor, Y.Y., Sundaramurthy, C.: Experience-based human capital and social capital of outside directors. J. Manag. **35**, 981–1006 (2009)

Lester, R., Hillman, A., Zardkoohi, A., Cannella, A.: Former government officials as outside directors: the role of human and social capital. Acad. Manag. J. **51**, 999–1013 (2008)

Macus, M.: Board capability; an interactions perspective on boards of directors and firm performance. Int. Stud. Manag. Organ. **38**(3), 98–116 (2008)

Mallin, C., Ow-Yong, K.: Corporate governance in small companies–the alternative investment market. Corp. Governance: Int. Rev. **6**(4), 224–232 (1998)

March, J.G.: Exploration and exploitation in organizational learning. Organ. Sci. **2**(1), 71–87 (1991)

Miller, D.: Stale in the saddle: CEO tenure and the match between organization and environment. Manag. Sci. **37**(1), 34–52 (1991)

Minichilli, A., Hansen, C.: The board advisory tasks in small firms and the event of crises. J. Manag. Governance **11**, 5–22 (2007)

Nelson, R., Winter, S.: An Evolutionary Theory of Economic Change. Belknap Press of Harvard University Press, Cambridge (1982)

Ocasio, W.: Institutionalized action and corporate governance: the reliance on rules of CEO succession. Adm. Sci. Q. **44**, 384–416 (1999)

Palacín-Sánchez, M.J., Pérez-López, C.: Los mercados alternativos bursátiles: una perspectiva regulatoria. Cuadernos de Economía **39**(109), 1–11 (2016)

Pérez-Calero, L., Villegas, M.D.M., Barroso, C.: A framework for board capital. Corp. Gov. **16** (3), 452–475 (2016)

Pugliese, A., Bezemer, P.-J., Zattoni, A., Huse, M., Van den Bosch, F.A.J., Volberda, H.W.: Boards of directors' contribution to strategy: a literature review and research agenda. Corp. Governance: Int. Rev. **17**, 292–306 (2009)

Rosenkopf, L., Nerkar, A.: Beyond local search: boundary-spanning, exploration, and impact in the optical disk industry. Strateg. Manag. J. **22**(4), 287–306 (2001)

Sirmon, D.G., Hitt, M.A., Ireland, R.D., Gilbert, B.A.: Resource orchestration to create competitive advantage: breadth, depth, and life cycle effects. J. Manag. **37**(5), 1390–1412 (2011)

Tian, J.J., Haleblian, J.J., Rajagopalan, N.: The effects of board human and social capital on investor reactions to new CEO selection. Strateg. Manag. J. **32**(7), 731–747 (2011)

Tuggle, C.S., Schnatterly, K., Johnson, R.A.: Attention patterns in the boardroom: how board composition and processes affect discussion of entrepreneurial issues. Acad. Manag. J. **53**(3), 550–571 (2010)

Van den Heuvel, J., Van Gils, A., Voordeckers, W.: Board roles in small and medium-sized family businesses: performance and importance. Corp. Governance: Int. Rev. **14**(5), 467–485 (2006)

Virany, B., Tushman, M.L., Romanelli, E.: Executive succession and organization outcomes in turbulent environments: an organization learning approach. Organ. Sci. **3**(1), 72–91 (1992)

Wincent, J., Anokhin, S., Boter, H.: Network board continuity and effectiveness of open innovation in Swedish strategic small-firm networks. R&D Manag. **39**(1), 55–67 (2009)

Winter, S.G. Four Rs of profitability: rents, resources, routines and replication. In: Montgomery, C.A. (ed.) Resource-Based and Evolutionary Theories of the Firm: Towards a Synthesis, pp. 147–178. Kluwer Academic Publishers, Norwell (1995)

Yar Hamidi, D.: Governance for innovation–board leadership and value creation in entrepreneurial firms. Doctoral dissertation, Halmstad University Press (2016)

Zahra, S.A., Filatotchev, I.: Governance of the entrepreneurial threshold firms: a knowledge based perspective. J. Manag. Stud. **41**, 885–897 (2004)

Zahra, S.A., Neubaum, D.O., Naldi, L.: The effects of ownership and governance on SMEs' international knowledge-based resources. Small Bus. Econ. **9**, 309–327 (2007)

Artificial Intelligence and Changing Paradigm in Healthcare

Domenico Marino[1]([⊠]) (iD), Antonio Miceli[2], and Giuseppe Quattrone[3]

[1] Università Mediterranea di Reggio Calabria, Reggio Calabria, Italy
dmarino@unirc.it
[2] Università di Messina, Messina, Italy
miceli@universitaericerca.it
[3] GTechnology, Modena, Italy
gquattrone1@gmail.com

Abstract. An important aspect that should be explored by the end of this decade is how the technologies of artificial intelligence, as applied in the health context, might ultimately improve the quality of the current system and whether the work done as part of the efforts now being made is optimised and sufficient to achieve new objectives. In particular, the ability to process large quantities of data will act as a catalyst, triggering an extremely high number of benefits in the health and wellness sector in terms of prevention, diagnosis and individual treatment.

1 Introduction

Health monitoring, crisis prevention and support for everyday activities represents an emerging field of application at a national level, with particular reference to fragile individuals, the elderly and people with chronic diseases. With this in mind, the prevention of functional decline and the treatment of physical frailty and cognitive weakness take on particular importance, as does the development of solutions for independent living, including by studying new diagnostic models and monitoring tools capable of providing for the clinical risk and at the same time reducing the related health and care costs.

An important aspect that should be explored by the end of this decade is how the technologies of artificial intelligence, as applied in the health context, might ultimately improve the quality of the current system and whether the work done as part of the efforts now being made is optimised and sufficient to achieve new objectives. In particular, the ability to process large quantities of data will act as a catalyst, triggering an extremely high number of benefits in the health and wellness sector in terms of prevention, diagnosis and individual treatment.

© Springer Nature Switzerland AG 2019
J. Gil-Lafuente et al. (Eds.): AEDEM 2017, SSDC 180, pp. 126–130, 2019.
https://doi.org/10.1007/978-3-030-00677-8_11

2   The Deficit Repayment Plans and the Italian Case

Despite the structural initiatives described above, the persisting deficits led the national authorities to intervene on spending levels with Budget Law 2005 (Article 1(180) of Law 311/04) which required Regional governments with an economic-financial imbalance to conclude an agreement with the Ministry of Health and the Ministry of the Economy and Finance. The agreement would set out the actions necessary to rebalance the Regions' budgets while ensuring the delivery of minimum standards of healthcare. This agreement led to a plan for the reorganisation, upgrading and strengthening of the Regional health service, which is called Deficit repayment plan (*Piano di Rientro*). Under the agreement, an initial phase focused on identifying the general and specific objectives for achieving the expected result. This was followed by actions and implementing decisions to launch the system reorganisation process and by subsequent monitoring based on a set of healthcare and economic-financial indicators.

This measure, with a roadmap and intermediate milestones established for each Region, translates operationally into streamlining of the service, which often involves further cuts to the number of beds, the closure of hospital departments with a small number of admissions, and the conversion or closure of hospitals considered not to be cost-effective.

Currently, the Regions with a Deficit Repayment Plan are mostly located in Southern Italy, plus Lazio and Piedmont (Fig. 1). The Region of Liguria and Sardinia have successfully passed the audit and have been granted the resources conditional on fulfilment of the conditions established in the Deficit Repayment Plan.

The strong deficits of most Southern Italian Regions clearly led the national Government to use invasive yet necessary instruments such as the above-mentioned agreements. However, it should be noted that the geographical areas subject to intervention on average seem to have low capacity to self-finance their healthcare expenditure compared with the other Regions of Central and Northern Italy. Therefore, any action to streamline their service will necessarily require significant cuts to the resources available. This notwithstanding, a purely economic-financial assessment would shift attention away from the main focus which should always be to plan the health service around the needs of citizens. These needs must be met at the appropriate standards as established by law, and ensuring adequate access to the service.

3   The Impact of Changing Paradigm in Healthcare

Achieving this paradigm shift in healthcare systems involves different focusing of care levels and healthcare provision channels. While the hospital would continue to play a key role in intensive care, community and home-based care would be the focus of the prevention effort and of monitoring to avert acute episodes.

This process, in which Italy is lagging behind other western European countries, requires increasingly holistic and personalised healthcare solutions. This approach focuses on systems for monitoring vital signs, on prevention and on patient well-being; it should be achieved by strategies ensuring quality and cost effectiveness.

Fig. 1. Regions with financial deficits - year 2012 *Source: Ministero della Salute*

The financial burden of healthcare delivery is on course to become unsustainable. On average, it absorbs 10.3% of the national GDP in Europe. In Italy, the figure is about 9.1%. This figure is set to rise in parallel with the steady increase in chronic diseases, which account for 75% of expenditure, also on account of the progressive ageing of the population. Projections show that the percentage of people over 65 will rise from 21% at present to 34% in 2051. Furthermore, the ratio of elderly people (over 65) to younger individuals (up to 64 years) is set to increase by more than 1/3 (nowadays there are 3 young people for every elderly person, in 2051 the ratio will be 1.9 younger persons per elderly person).

As concerns the commitments made with the Italian Digital Agenda for the eHealth service, between the present time and 2020 public funding of healthcare should rise from EUR 2 to EUR 7.8 billion, reaching in 2020 a total estimated commitment between EUR 9.5 and 10.2 billion. Part of these funds should be used to develop Telemedicine services.

Looking at the growth of demand for home care (both public and private), this coincides largely with the "growth in telemedicine services".

Starting with Hippocrates, basing medicine on the observation of events has long been the guiding epistemological criterion of the healthcare profession.

This criterion evolved as medicine progressed, resulting in the formula of *Evidence-Based Medicine (EMB)*, which can be defined as "the process of systematic search, assessment and use of the results of contemporary research as a basis for clinical decisions" or also as "the use of mathematical estimates of risks, benefits and damage derived from high-quality studies conducted on population samples to support the clinical decision-making process in diagnosis or in the management of individual patients". The possibility of using big data and artificial intelligence has had a strong impact on this epistemological assumption of present-day clinical practice. The use of big data and artificial intelligence is ushering in a type of medicine based on elements that are not apparent to human doctors but can be extracted by using big data and deep learning techniques, due to the ability of computers to cover and process a far larger amount of information than a human being.

Ordinary tools of analysis, although not yet in use in certain areas of the country, are becoming obsolete. This requires rapid adjustments also to the regulatory framework. The Gelli law on medical liability is based on the by now outdated concept of Evidence-Based Medicine and fails entirely to consider the "non-evident evidence" of big data and artificial intelligence.

It is therefore necessary to update rapidly the legislation relating to evidence-based medicine to make it compatible with predictive medicine. This must be accompanied by reform and re-engineering of the health service programming cycle.

Today, by using big data and deep learning techniques we can deliver effective preventive medicine long before the onset of symptoms. For chronic and degenerative diseases, this provides a significant advantage. Instant access to the entire set of data makes it possible to plan evolution of the clinical presentation by means of algorithms supporting decision-making, improving the overall efficiency of the process. The overall process is constructivist and is aimed at delivering significant benefits in terms of patients' treatment and care.

The diagnostic and care model also based on the patient's personalised electronic medical record will respond to the demand for increasingly effective, efficient and high-quality diagnosis, prognosis and treatment services. A good trade-off between quality of service and implementation costs can be achieved via the application of innovative technologies, systems and procedures for management of the clinical process, based on an e-Health Service Management logic. Creation of the electronic medical record, constantly updated with data from remote monitoring will favour very early diagnosis of many diseases, the identification of risks and the remote delivery of treatment and care. Health status monitoring, prevention of acute episodes and support in daily life are all emerging areas for e-health services, in particular for fragile and elderly individuals and people with chronic diseases.

This revolution can help to cut significantly the costs of healthcare, by reducing sharply the number of acute cases, preventing the development of many chronic diseases and delivering tele assistance and telemedicine. (3)

To ensure that this revolution is achieved, the regulatory framework, health policies and clinical approaches to diseases must be adjusted as appropriate. A new, "4.0" generation of doctors must be trained, and much remains to be done in this area.

4 Conclusion

The analysis performed on the Deficit repayment plan has shown a degree of imbalance between the Regions in terms of resources allocated and of the associated volumes of healthcare activity produced. In particular, study of the structure of the hospital services available, compared with the demand for healthcare shows a mismatch between demand and supply, which is only partly explained by the different geographical and population sizes of the different Regions. The creation of a telemedicine and remote monitoring system, will offer support for the long-term care of/provision for diabetic diseases by: (i) guaranteeing continuity of care at a hospital and regional level, (ii) integrating social and healthcare activities, (iii) favouring the continuation of such activities within the patient's own living environment for as long as possible and (iv) improving the patient's quality of life, in addition to providing improved support for diagnosis and treatment.

This approach is consistent with both international and national strategies for innovation, above all since it is developing new decision-making paradigms. As the result of the instantaneous access to the entire data set, provision can be made for the development of the health record through decision-making support algorithms, which make the entire process more efficient.

This revolution can help to cut significantly the costs of healthcare, by reducing sharply the number of acute cases, preventing the development of many chronic diseases and delivering tele assistance and telemedicine.

Reference

Marino, D., Quattrone, G.: Mobilità sanitaria, prime valutazioni (2018, forthcoming)

Assessment Evolution: Introduction of Experiential Learning, Use of ICT and Influence on Academic Results and Performance

Verónica Baena-Graciá[✉], Miriam Jiménez-Bernal,
and Elisabet Marina-Sanz

Universidad Europea de Madrid, Madrid, Spain
veronica.baena@universidadeuropea.es

Abstract. Innovation in Higher Education includes not only the implementation of active methodologies, inclusion of ICT (Information and Communication Technologies) and a changing teacher role, but also the development of new tools and techniques for assessing content acquisition and skills. In this paper, the authors present a preliminary study on the correlation between the implementation of active methodologies, the development of new tools to assess content acquisition and skills and the academic results in several courses belonging to the Degrees in Pre-primary and Primary Education. Modalities include blended-learning and traditional face-to-face learning, and the groups of students vary from less than 10 to 40.

1 Introduction

Adaptation to the European Space for Higher Education and the apparition of new degrees imply not only modifications in any syllabus, what involves a complete reorganization and restructuration, but also a significant change regarding the teaching and learning methods used. With this educational framework in mind, the concept of competence or skills is considered to be intrinsically linked to professional action and performance, understanding that the development of this performance requires the participation of diverse strategies and resources that depend on the Higher Education institutions as well as on external partners, such as companies, associations, NGOs and collaborating entities (Grimaldo Moreno and Arevalillo-Herráez 2011).

Also, this new educational scenario requires rethinking the assessment procedures —entailing activities that are significant in real contexts and practical exercises that reflect the students' professional development, as well as a great urgency for a methodological change (De Miguel Díaz 2006). In this sense, the assessment systems

This research was supported in part by funds received from the David A. Wilson Award for Excellence in Teaching and Learning, which was created by the Laureate International Universities network to support research focused on teaching and learning. For more information on the award or Laureate, please visit www.laureate.net.

© Springer Nature Switzerland AG 2019
J. Gil-Lafuente et al. (Eds.): AEDEM 2017, SSDC 180, pp. 131–142, 2019.
https://doi.org/10.1007/978-3-030-00677-8_12

that have been traditionally used in Higher Education have mostly emphasized theoretical knowledge. Assessment in this stage is a curricular element that, after the university transformation, is suffering relevant modifications from a teacher centered model—or learning assessment—to a student centered model—or assessment for learning—(Valverde and Ciudad 2014).

It is of the utmost relevance to remark that the literature indicates the importance of the inclusion among new assessment methods those ones that involve students in a self-regulated process, and that put into practice self-assessment criteria. This assessment oriented towards learning promotes and maximizes the opportunities of students to learn, developing competences and skills that will be significant and highly valued in their professional performance.

As mentioned, the change in the paradigm caused by the European Space for Higher Education means a revolution and a significant renewal of the principles and values supporting the teacher's role in Higher Education, what is shown not only in assessing aspects and changes, but also in the design of new teaching modalities focused on the development of skills (De Miguel Díaz 2006). Thus, this pedagogical revolution must intend to overcome the traditional dichotomy between theoretical and practical methods and classes, introducing new organizational models (workshops, teamwork, mentoring, autonomous work and external internships, for instance) and remarking the relevance of the autonomous work as the main objective of the election and selection of the teaching methods.

Havind said that, we can observe how several empirical studies indicate that there is a significant improvement in academic results of students, achieving higher quotes of attendance to classes and to exams—through the use of new assessment methods combined with a methodological diversity—in comparison with those obtained with more traditional teaching methods (Grimaldo Moreno and Arevalillo-Herráez 2011; Palazón et al. 2011; Arribas 2012). In conclusion, the use of active methodologies and a formative assessment system—where the implication of students, teamwork, mark distribution and self-assessment are a priority—seems to foster quality learning and a more satisfactory experience for students.

2   Objectives and Methodology

The possibility to train students in an integral way, as citizens as well as professionals, increasing their knowledge and contributing to the development of competences, skills and values is always present in Higher Education teachers' performance. Thus, positive academic results and improvement are expected when modifications are implemented regarding methodology, materials or assessment. This paper aims, then, at demonstrating the evolution of active methodologies towards Experiential Learning and learning by doing and, consequently, the necessary changes in the assessment of students, as well as their correlation with their academic performance and results.

In the following sections, the methodology used will be presented, together with a description of the experience—involving teachers' guides, where training activities, methodologies and assessment tools are explained. The academic results will be discussed and some preliminary conclusions will be established.

The methodology used by the research group is a mixed one, where qualitative and quantitative tools are employed to obtain measurable results that will be complemented with descriptive qualitative data. Thus, we will be allowed to establish a relation between the quantitative results obtained and social processes, such as globalization—associated with internationality—and skills development.

3 Description of the Experience

3.1 Brief Analysis of Modifications in the Teachers' Guides

In this section we will briefly analyze the teachers' guides from three different courses: Communication Skills (degree in International Business and International Relations), Didactics in Language and Literature and School Library (degree in Primary Education). All of the abovementioned courses are taught in the School of Social Sciences and Communication are mandatory within their syllabuses and have an estimated dedication of 150 h by students, corresponding to 6 ECTS credits. The first one is taught in English and in a traditional face-to-face modality, while the ones belonging to the degree in Education are taught in English and Spanish, not only face-to-face, but also in a blended learning modality.

Regarding Didactics in Language and Literature, contents and transferrable competences and skills have been undergone significant changes in general terms (interculturality in the classroom, language functions, Spanish literature, literary genres and didactic usage of Literature), even though their names have been slightly modified (social responsibility, self confidence, communication skills, interpersonal communication, teamwork and initiative or entrepreneurship). However, concerning the methodologies used and assessment, modifications are noteworthy.

The School Library course, as in the previous case, contents and skills have not suffered any changes, but Experiential Learning and Service-Learning, active methodologies based on experience, were introduced in the academic year 2014/15. The difficulties and differences, then, are based on the methodologies used and the characteristics of the groups, which tend to be reduced in the face-to-face modality (around seven students) and blended learning (up to forty students).

The methodologies used in Communication Skills during the first year (2015/16) was mainly based on Dialogical Learning and flipped classrooms, while Experiential Learning was introduced during this academic year (2016/17). The course was taught in a face to face modality, and the forty students belonged to the "digital native" generation. Some of them—50% approx.—were Spanish native speakers in their first year and the rest of them were international students from different countries where Spanish is not spoken, so the background, culture, language and previous experience of all of them were diverse and enriching.

3.2 Assessment: Tools and Procedures

During the first academic year, Didactics in Language and Literature was focused on teaching literature in a more traditional way, lectures being the most commonly used methodology, even though many practical activities and text analysis were carried out during the face to face sessions. The use of the virtual platform was, as well, much more limited then and the assessment consisted of a theoretical and practical test—30% of the final mark, a portfolio-40%, an essay and a presentation on a specific topic—20%—and the assessment of the student's implication through the attendance and participation in the face to face sessions—10% (Table 1).

Table 1. Assessment in didactics in language and literature

Academic year	Tools and procedures	Percentages in the final mark
2012/13	Test	30%
	Portfolio	40%
	Essay and oral presentation	20%
2013/14	Prueba de conocimiento	30%
	Portfolio (blog)	20%
	Essay and Reading report	20%
	Literary work	15%
2014/15	Test	30%
	Portfolio	15%
	Literary work	10%
	Essay and Reading report	15%
	Service Learning project	30%
2015/16	Oral test	15%
	Written test	25%
	Portfolio (face to face activities)	20%
	Portfolio (online activities)	20%
	Service Learning project	20%
2016/17	Oral test	15%
	Written test	25%
	Portfolio	20%
	Essay and reading report	10%
	Microprojects	20%
	Audiovisual resources creation	10%

As the figure shows, during the second academic year the modifications consisted of the inclusion of various training activities that had a correspondence in assessment and assessment tools. Creativity and communication skills, highly valued not only in educational settings but also in professional ones, as well as the ability to apply theoretical knowledge to practice—regarding literature in this case, were assessed by means of the creation of a literary text and the participation in a literary contest selected by the students. The significance of reading was increased in the final mark through a reading report that ensured the selection of and work on a book chosen by them.

During the academic year 2014/15 the Service Learning (SL) methodology was introduced by means of a project that was included as part of the assessment—30% of the final mark, taking into account the estimated number of hours of dedication to the project by students. In this activity, a school from a deprived zone in Madrid and with a high diversity collaborated as an external partner. Thus, a real situation was offered to students, who had the opportunity to detect the real needs of the partner and to provide the, with a series of pedagogically relevant solutions. Also the test, participation in face to face activities and a literary contest, as well as the essay on a book were maintained, even though they were more specific and concise.

The test was divided, in the academic year 2015/16, into an oral and a written part, in order to foster a direct assessment of the oral and written communication skills by using a rubric. The autonomous work of the student outside the classroom was considered as important as the face to face work on activities. More rubrics were included to assess all the activities, including the SL project, where evaluation was carried out by the teacher, by the research group, by the external partner and by students—through self-assessment and peer-assessment. Thus, assessment relied not only on the acquisition of contents, but also on the development of skills and competences such as application of theoretical knowledge to practical situations, planning and time management, flexibility and teamwork.

Finally, during this academic year, 2016/17, several activities based on Experiential Learning and Service Learning have been carried out in the face to face modality, including a microproject with students of Spanish as a Foreign Language in Germany and a microproject in collaboration with an NGO—both of them were part of the 20% devoted to microprojects in the final mark.

The test kept its importance and the division into an oral and a written part, and the specifications about the essay on a Children Literature book and the portfolio were slightly modified. Instead of being a mere collection of activities carried out during the term, the portfolio turned into a space for reflection upon the learning process and the relation between knowledge and the students' future professional career. The introduction of audiovisual resources in the syllabus design allows the teacher to assess competences such as teamwork, application of theoretical knowledge to practice, initiative and entrepreneurship and management of Information and Communication Technologies, as part of the digital skills required to any teacher nowadays.

Rubrics have been introduced step by step as assessment tools to evaluate the acquisition of contents and the development of skills and competences in projects, oral tests and training activities (essays, reports, written essays, debates and so on). Furthermore, tools have evolved to a mixed evaluation—that is, hetero-, self- and peer-assessment, where students evaluate their own work and implication and the ones of their partners by means of rubrics and reflective diaries. Thus, the level of skills development has been checked concerning teamwork, social responsibility, planning and time management, flexibility, communication skills, creativity and application of theoretical knowledge to practice (Table 2).

Concerning the School Library course, introduction of Experiential Learning—as reflected on the assessment item related to the participation in a project in collaboration with educational centers and NGOs—allows some continuity in assessment tools, where rubrics play a main role, and in percentages in the final mark. Thus, the test (30–40%), the virtual and face to face activities (debates, essays and so on) and the project to detect needs and suggest possible solutions in terms of feasible actions for the

Table 2. Example of a peer-assessment rubric

Mark (0 = absolutely disagree, 10 = absolutely agree)	
The resource created is appropriate for the Primary Education stage	
The resource can be easily related to Language and Literature	
The resource is creative	
I would use the resource in my classroom	
Comments (justification for the marks given and other comments)	

collaborating entity (20–30%) have remained an essential and immutable part of the course.

It is true that some competences and skills can be easily evaluated observing the development of many of the activities—for instance, communication skills in essays and debates or planning and time management, which can be associated with deadlines and tasks and their fulfillment in the right order and within the established period of time. However, some other competences can be more difficult to be assessed from an objective point of view, independently from the perception of the group or the individual. In this sense, for instance, teamwork has been assessed through the use of items such as "the student is able to assume different roles" or "the student shows the ability to negotiate and solve conflicts within the group". A correct description of the item promotes the right use of rubrics in self- and peer-assessment, and avoids the controversy of students grading other classmates just for personal reasons and underestimating or overestimating their own abilities.

Rubrics were not only introduced in order to observe and check the development of skills thanks to the application of different items, but also because, given the fact that this course was taught in a language other than the students' mother tongue, it was necessary to integrate both language and content in the evaluation. The following is an example of a basic rubric used in the peer-assessment part of one of the training activities, which referred to the analysis and evaluation of an audiovisual resource (Table 3).

Regarding the Communication Skills course, the assessment tools and procedures were more specified and detailed during the second year. Thus, the test carried out with the first group—divided into an oral part and a written part—was maintained, since it allowed the observation of the development of communication skills as well as the acquisition of contents. The online activities, which in the first year included the participation in forums and the self-assessment tests, included this time discussions and debates on movies and Experiential Learning teamwork activities in collaboration with students from other courses. Finally, the face to face activities, besides debates and written reports, included also research and role plays, what facilitated the assessment of negotiation skills, among others. Rubrics were used and were made available for students for every activity.

Table 3. Example of a basic rubric for peer-assessment

	Content	Language
Title and brief description of the resource	The brief description evidences the content of the resource in a sufficient way	The vocabulary and structures used are appropriate and grammatically correct
Authorship and updates	The author is indicated and reliable and there is evidence of documentation on the frequency of updating	There is evidence of the student mastering the process of documentation through the use of adequate keywords and descriptors
Adequacy of the contents for the selected educational level or age	There is a justification of the level of adequacy of the contents and its correspondence with the needs of the selected educational level or age	The vocabulary and structures used are appropriate and grammatically correct
Accessibility of the resource (according to the impairments and digital divides studied in class)	There is a justification of the level of accessibility of the resource for people with certain impairments	There is evidence of the student understanding the concept "accessibility"
Global mark and justification	The global mark is sufficiently justified, with reasonable arguments and a description of the pros and cons of the evaluated resource	The vocabulary and structures used are appropriate and grammatically correct

3.3   Methodologies: Experiential Learning and Service-Learning

Methodologies utilized in the last courses involve, as the assessment tools in the described courses show, Experiential learning and flipped classroom, among others. Cooperative Learning and Dialogical Learning are also present in the activities developed during the face to face sessions, as well as in the online part of the subjects carried out through the virtual platform.

The Theory of Experiential Learning emphasizes the essential role of experimentation and experiences in the learning process of any individual. Its origins have their intellectual roots in Dewey's philosophical pragmatism, Lewin's social psychology and Piaget's epistemological genetics on the cognitive development process (Kolb et al. 2000). Experiential Learning, specifically, allows the application of rubrics and the observation of learning processes among students, given an appropriate division of students according to the number of them participating in the course.

In this sense, any project in collaboration with other students and external partners can contribute to the assessment process and to an improvement of the results, besides being still an innovation in the degrees in Education, International Business and International Relations, by virtue of the inclusion of new items, new partners and interdisciplinary models. In contrast with past decades, where internships were the only contact with the real professional world—and bearing in mind that this situation did not always occur in the best possible conditions for the student, these projects, adequately framed in different course syllabuses, contribute to the empowerment of Higher Education teachers, who have lost their role of experts merely transmitting contents. Teachers possess, thus, a new tool to allow students apply theoretical knowledge to practical settings before actually being part of the professional world.

It must also be noticed that, in spite of the difficulties that the blended learning modality introduces in the participation in SL activities and Experiential Learning projects, both methodologies have been implemented and used in training activities, associated or not with a specific assessment tool or established percentage in the final mark.

In order to illustrate this, we will briefly include the description of the project in which students attending the School Library course shall participate. Students, working in teams, need to complete the following tasks: firstly, document themselves about the legislation related to school libraries in Spain; secondly, identify an educational center with a school library that requires some actions to be improved and updated, in accordance with international guidelines and current legislation; thirdly, describe the needs detected and, finally, suggest and define specific actions that could be taken.

The learning outcome, in terms of production, is a document including the following sections: documentation—school library requirements, needs detection, description of the selected case, objectives of the project, proposals for actions and activities, and a reflective diary with several entries where the participation of each member of the team are detailed, together with a personal reflection upon the learning. This documents needs to be approved by all members of the team, thus supporting self-assessment and peer-assessment, avoiding incoherence and facilitating the assessment of teamwork by the teacher, also in the blended learning modality, where observation is more difficult or even impossible.

The flipped classroom, the use of the virtual platform as a support tool for face to face groups—besides its relevance in blended learning, and the inclusion of activities and student productions—audiovisual resources, presentations and so on—is a good example of how ICTs are used in Higher Education, independently from the degree and modality.

3.4   Academic Results: Historical Records

The following figures show the historical records of the academic results in the School Library and Didactics in Language and Literature courses, presenting the total number of students, as well as the total of students who passed or failed, and the average mark of the ones who passed. In the event of several groups of the same course during the same academic year, the group shall be named adding to the academic year the corresponding letter of the alphabet, starting from "a", and "S" will be used to indicate the blended learning modality. Those students who decided to quit the course are not included and only the ordinary exam period is analyzed (Tables 4 and 5).

After observing the data, we may conclude that the variations in the academic results are not significant between different groups and that there is not, apparently, any

Table 4. Marks: School library course

Group (academic year)	Number of students (sample)	Pass	Fail	Average mark (passed)
2012–2013 a	11	11	0	8.9
2012–2013 b S	40	40	0	8
2012–2013 c	13	11	2	8.9
2012–2013 d S	17	17	0	7.8
2012–2013 e S	31	30	1	7.9
2013–2014 a	7	7	0	7.7
2013–2014 b S	37	34	3	8.3
2013–2014 c S	40	38	2	8.5
2014–2015 a	8	7	0	8.5
2014–2015 b S	36	36	0	8.4
2015–2016 a S	21	21	0	8.6
2015–2016 b	10	6	4	7.3
2016–2017 a S	38	38	0	8.6

Table 5. Marks: Didactics in language and literature course

Group (academic year)	Number of students (sample)	Pass	Fail	Average mark (passed)
2012–2013	8	8	0	8
2013–2014	10	10	0	8
2014–2015 a S	41	36	5	8.3
2014–2015 b S	33	29	2	8.3
2015–2016	19	12	5	7.5
2016–2017 a S	42	41	0	8
2016–2017 b	11	6	3	7.6

improvement after the inclusion of active methodologies. However, it is necessary to remark, in the first place, that most of the students who failed were due to cases of plagiarism or the lack of commitment, that is, the non participation or delivery of all the mandatory activities.

Also, it could be derived from the data that there is not any correlation between the application of several active methodologies and assessment tools and a significant improvement of the academic results. Nevertheless, if we consider the fact that the average marks do not significantly vary from one modality to another, independently from the number of students involved, we could suggest that these actions were important and influenced the academic performance of students, providing that the increase in the number and quality of assessment tools, as well as the introduction of skills in the rubrics, could imply a noticeable improvement in the evaluation quality, which is far more rich and complete, as well as in inclusiveness.

Some variations can be appreciated between face to face and blended learning modalities, given the fact that in Didactics in Language and Literature the average results obtained by students are better in the blended learning modality than in the face to face one, while any variations in the tendency in the School Library course are just observed after the introduction of Service Learning, since 2013/14. As it was mentioned before, blended learning can cause some difficulties when implementing active methodologies such as Experiential Learning, but the figures presented above indicate that these difficulties also appear in face to face modalities. In fact, the use of this methodology fostered an improvement in the academic results blended learning groups.

Finally, the use of ICTs tends to be better accepted by students in blended learning modalities, since most of the autonomous work of the student is done through the virtual platform. Students read the notes, complete self-assessment tests and participate in online discussions and debates, among other things, and attend virtual workshops frequently. Besides developing their digital skills, this context fosters the use of different competences and, if we observe the academic results shown in the previous figures, we may state that the use of ICTs—which involves multimodal materials, online activities and the promotion of active methodologies such as the flipped classroom—contributes to the improvement of students performance.

3.5 Students' Opinions

We would like to briefly introduce and discuss some of the opinions collected from students in the Communication Skills course after the implementation of Experiential Learning and the increasing use of ICTs (Table 6):

Table 6. Students' opinions

Topic	Opinions
Use of ICTs	'Things are more technological here, it is more advanced at that level in comparison with Venezuela. They are more prone to read books and those [digital] boards do not exist there. And since the degree is international, obviously, it has a little more international range...' (A2)
Active methodologies	'For me, I'm even more controlled than when I attended the High School, for instance, the teacher came, gave a lecture and you knew that in the end, before Christmas, you had an exam. Here we have tasks, self-assessment, I don't know. At the end you are more conscious of how you are progressing in the course during the whole year. For me, that is positive' (A6)
	'And, for instance, we have a control, everything is on the web, every mark. We have a lot of tasks, exams...it's not like university here in Spain, for instance, where you have an exam at the end of the course and that's it' (A1)

The answers seem to confirm that active methodologies serve as a means for transforming students into the main actors in the process. The student centered approach, then, and continuous assessment, involving different methodologies and training activities is, apparently, highly valued in comparison with traditional methods where only content acquisition was relevant.

4 Discussion, Conclusion and Implications

The European Space for Higher Education implies deep and diverse changes. The main one is related to the challenge of establishing students as the focus of their own training processes (Baena and Padilla 2012). In this next context, modifications that have been occurring in Higher Education in Spain -and their relevant effects on the conditions of a more and more competitive professional market- make it necessary for universities to implement deeper changes in the required adaptations for their training syllabuses and teaching methodologies in the new Bachelor's and Master's degrees (Gibbs 2007; Barber 2008).

As we have observed in previous sections, this work confirms the evolution in methodologies towards an experiential learning and, consequently, the modifications in students' assessment, as well as its correlation with academic results and performance. Also, results obtained show improvements in the academic performance of those students participating in the experiences described here and a positive perception of the changes included in the different courses. We can state, then, that the introduction of new methodologies and the inclusion of various elements in the assessment process contribute to a better consideration of content acquisition and skills and values development.

Those assertions are coherent with the literature about Generation and (GY) or Millennials (students born between 1980 and 1994). This generation is characterized by being the best prepared for teamwork, among other salient skills. Furthermore, they have grown up in a highly computerized context, so they tend to be very comfortable using ICTs and it is easier for them to interact and cooperate with individuals from different parts of the world, and to obtain information in a quick way.

It can be concluded, then, that the University shall take into account that students need not only a deep knowledge of the contents in the courses they are attending, but also the development of certain skills and competences that will help them adapt and perform efficiently in a working environment that is competitive, complex and always changing, (Baena and Padilla 2012). For this reason, the design of activities that foster experiential learning and that are adapted to the needs and interests of the GY is essential.

References

Arribas, J.M.: El rendimiento académico en función del sistema de evaluación empleado. RELIEVE. Revista Electrónica de Investigación and Evaluación Educativa (2012). http://www.redalyc.org/pdf/916/91624440003.pdf

Baena, V., Padilla, V.: Refuerzo and desarrollo de competencias mediante la elaboración de una campaña real de marketing: la FormulaUEM. REDU. Revista de Docencia Universitaria. Número monográfico sobre 'buenas prácticas docentes en la enseñanza universitaria' **10**, 199–214 (2012)

Barber, M.: A formula for great teaching. Times Educ. Suppl. **48**, 19–29 (2008)

De Miguel Díaz, M. (ed.), Alfaro Rocher, I.J., Apodaca Urquijo, P., Arias Blanco, J.M., García Jiménez, E., Lobato., Fraile, C., Pérez Boullosa, A.: Modalidades de enseñanza centradas en el desarrollo de competencias orientaciones para promover el cambio metodológico en el Espacio Europeo de Educación Superior, pp. 159–172. Ediciones Universidad de Oviedo (2006). http://www.academia.edu/5817175/Modalidades_de_Ense%C3%B1anza_Centradas_en_el_Desarrollo_de_Competencias_Orientaciones_para_Promover_el_Cambio_Metodol%C3%B3gico_en_el_Espacio_Europeo_de_Educaci%C3%B3n_Superior

Gibbs, P.: Editorial. J. Bus. Res. **60**(9), 925–926 (2007)

Grimaldo-Moreno, F., Arevalillo-Herráez, M.: Metodología docente orientada a la mejora de la motivación and rendimiento académico basada en el desarrollo de competencias transversales. IEEE-RITA **6**(2), 70–77 (2011)

Kolb, D.A., Boyatzis, R.E., Mainemelis, C.: Experiential learning theory: previous research and new directions. In: Sternberg, R.J., Zhang, L.F. (eds.) Perspectives on cognitive, learning, and thinking styles. Lawrence Erlbaum, Mahwah (2000)

De Los, P.-P., Cobos, A., Gómez-Gallego, M., Gómez-Gallego, J.C., Pérez-Cárceles, M.C., García, G.: Relación entre la aplicación de metodologías docentes activas and el aprendizaje del estudiante universitario. Bordón **63**(2), 27–40 (2011)

Valverde, V., Ciudad, A.: El uso de e-rúbricas para la evaluación de competencias en estudiantes universitarios. Estudio sobre fiabilidad del instrumento. REDU: Revista de Docencia Universitaria. **12**(1) (2014). https://dialnet.unirioja.es/servlet/articulo?codigo=4691792

The Dimensions of Service Quality

Carlos del Castillo Peces[✉], Carmelo Mercado Idoeta,
Miguel Prado Román, and Cristina del Castillo Feito

Universidad Rey Juan Carlos and European Academy of Management
and Business Economics (AEDEM), Madrid, Spain
carlos.delcastillo@urjc.es

Abstract. The quality of the service perceived by the customers is a key element to increase their fidelity. On the other hand, the private security sector is of great socio-economic importance both globally and in Spain. The main objective of this work is to analyze the dimensions of quality of service valued by customers, as well as their relative importance. For this purpose, a questionnaire has been drawn up and sent to a sample of 3000 clients, and 216 valid responses have been obtained. The results show that all the dimensions of technical and functional quality influence the perception of quality of service, emphasizing to a greater extent the aspects related to attitudes and behavior, and to a lesser extent with respect to accessibility and flexibility. Likewise, these results can be affected by the number of years with the same service provider, but not by the size of the company receiving them.

Keywords: Quality of service · Perceived quality · Technical quality
Functional quality · Quality dimensions · Private security

1 Introduction

Customer loyalty is one of the most important business concerns since keeping a long relationship with clients is essential to assure its place in the market (Castillo et al. 2013). In this context, modern management of quality is a key strategy to achieve competitive advantages based on satisfied and loyal customers and better results (San Miguel et al. 2009). This affects both tangible products and services. In fact, the focus on loyalty is first developed for services (Cobo and González 2007) since this sector became one of the most important of the world in terms of activities as well as for number of employees (Martinez 2013). This explains why service quality becomes a key factor for any business success (Ladhari 2009).

Private security services are one of the key activities included in the services sector. From an economical perspective, this sector turned over €34.5 billion in Europe with more than 2 million security guards (COESS 2014) and over €3.5 billion in Spain with almost 80 thousand guards in 2015 (APROSER 2016). From a social point of view, society is turning more risky where industries and citizens as well as public power, are focusing on crime, insecurity and further similar activities (Cools 2013). Therefore, service quality belongs to the most relevant aspects in mix marketing developed by industries in this sector (DBK 2014).

J. Gil-Lafuente et al. (Eds.): AEDEM 2017, SSDC 180, pp. 143–157, 2019.
https://doi.org/10.1007/978-3-030-00677-8_13

The main purpose of this paper is to analyze the dimensions in service quality from the customer perspective in private security sector in Spain as well as its relative value. On the other hand, we will also analyze how these dimensions can be influenced by the size of the company receiving the services as well as the number of years keeping its current private security service provider.

Many studies about this sector have been developed in Spain during last years, however, they had legal and administrative approaches (Vigara 2007; Gutiérrez 2009) or they are related to defense industry (Rodriguez 2012) or they had a military approach (Laborie 2011) instead of focusing on business management. This paper pretends to explain the customer's behavior in relation to these services helping other companies to manage their relationships with clients so they can be longer and more effective.

2   Theoretical Framework

Parasuraman et al. (1985) stated that performance measurement of service quality occurs when comparing customer expectations with results, which means analyzing both the outcome and the process of providing the service. In order to determine the service quality and before analyzing customer perceptions, it is necessary to know what aspects will be used to evaluate the service (Grönroos 1994). Service quality is subjective since different customers can have different perceptions based on their expectations. That is the reason why organizations should pay special attention on what elements are considered by clients when evaluating service quality (Laguna and Palacios 2009; Omar 2011; Shpëtim 2012).

There are different opinions about what dimensions have an impact when evaluating service quality. Grönroos (1984) compares expected service with experienced service in order to calculate the total service quality perceived. On one hand, the experienced service results from both dimensions: its technical quality (what is received) and its functional quality (how it is received) and they are influenced by the organization's corporate image. On the other hand, the expected service can be affected by several factors like marketing communication, word of mouth recommendations, reputation and customer's needs.

Both works developed by Grönroos (1984), Parasuraman et al. (1985) defined a scale for measuring service quality called SERVQUAL built on the expectancy-disconfirmation paradigm and applied to ten different dimensions whose later were reduced to five (tangible items, reliability, responsiveness, assurance and empathy) measured through 22 items.

The right model to predict quality perception in services with close contact between client and company is the technical and functional quality model (Lovelock 1996; Lassar et al. 2000). Private security services have a high contact since 70% of business turnover in this sector in 2015 resulted from surveillance and funding transportation services where there is a high contact between service provider (guard) and customer (APROSER 2016).

Therefore, the previously mentioned Grönroos (1984) service quality model will be used in this investigation with its three main items: technical quality (service produced), functional quality (how it is delivered) and the corporate image (acts as a filter that helps to increase or reduce the impact of bad services).

Markets with mainly industrial clients do not have an emotional connection between service provider and customer as strong as in consumer markets, and that is why corporate image does not play such an important role (Davis et al. 2008). In private security sector, private clients represent 4.45% of demand (APROSER 2016) so the biggest customer group is composed of institutional clients and public administrations whose service provider selection processes are based on benefit-cost results and not so much emotional.

According to the above mentioned and previous investigations (Ruiz et al. 1995; Lassar et al. 2000), this work is going to be exclusively focused on the two big items of service quality: the service delivery (technical quality) and the interactions between service provider and customer (functional quality). On the other hand, other studies defend the influence of all quality service dimensions on quality perception (Culiberg and Rojsek 2010; Nimako 2012; Suneeta and Koranne 2014; De Keyser and Lariviere 2014), and that is why all quality dimensions will be considered.

3 Methodology

3.1 Sample

Surveillance services represent 62% of total €3,481 million turnover in 2015 (APROSER 2016), funding transportation represents the 8% and the static security systems (alarms) belong to the remaining 30%. The first two activities include inherent features of services (intangibility, ownership, heterogeneity and perishability) and the security systems are not considered services but tangible products that can be bought or sold. Because this work's purpose is to investigate the service quality in this sector, only surveillance activities and funding transportation will be analyzed as since, as already mentioned, they represent the 70% of total.

Thus, services provided to private clients will be excluded in this paper, since they are barely affected by the previously mentioned activities (surveillance and funding transportation) and because this sector only represented the 4.45% of total turnover in 2015 (APROSER 2016).

On the other hand, according to annual statistics on employment of the Ministry of Interior, there were 359,402 active contracts related to surveillance services and funding transportation with different clients in 2015 so this amount can be understood as this job population size.

We could count on Stanley Security Solutions collaboration, in order to see a representative sample of clients receiving these services. This organization is one of the first ten companies in terms of turnover in this sector that has more than 23,000 clients all over the country through its 26 offices whose 3,000 were selected considering two conditions: only clients related to surveillance and funding transportations services and the activity breakdown had to be similar to the one demanded on private security markets in Spain (APROSER 2016).

3.2   Data Collection and Investigation Tool

Online surveys have been applied because of the advantages they afford (Sánchez et al. 2009). In order to evaluate all items of each service quality dimension, a Likert scale in ascending order from 1 to 5 has been developed. The fact sheet of this investigation is detailed below (Table 1):

Table 1. Investigation fact sheet

Universe	Clients who contract private security services in Spain
Scope	Spain
Population size	359,402 clients
Sample size	3,000 clients
Sampling method:	
• **Private security sector clients:**	Convenience sampling
• **Stanley Security Solutions clients:**	Random probability sampling
Data collection method	Online structured questionnaire
Questionnaires received	216
Sampling error	
• **Private security sector clients:**	Not applicable
• **Stanley Security Solutions clients:**	6,5%
Confidence level	
• **Private security sector clients:**	Not applicable
• **Stanley Security Solutions clients:**	95% (z=1,96; p=q=0,5)
Data collection period	From September 2015 to December 2015

3.3   Construct Measurement

Measuring scale of technical quality. Grönroos (1988) differentiates two dimensions for this construct (professionalism and skills) based on the following items (Table 2):

Table 2. Technical quality items

Technical quality		
Professionalism		
Name	Item	Source
T-CONOC	Employees have the technical knowledge to realize the contracted services	Ennew and Binks (1999), Aldlaigan and Buttle (2002)
T-MEDIO	Current provider counts on modern and proper technical resources	Aldlaigan and Buttle (2002)
Skills		
Name	Item	Source
T-APARI	Employees have an appropriate and professional appearance	Ennew and Binks (1999), Aldlaigan and Buttle 2002
T-DESTR	Employees have the required skills and abilities to do the contracted services correctly	Aldlaigan and Buttle (2002)

Measuring scale of Functional Quality: Some of the items considered are similar to those used in previous studies that applied the SERVQUAL Model since it is focused on what Grönroos called functional quality (Caruana 2002) (Table 3).

Table 3. Functional quality items

Functional quality		
Behavior and attitude		
Name	Item	Source
F-INTER	When there is a problem, employees show a sincere interest to solve it	Alén and Fraiz (2006), Nuviala et al. (2011), Omar (2011), Díaz et al. (2012), Shpëtim (2012)
F-DISPO	Employees are always willing to help clients and answer their questions	Miguel and Flórez (2008), Bravo et al. (2011); Jemmasi et al. (2011)
F-COMPO	Employee's behavior exhibits professionalism to their clients	Alén and Fraiz (2006), López and Serrano (2010), Bravo et al. (2011)
F-TRATO	Clients always get a kind and caring attention	Miguel and Flórez (2008), Bravo et al. (2011), Nuviala et al. (2011), Díaz et al. (2012)

(continued)

Table 3. (*continued*)

Accessibility and flexibility		
Name	Item	Source
F-HORAR	Customer service hours are scheduled according to their needs	Miguel and Flórez (2008), Omar (2011), Nuviala et al. (2011), Díaz et al. (2012)
F-LOCAL	Customer service centers are located according to client's requirements	Miguel and Flórez (2008), López and Serrano (2010), Bravo et al. (2011)
F-NUMER	The number of customer service centers is decided according to client's needs	Díaz et al. (2012)

Reliability		
Name	Item	Source
F-PLAZO	Current service providers always meet agreed deadlines	Caruana (2002), Nuviala et al. (2011), Omar (2011), Jemmasi et al. (2011), Correia and Miranda (2012)
F-PRIME	Current providers do the service correctly from the very beginning	Caruana (2002)
F-CLARA	Information given by current service providers is clear and reliable	Lassar et al. (2000), López and Serrano (2010), Correia and Miranda (2012)

Recovery		
Name	Item	Source
F-CORRE	When clients report an incident, their current service provider takes the required measures to solve it	Aldlaigan and Buttle (2002), Cambra et al. (2011)

4   Results

The first questions in the online questionnaire aimed to define the type of each participating client. The following results were obtained:

In view of the results, it can be confirmed the sample of clients mainly represents the sector since the activity sector classification is very similar to the total market classification in 2015 (APROSER 2016). Regarding the turnover size, and according to the European Commission Recommendation 2003/361, 29% of the sample consists of big companies (over 50 million euros turnover), 11% of medium-sized companies (between 10 and 50 million euros) and the remaining 60% are small-sized (under 10 million euros).

In relation to the number of years as a client, over 70% of them are keeping their current service provider. This allows us to trust the results since they are based on long-term experiences where different situations have been faced.

Table 4. Description of participating companies

Features	Categories	% total sampling
Activity sector	Industry and commercial sector and other services	72
	Public administration	17
	Financial companies	11
Classification by client's turnover	Under 2 million euros	39
	Between 2 and 10 million euros	21
	Between 10 and 50 million euros	11
	Over 250 million euros	29
Number of years with the current private security service provider	Under 2 years	29
	Between 2 and 4 years	13
	Between 4 and 6 years	19
	Over 6 years	39

4.1   Dimensions of Service Quality

Regarding the most important dimensions in service quality, the descriptive statistics that better represents all items of each construct are exposed in the following table:

As Table 5 shows, all considered items for the two dimensions of technical quality, i.e. professionalism and skills, are both very relevant since their means were higher than 4 in Likert scale and the range is limited (from 4.10 to 4.34) and it is even lower if the item related to the employees appearance at 4.10 is excluded. Finally, the average of all means obtained by all four items of technical quality is 4.26.

Table 5. Descriptive statistics of technical quality

	N	Minimum	Maximum	Mean	Standard deviation
T_CONOC	216	1	5	4.32	.995
T_MEDIO	216	1	5	4.27	.848
T_APARI	216	1	5	4.10	.897
T_DESTR	216	1	5	4.34	.690
N valid (by list)	216				

Table 6 proves a bigger spread among items of the dimensions of functional quality, being Behavior and Attitudes the most valued dimension (its items obtained means between 4.25 and 4.50). The next place is taken by Reliability and Recovery, whose items scored between 4.11 and 4.25, thus as well as in technical quality, items are homogeneous. Finally the items related to Accessibility and Flexibility are the only that scored under 4 in the scale and even with a bigger spread (values oscillate between 3.49 and 3.99). The average mean reached by all eleven items in functional quality is 4.13.

Table 6. Descriptive statistics of functional quality

	N	Minimum	Maximum	Mean	Standard deviation
F_INTER	216	1	5	4.50	.852
F_DISPO	216	1	5	4.44	.822
F_COMPO	216	1	5	4.44	.833
F_TRATO	216	1	5	4.25	.852
F_HORAR	216	1	5	3.99	1.083
F_LOCAL	216	1	5	3.52	1.048
F_NUMER	216	1	5	3.49	.925
F_PLAZO	216	1	5	4.11	1.084
F_PRIME	216	1	5	4.25	.952
F_CLARA	216	1	5	4.24	.939
F_CORRE	216	1	5	4.21	.926
N valid (by list)	216				

Table 7 shows a global ranking of all indicators including those related to dimensions in both technical and functional quality.

Table 7. Descriptive statistics (global ranking)

	N	Minimum	Maximum	Mean	Standard deviation
F_INTER	216	1	5	4.50	.852
F_DISPO	216	1	5	4.44	.822
F_COMPO	216	1	5	4.44	.833
T_DESTR	216	1	5	4.34	.690
T_CONOC	216	1	5	4.32	.995
T_MEDIO	216	1	5	4.27	.848
F_PRIME	216	1	5	4.25	.952
F_TRATO	216	1	5	4.25	.852
F_CLARA	216	1	5	4.24	.939
F_CORRECT	216	1	5	4.21	.926
F_PLAZO	216	1	5	4.11	1.084
T_APARI	216	1	5	4.10	.897
F_HORAR	216	1	5	3.99	1.083
F_LOCAL	216	1	5	3.52	1.048
F_NUMER	216	1	5	3.49	.925
N valid (by list)	216				

The most important dimension of service quality is Behavior and Attitudes of employees (its three items take the first three positions), followed by the dimensions in technical quality (Professionalism and Skills), with the exception of employee's appearance, that is right before the dimension Accessibility and Flexibility in functional quality that takes the last place.

In relation to the second purpose of this investigation, it has also been analyzed if results are influenced by the size of the company receiving the service and as well as by the number of years keeping the current private security provider. To do so, we analyzed the variance, also called ANOVA, considering the results obtained from the different groups of companies according to prior Table 4: 4 groups of companies based on their turnover and other 4 groups based on the number of years they have kept their current private security provider.

With respect to the size, Tables 8 and 9 show ANOVA's results for dimensions of both technical and functional quality:

Table 8. ANOVA (Technical quality)

		Sum of squares	df	Mean square	F	Sig.
T_CONOC	Between groups	3.929	3	1.310	1.328	.266
	Within groups	209.029	212	.986		
	Total	212.958	215			
T_MEDIO	Between groups	2.588	3	.863	1.205	.309
	Within groups	151.838	212	.716		
	Total	154.426	215			
T_APARI	Between groups	.672	3	.224	.276	.843
	Within groups	172.286	212	.813		
	Total	172.958	215			
T DESTR	Between groups	1.060	3	.353	.740	.530
	Within groups	101.269	212	.478		
	Total	102.329	215			

Table 9. ANOVA (Functional quality)

		Sum of squares	df	Mean square	F	Sig.
F_INTER	Between groups	3.346	3	1.115	1.549	.203
	Within groups	152.649	212	.720		
	Total	155.995	215			
F_DISPO	Between groups	.926	3	.309	.453	.715
	Within groups	144.407	212	.681		
	Total	145.333	215			
F_COMPO	Between groups	9.147	3	3.049	4.611	.004
	Within groups	140.187	212	.661		
	Total	149.333	215			
F_TRATO	Between groups	1.500	3	.500	.686	.561
	Within groups	154.495	212	.729		
	Total	155.995	215			
F_HORAR	Between groups	4.636	3	1.545	1.325	.267
	Within groups	247.345	212	1.167		
	Total	251.981	215			

(continued)

Table 9. (*continued*)

		Sum of squares	df	Mean square	F	Sig.
F_LOCAL	Between groups	4.342	3	1.447	1.325	.267
	Within groups	231.584	212	1.092		
	Total	235.926	215			
F_NUMER	Between groups	1.715	3	.572	.665	.575
	Within groups	182.244	212	.860		
	Total	183.958	215			
F_PLAZO	Between groups	3.157	3	1.052	.895	.445
	Within groups	249.394	212	1.176		
	Total	252.551	215			
F_PRIME	Between groups	3.520	3	1.173	1.299	.276
	Within groups	191.475	212	.903		
	Total	194.995	215			
F_CLARA	Between groups	3.045	3	1.015	1.154	.328
	Within groups	186.436	212	.879		
	Total	189.481	215			
F_CORRE	Between groups	8.789	3	2.930	3.541	.016
	Within groups	175.414	212	.827		
	Total	184.204	215			

Tables 8 and 9 prove the four groups of companies evaluated according to its size had similar results about dimension's relevance in service quality since only two of the 15 items considered had a lower than 0.05 significance level (p) (Behavior and Attitudes and Recovery dimensions, both in functional quality).

Table 10. ANOVA (Technical quality)

		Sum of squares	df	Mean square	F	Sig.
T_CONOC	Between groups	12.143	3	4.048	4.273	.006
	Within groups	200.816	212	.947		
	Total	212.958	215			
T_MEDIO	Between groups	5.790	3	1.930	2.753	.044
	Within groups	148.636	212	.701		
	Total	154.426	215			
T_APARI	Between groups	1.698	3	.566	.701	.553
	Within groups	171.260	212	.808		
	Total	172.958	215			
T_DESTR	Between groups	3.301	3	1.100	2.355	.073
	Within groups	99.028	212	.467		
	Total	102.329	215			

In relation to the number of years keeping the same private security provider, Tables 10 and 11 show ANOVA's results as well as for dimensions in technical and functional quality:

Table 11. ANOVA (Functional quality)

		Sum of squares	df	Mean square	F	Sig.
F_INTER	Between groups	19.155	3	6.385	9.892	.000
	Within groups	136.840	212	.645		
	Total	155.995	215			
F_DISPO	Between groups	23.649	3	7.883	13.734	.000
	Within groups	121.684	212	.574		
	Total	145.333	215			
F_COMPO	Between groups	16.125	3	5.375	8.554	.000
	Within groups	133.208	212	.628		
	Total	149.333	215			
F_TRATO	Between groups	9.217	3	3.072	4.438	.005
	Within groups	146.778	212	.692		
	Total	155.995	215			
F_HORAR	Between groups	13.567	3	4.522	4.021	.008
	Within groups	238.415	212	1.125		
	Total	251.981	215			
F_LOCAI.	Between groups	10.717	3	3.572	3.363	.020
	Within groups	225.209	212	1.062		
	Total	235.926	215			
F_NUMER	Between groups	8.425	3	2.808	3.392	.019
	Within groups	175.533	212	.828		
	Total	183.958	215			
F_PLAZO	Between groups	25.786	3	8.595	8.036	.000
	Within groups	226.765	212	1.070		
	Total	252.551	215			
F_PRIME	Between groups	21.121	3	7.040	8.584	.000
	Within groups	173.875	212	.820		
	Total	194.995	215			
F_CLARA	Between groups	15.766	3	5.255	6.414	.000
	Within groups	173.715	212	.819		
	Total	189.481	215			
F_CORRE	Between groups	14.777	3	4.926	6.163	.000
	Within groups	169.427	212	.799		
	Total	184.204	215			

Regarding the period keeping the current provider, the last two tables show there are big differences between the 4 groups of companies considered in the results about dimension's relevance in service quality since 14 of 15 items have a lower than 0.05 significance level (p): it means in all items of dimensions in functional quality and almost all in technical quality with the exception of employee's appearance which, as mentioned before, was the less valued item in technical quality.

5   Discussion, Conclusion and Implications

The main purpose of this work was to analyze the dimensions in service quality perception as well as its relative significance in the sector of private security services in Spain. In addition to this, it was important to analyze if results were affected by the size of the company receiving the service as well as by the number of years keeping the same service provider of private security.

Regarding the first purpose of this paper, results show that all dimensions in service quality are relevant for the client's perception of quality in this sector (the lowest mean obtained is 3.49 based on a Likert scale from 1 to 5). This result is similar to those obtained in previously mentioned investigations (Culiberg and Rojšek 2010; Nimako 2012; Suneeta and Koranne 2014; De Keyser and Lariviere 2014).

However, results also show a different relevance relative to the before mentioned dimensions. The most valued dimension is Behavior and Attitudes related to functional quality (its three items lead the developed ranking and scored a mean between 4.44 and 4.50). Considering there is a close and direct contact between employee and client in this kind of services, no wonder the most important thing for clients is to feel employees really care and try to solve their problems, to be treated correctly and kindly and to get their doubts answered but also with professionalism.

It is also important to highlight the dimensions in technical quality (Professionalism and Skills). Excluding the item of employee's appearance (employees normally use uniforms in this sector), these dimensions reached means between 4.27 and 4.34. This is mainly because when clients need these services they usually are in a risk situation and they need to be assisted by qualified personnel and solve this situation as soon as possible.

Regarding the less valued dimensions, it is worth to mention the Accessibility and Flexibility in functional quality, whose items close the ranking with means between 3.49 and 3.99. Generally private security services are always provided where the clients need, so the service provider is the one moving where the client requires the service, thus the service provider location does not affect the client's quality perception.

Regarding the second purpose of this investigation, i.e. the possible influence of the size of the company receiving the service and of the number of years keeping the same service provider on the dimensions in service quality, results prove the size is not relevant for the valuation of the considered dimensions in technical and functional quality however the number of years with the same service provider does have an impact on this valuation.

This investigation has certain limitations that must be highlighted. First, its scope is limited to a specific area (Spain) so our conclusions could not be the same in other sector or countries. In addition to this, all private security services are included in this work, so the results could change in case of focusing the study on a specific type of service.

Thus, based on the obtained results, new lines of research can be opened for further investigations that will help to have a deeper knowledge about a so dynamic and important but also unknown sector. Two of these lines come from the already mentioned limitations, i.e. the investigation could be focused on a specific service of this sector as well as other geographic areas, so that comparative analyses based on the obtained outcome could be developed. On the other hand, the impact that has the number of years with the same service provider on the valuation of the dimensions in service quality could be analyzed deeper. Besides, introducing new quality variables in the model like customer's loyalty and satisfaction could complete this chapter.

References

Aldlaigan, A.H., Buttle, F.A.: SYSTRA-SQ: a new measure of bank service quality. Int. J. Serv. Ind. Manag. 13(4), 362–381 (2002)

Alén, M.E., Fraiz, J.A.: Evaluación de la relación existente entre la calidad de servicio, la satisfacción y las intenciones de comportamiento en el ámbito del turismo termal. Revista Europea de Dirección y Economía de la Empresa 15(3), 171–184 (2006)

APROSER, Asociación Profesional de Compañías Privadas de Servicios de Seguridad: El sector de la seguridad privada en España. Estudio económico 2015 (2016). www.aproser.es. Consultado el 11 de Mayo de 2016

Bravo, R., Matute, J., Pina, J.M.: Efectos de la imagen corporativa en el comportamiento del consumidor. Un estudio aplicado a la banca comercial. INNOVAR, Revista de Ciencias Administrativas y Sociales 21(40), 35–51 (2011)

Cambra, J., Ruiz, R., Berbel, J., Vázquez, R.: Podemos fidelizar clientes inicialmente insatisfechos. Revista de Ciencias Sociales 17(4), 643–657 (2011)

Caruana, A.: Service loyalty: the effects of service quality and the mediating role of customer satisfaction. Eur. J. Mark. 36(7/8), 811–828 (2002)

Castillo, V., Cortés, G., Muñoz, G.T., Catillo, S.I.: Análisis estadístico sobre el seguimiento de clientes y su valoración por las empresas de la región de Córdoba-Orizaba. Revista de la Ingeniería Industrial 7(1), 51–57 (2013)

Cobo, F., González, L.: Las implicaciones estratégicas del marketing relacional: fidelización y mercados ampliados. Anuario Jurídico y Económico Escurialense 40, 543–568 (2007)

COESS: Private Security in Europe: facts and figures (2014). http://www.coess.org/newsroom.php?page=facts-and-figures. Consultado el 16-11-2015

Cools, M.: El valor socioeconómico añadido de los servicios de seguridad privada en Europa. Cuarto libro Blanco. Secretariado General de la Confederación Europea de Servicios de Seguridad, Wemmel (2013)

Correia, S.M., Miranda, F.J.: DUAQUAL: calidad percibida por docentes y alumnos en la gestión universitaria. Cuadernos de Gestión 12(1), 107–122 (2012)

Culiberg, B., Rojšek, I.: Identifying service quality dimensions as antecedents to customer satisfaction in retail banking. Econ. Bus. Rev. 12(3), 151–166 (2010)

Davis, D., Golicic, S., Marquardt, A.: Branding a B2B service: does a brand differentiate a logistics service provider? Ind. Mark. Manag. **37**(2), 218–227 (2008)

DBK: Informe de compañías de seguridad. 19ª Edición. DBK Informa, Madrid (2014)

De Keyser, A., Lariviere, B.: ¨How technical and functional service quality drive consumer happiness: moderating influences of channel usage. J. Serv. Manag. **25**(1), 30–48 (2014)

Díaz, I.M., Verdugo, M., Palacios, B.: Calidad percibida por el espectador de fútbol. Revista de Psicología del Deporte **21**(1), 25–33 (2012)

Ennew, C.T., Binks, M.R.: Impact of participative service relationships on quality, satisfaction and retention: an exploratory study. J. Bus. Res. **46**(2), 121–132 (1999)

Grönroos, C.: A service quality model and its marketing implications. Eur. J. Mark. **18**(4), 36–44 (1984)

Grönroos, C.: Service quality: the six criteria of good perceived service quality. Rev. Bus. **3**, 10–13 (1988)

Grönroos, C.: Marketing y gestión de servicios: la gestión de los momentos de la verdad y la competencia en los servicios. Ediciones Díaz de Santos, Madrid (1994)

Gutiérrez, A.G.: Aspectos jurídicos de la investigación privada. Cuadernos de criminología: revista de criminología y ciencias forenses **7**, 14–21 (2009)

Jemmasi, M., Strong, K.C., Taylor, S.A.: Measuring service quality for strategic planning and analysis in service firms. J. Appl. Bus. Res. **10**(4), 24–34 (2011)

Laborie, M.A.: La privatización de la seguridad. Las empresas militares y de seguridad privada en el entorno estratégico actual. Tesis Doctoral, Universidad Nacional de Educación a Distancia, Madrid (2011)

Ladhari, R.: A review of twenty years of SERVQUAL research. Int. J. Qual. Serv. Sci. **1**(2), 172–198 (2009)

Laguna, M., Palacios, A.: La calidad percibida como determinante de tipologías de clientes y su relación con la satisfacción: Aplicación a los servicios hoteleros. Revista Europea de Dirección y Economía de la Empresa **18**(3), 189–210 (2009)

Lassar, W.M., Manolis, C., Winsor, R.D.: Service quality perspectives and satisfaction in private banking. J. Serv. Mark. **14**(3), 244–271 (2000)

López, M.C., Serrano, A.M.: ¨Dimensiones y medición de la calidad de servicio en empresas hoteleras. Revista Colombiana de Marketing **2**(3), 1–13 (2010)

Lovelock, C.: Services Marketing. 3ʳᵈ edn. Prentice Hall, Upper Saddle River (1996)

Martínez, R.: Relación entre calidad y productividad en las PYMEs del sector servicios. Publicaciones en Ciencias y Tecnología **7**(1), 85–102 (2013)

Miguel, J.A., Flórez, M.: Calidad de servicio percibida por clientes de entidades bancarias de Castilla y Leon y su repercusión en la satisfacción y lealtad de la misma. Pecunia, Monográfico, pp. 105–128 (2008)

Nimako, S.G.: Linking quality, satisfaction and behaviour intentions in Ghana's mobile telecommunication industry. Eur. J. Bus. Manag. **4**(7), 1–17 (2012)

Nuviala, A., Tamayo, J.A., Fernández, A., Pérez, J.A., Nuviala, R.: Calidad del servicio deportivo en la edad escolar desde una doble perspectiva. Revista Internacional de Medicina y Ciencias de la Actividad Física y el Deporte **11**(42), 220–235 (2011)

Omar, K.: Interrelations between service quality attributes, customer satisfaction and customer loyalty in the retail banking sector in Bangladesh. Int. J. Bus. Manag. **6**(3), 12–36 (2011)

Parasuraman, A., Zeithaml, V.A., Berry, L.L.: A conceptual model of service quality and its implications for future research. J. Mark. **49**(4), 41–50 (1985)

Rodriguez, P. (2012): Evolución y futuro de las empresas de seguridad privadas y defensa en España. Tesis Doctoral, Universidad Autónoma de Madrid, Madrid

Ruiz, A., Vázquez, R., Diaz, A.: La calidad percibida del servicio en establecimientos hoteleros de turismo rural. Papers de Turisme **19**, 17–33 (1995)

San Miguel, E., Heras, I., Elgoibar, P.: Marketing y gestión de la calidad: dos enfoques convergentes. Revista de Dirección y Administración de Empresas **16**, 83–109 (2009)

Sánchez, J., Muñoz, F., Montoro, F.J.: ¿Cómo mejorar la tasa de respuesta en encuestas on line? Revista de Estudios Empresariales, Segunda época **1**, 45–62 (2009)

Shpëtim, C.: Exploring the relationships among service quality, satisfaction. Trust and store loyalty among retail customers. J. Compet. **4**(4), 16–35 (2012)

Suneeta, B., Koranne, S.: ¨Conceptual study of relationship between service quality and customer satisfaction. Int. Res. J. Soc. Sci. **3**(2), 9–14 (2014)

Vigara, J.: La seguridad privada en España: el sistema jurídico y administrativo y su necesaria evolución. Tesis Doctoral, Universidad Europea de Madrid, Madrid (2007)

Relationship Between Innovation Process and Innovation Results: An Exploratory Analysis of Innovative Peruvian Firms

Jean Pierre Seclen Luna[✉]

Department of Management, Pontifical Catholic University of Peru, Lima, Peru
jseclen@pucp.pe

Abstract. Innovation is an unavoidable competitive strategy in the current economic context. In the literature it is mentioned that innovation is a process that is not by itself perfectly defined and limited in its different phases, being a set of interrelated and complex activities to systematize. Innovation, insofar as it is considered as a structured strategic process, could influence on their innovation results. This article aims to advance knowledge and identify which variables of an innovation process can better explain their innovation results. A group of innovative Peruvian companies that have been beneficiaries of public aid to develop innovation projects was studied. Evidence shows us that there is a relationship between the innovation process and innovation results, and there are variables in the innovation process that influence these results more than others. A quantitative methodology was used based on a proposed econometric model.

1 Introduction

The innovation is a phenomenon that has been studied for decades worldwide from different perspectives contributing to the comfort and the quality of life of the persons (Dodgson et al. 2014). Despite it, in Peru there are few empirical evidence that has approached this question, principally from the perspective of innovation processes (Seclen 2017).

The Peruvian Ministry of Production comes promoting the competitiveness of the Peruvian companies for a decade. One way to achieve these goals is through the financing of innovative projects to develop a new product, however, their impacts are not well known (Seclen and Ponce 2017).

In this research, we assume that the companies' beneficiaries of these aids public are those that come realizing innovations and, therefore, they are more sensitized on the innovation process. In this context, we try to know the relations that exist between the innovation process and the innovation results of these companies by using econometrics regressions. It is an attempt to advance in the knowledge on what variables of the innovation process can explain better the innovation results, understood these as outputs obtained after having realized activities of innovation (Acs and Audrestch 2010; OECD 2005).

The structure of the paper is as follows: first, we establish a theoretical framework about the importance of innovation process and innovation results. Following this, we

© Springer Nature Switzerland AG 2019
J. Gil-Lafuente et al. (Eds.): AEDEM 2017, SSDC 180, pp. 158–171, 2019.
https://doi.org/10.1007/978-3-030-00677-8_14

carry out an empirical study to test these assumptions and present the results. Finally, we interpret the relationship between the innovation process and innovation results.

2 Conceptual Framework

At present, there is no doubt that the innovation is an important source of growth and a fundamental factor in the companies that demonstrate to be more competitive (Lam 2010). Even, in some cases, this supposes an unavoidable factor for the companies' survival (Seclen and Barrutia 2013).

To understand the phenomenon of the innovation, it is very important to use a wide definition of innovation. In this research, we take as a reference the Oslo Manual (OECD 2005). In this way, based on the manual, the innovation can be understood as "the process across which a company independently of their size improves or make new products, processes, ways of commercializing and realizing organizational changes, to generate competitive sustainable advantages over the time and assure their survival, respecting the environment and the society" (Seclen 2014, p. 152).

According with the previous definition, the innovation results can be understood as those outputs obtained after having realized activities of innovation (Acs and Audrestsch 2010; OECD 2005). In concrete, the four principal innovation results are: product innovation (on reliability, performance, characteristics, components, materials, design, etc.), process innovation (use of new methods of production, acquisition of machineries, use of management systems, etc.), organizational innovation (new ways to manage the human talent, changes in the structure organizational, formation of strategic alliances, etc.) and innovation in commercialization (implementation of new methods of marketing that answers better to the needs of their clients, like a new channels of sales, prices, etc.).

On the other hand, as we know it is widely accepted that the innovation is a complex and uncertain process in which many actors and factors interact, and their integration does not turn out to be automatic (COTEC 1998). Therefore, the activities that they lead to the innovation have to be structured and systematized (Tidd et al. 2005). In last instance, the management of the innovation implies coordinating between diverse areas or departments of the organization, in such a way that it could be understood as a process or a project, depending on the situation and context of the company (Bessant and Tidd 2007; Gaubinger et al. 2014).

Many investigators deny the existence of a better way of managing the innovation and affirm that there does not exist a "unique" model of innovation that the companies follow or an approach that can be generalized, because the companies are heterogeneous (they possess different routines, competences and capabilities) and they can apply different strategies to manage their innovation process. In any case, a wide variety of investigators have proposed their own model of the innovation process. From this way, we could see that many models present some differences and share points in common (Sattler 2011). For example, in the "Stage-Gate" Model (Cooper 2008), the Process Model of Integrated Innovation Management (Gaubinger et al. 2014) or the Model UNE 166002 on Management of the R&D&i (AENOR 2002), we can appreciate that the innovation is a process that transforms specifics inputs in outputs. In this way, from the mentioned models we characterize a "standard and basic innovation process" that tries neither, integrate nor to generalize, all the existing models, but, it has as intention to explain clearly in a simple way the main activities that must develop in each of the phases of innovation process suitable (Fig. 1).

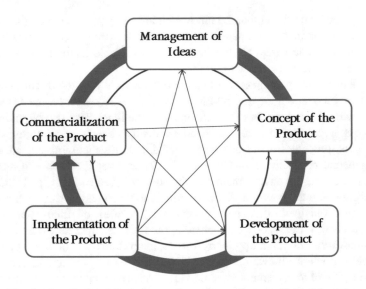

Fig. 1. Standard and basic innovation process. Source: Own elaboration

In definite, this research considers that the innovation process is made up of five phases that can be simultaneous and related: the management of ideas, the concept of the product, the development of the product, the implementation of the product and the commercialization of the product.

Without intentions of being very exhaustive, it is important to mention that the innovation process must begin with a strategic reflection where the process is planned and assigns the staff that will dedicate in an exclusive way to the mentioned process of innovation. Normally, the first phase includes the identification of opportunities from the generation and preliminary evaluation of ideas that they are potentially generating of commercial benefits. In the second phase, as soon as the opportunities have been detected and, in addition, the ideas have been selected of suitable form, the evaluation of their potential economic success on their target market is key. For it, the functional and economic-financial analysis is fundamental (Dornberger et al. 2012).

Following the evaluation of the feasibility of these ideas related to the market and the technology, the third phase carries to the technological accomplishment or materialization of the new ideas, where the basic and applied research, as well as the prototype of the product acquire great relevancy.

The fourth phase includes the implementation that begins with the tests of production where flexibility is needed and the costs of production according the demand of the market. Of this form, the capacity of the company to produce a good or service is a fundamental factor. The fifth phase, the commercialization of the product is characterized for introducing the product on the market in opportune way (AENOR 2002; Bessant and Tidd 2007; Cooper 2008; Gaubinger et al. 2014; Gebauer 2011; Kotler et al. 2008; Sattler 2011). Finally, it is necessary to evaluate the product after their introduction to the market, to identify threats and opportunities.

The research model it is described as follows: first, we take the idea that the innovation process has influence on the innovation results (Fig. 2). In this way, the quantity of good ideas that are generated from the interior of the organization, the time that is delayed in capturing the external ideas to the organization, the time in which it is delayed in conceptualizing, developing, and commercialize the product, as well as hiring experts for the innovation process, might be some key indicators that explain the innovation results (Table 1).

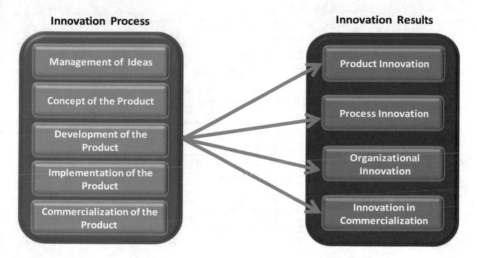

Fig. 2. Relationship between the innovation process and innovation results. Source: Own elaboration

Table 1. Determination variables of innovation results

Phase	Variable	Indicator
Management of ideas	Good ideas that are generated within the organization	Number of good ideas that were generated in the organization in a year
	Hiring an expert in ideas generation and selection techniques	0 = No 1 = Yes
	Good ideas that come from outside the organization	Number of good ideas captured from outside the organization in a year
	Time that it takes to capture external ideas	Time measured in months

(*continued*)

Table 1. (*continued*)

Phase	Variable	Indicator
Concept of the product	Evaluation of the viability of the product	Number of concepts carried out in a year
	Time that it takes to make the concept of the product	Time measured in months
	Hiring an expert in techniques for the concept of the product	0 = No 1 = Yes
Development of the product	Products that have been developed in the organization	Number of new products developed by the organization in a year
	Time that it takes to develop a new product	Time measured in months
	Hiring an expert in techniques to develop products	0 = No 1 = Yes
Implementation of the product	Products that have been implemented in the organization	Number of new products implemented by the organization in a year
	Time that it takes to implement the product	Time measured in months
	Hiring an expert in techniques to implement the product	0 = No 1 = Yes
Commercialization of the product	Time that it takes to launch and introduce a new product to the market	Time measured in months
	Hiring an expert in techniques to commercialization the product	0 = No 1 = Yes

Source: Own elaboration

3 Methodology

This research is a part of project "Guide for comprehensive assessment of innovation management" that is developed by the Research Group of Innovation Management of the Pontifical Catholic University of Peru. The aim of this investigation is to know the relations that exist between the innovation process and the innovation results. Therefore, the research has an exploratory and correlational scope, since we identify those variables that condition the probability of obtaining innovation results.

We study the Peruvian companies that were financed by public funds of the Program "PIPEI, PITEI, PIMEN and PITEA" in the period from 2013 to 2015 and that finished their respective projects of innovation. The population is shaped by 107 companies and the sample was 84 companies (Table 2). The information was collected between April and July 2017. The surveys were sent via online and by mail, addressed to the manager of the company or the director of the area of R&D, obtaining 90 questionnaires answered, but finally 84 surveys were valid, reaching a response rate of 78%.

Table 2. Sample composition

Industry/Size	Micro	Small	Medium	Large	Total
ICTs	6	8	1	2	17
Metalworking	2	3	3	1	9
Wood	1	2	0	0	3
Transports	1	1	1	0	3
Business Consulting	5	5	0	0	10
Agriculture	8	7	1	1	17
Ceramic	1	2	0	0	3
Surgical Equipment	1	2	0	0	3
Engineering	2	3	0	0	5
R&D	5	2	0	0	7
Others	1	3	1	2	7
Total	33	38	7	6	84

Source: Own elaboration

4 Results

Econometric regressions were performed using the multiple linear regression model by the statistical package SPSS. The correlation in the statistical causal sense focuses on the conditioning variables of the innovation process for the generation of innovation results (dependent variable is ordinal and positive). The econometric model proposed is the following:

$$Y = \beta 0 + \beta 1(X_1) + \beta 2(X_2) + \ldots + \beta k(Xk) + \epsilon$$
$$\begin{aligned} Y = {} & \beta 0 + \beta 1(X_1) + \beta 2(X_2) + \beta 3(X_3) + \beta 4(X_4) + \beta 5(X_5) + \beta 6(X_6) \\ & + \beta 7(X_7) + \beta 8(X_8) + \beta 9(X_9) + \beta 10(X_{10}) + \beta 11(X_{11}) \\ & + \beta 12(X_{12}) + \beta 13(X_{13}) + \beta 14(X_{14}) + \beta 15(X_{15}) \end{aligned}$$

Where:

Y = Likelihood of generating an innovation
X_1 = Good ideas that are generated within the organization
X_2 = Hiring an expert in ideas generation and selection techniques
X_3 = Good ideas that come from outside the organization
X_4 = Time that it takes to capture external ideas
X_5 = Evaluation of the viability of the product
X_6 = Time that it takes to make the concept of the product
X_7 = Hiring an expert in techniques for the concept of the product
X_8 = Products that have been developed in the organization
X_9 = Time that it takes to develop a new product
X_{10} = Hiring an expert in techniques to develop products
X_{11} = Products that have been implemented in the organization
X_{12} = Time that it takes to implement the product

X_{13} = Hiring an expert in techniques to implement the product
X_{14} = Time that it takes to launch and introduce a new product to the Market
X_{15} = Hiring an expert in techniques to commercialization the product

First, the main characteristics found are: 61% of the companies studied are family business, 54% do not belong to an industrial association and on average have 56 workers. It should be noted that the largest proportion of family businesses are microenterprises and small businesses, while the largest proportion of companies that belong to an industrial association are medium and large companies. We also found that the companies studied mainly focused on developing product innovations, followed by processes innovations, innovations in commercialization and finally, organizational innovations.

Second, Table 3 shows the adjustment of the model, where the multiple correlation coefficients and the coefficient of determination are more than 0.5 for the case of process innovation and innovation in commercialization, so that the variables considered in these models could explain better their results. However, the coefficient of determination for the case of product innovation and organizational innovation are less than 0.5 so they relatively could explain their results. In any case, the test Durbin-Watson presents values between 1.5 and 2.1, so we assume that the residuals are independent.

Table 3. Model adjustment for each type of innovation

Model	R	R square	R square adjusted	Standard estimation error	Durbin-Watson
Product innovation	.651	.424	.265	3.058	1.554
Process innovation	.858	.737	.664	1.715	1.870
Organizational innovation	.584	.341	.159	1.037	2.013
Innovation in commercialization	.942	.887	.856	1.821	2.375

Source: Own elaboration

On the other hand, Table 4 shows the ANOVA of the model, which shows the value of the statistic F and their critical level (Sig.). All this confirms the robustness of the model and the results.

Table 4. Model ANOVA for each type of innovation

Regression	Sum of squares	Gl	F	Sig.
Product innovation	361.891	15	2.186	.014
Process innovation	505.717	15	9.612	.000
Organizational innovation	35.865	15	2.137	.016
Innovation in commercialization	1321.013	15	9.413	.000

Source: Own elaboration

Finally, we will focus on the coefficients of the model, where we can assess the relative importance of the independent variables within the equation, from critical level of less than 0.05 for each innovation result. In this way, in Table 5 we can see that the variables that determine product innovation are: the good ideas that are generated within the organization in a year (on average companies generate 4 good ideas), the hiring of experts for the development of new products (43% of companies hire specialists to develop new products), and the number of products that have been implemented in a year (on average, companies implement 2 new products).

Table 5. Coefficients of the model for product innovation

Model	Non-standardized coefficients		Standardized coefficients	T	Sig.
	B	Standard error	Beta		
(Constant)	2.788	0.739		3.776	0.000
Good ideas that are generated within the organization	0.286	0.119	0.593	2.395	0.019
Hiring an expert in ideas generation and selection techniques	−0.206	1.408	−0.028	−0.14	0.884
Good ideas that come from outside the organization	0.054	0.106	0.072	0.509	0.612
Time that it takes to capture external ideas	−0.155	0.122	−0.179	−1.27	0.208
Evaluation of the viability of the product	0.108	0.158	0.154	0.684	0.497
Time that it takes to make the concept of the product	0.121	0.207	0.114	0.583	0.562
Hiring an expert in techniques for the concept of the product	−1.364	1.872	−0.186	−0.72	0.469
Products that have been developed in the organization	−0.087	0.324	−0.122	−0.26	0.790
Time that it takes to develop a new product	0.027	0.157	0.038	0.173	0.863
Hiring an expert in techniques to develop products	2.678	1.323	0.374	2.024	0.047
Products that have been implemented in the organization	0.610	0.284	0.922	2.145	0.036
Time that it takes to implement the product	−0.088	0.137	−0.146	−0.63	0.525
Hiring an expert in techniques to implement the product	−1.147	1.350	−0.159	−0.85	0.399
Time that it takes to launch and introduce a new product to the Market	0.007	0.139	0.010	0.052	0.958
Hiring an expert in techniques to commercialization the product	0.302	1.143	0.040	0.264	0.793

Source: Own elaboration

On the other hand, in Table 6 we can see that the variables that condition the process innovation are: the good ideas that are generated within the organization in a year (on average companies generate 4 good ideas), the good ideas captured from the outside of the organization in a year (on average companies capture 2 good ideas), the number of new products that are conceptualized in a year (on average, companies make the concept of 3 good ideas) and the time to implement the new product (on average companies take 7 months).

Table 6. Coefficients of the model for process innovation

Model	Non-standardized coefficients		Standardized coefficients	T	Sig.
	B	Standard error	Beta		
(Constant)	0.822	0.416		1.974	0.053
Good ideas that are generated within the organization	0.165	0.067	0.412	2.452	0.017
Hiring an expert in ideas generation and selection techniques	0.013	0.794	0.002	0.016	0.987
Good ideas that come from outside the organization	0.287	0.060	0.461	4.816	0.000
Time that it takes to capture external ideas	−0.054	0.069	−0.076	−0.78	0.433
Evaluation of the viability of the product	0.308	0.089	0.531	3.464	0.001
Time that it takes to make the concept of the product	−0.034	0.117	−0.038	−0.28	0.774
Hiring an expert in techniques for the concept of the product	−0.947	1.055	−0.155	−0.89	0.373
Products that have been developed in the organization	−0.019	0.183	−0.033	−0.10	0.917
Time that it takes to develop a new product	−0.021	0.089	−0.036	−0.23	0.813
Hiring an expert in techniques to develop products	−1.384	0.746	−0.233	−1.85	0.068
Products that have been implemented in the organization	0.088	0.160	0.161	0.551	0.583
Time that it takes to implement the product	−0.254	0.770	−0.210	−1.79	0.048
Hiring an expert in techniques to implement the product	0.567	0.761	0.095	0.745	0.459
Time that it takes to launch and introduce a new product to the Market	0.053	0.078	0.088	0.683	0.497
Hiring an expert in techniques to commercialization the product	0.176	0.644	0.029	0.274	0.785

Source: Own elaboration

Likewise, in Table 7 we appreciate that the variables that condition organizational innovation are: the time to capture ideas from outside the company (on average companies take 3 months), the time to develop the new product (on average companies are delayed 5 months) and the time to implement the new product (on average companies take 7 months).

Table 7. Coefficients of the model for organizational innovation

Model	Non-standardized coefficients		Standardized coefficients	T	Sig.
	B	Standard error	Beta		
(Constant)	1.052	0.235		4.473	0.000
Good ideas that are generated within the organization	0.029	0.038	0.192	0.775	0.441
Hiring an expert in ideas generation and selection techniques	−0.528	0.448	−0.228	−1.17	0.243
Good ideas that come from outside the organization	0.014	0.034	0.059	0.420	0.676
Time that it takes to capture external ideas	−0.087	0.039	−0.319	−2.25	0.028
Evaluation of the viability of the product	0.065	0.050	0.291	1.287	0.202
Time that it takes to make the concept of the product	−0.008	0.066	−0.025	−0.12	0.901
Hiring an expert in techniques for the concept of the product	0.414	0.596	0.178	0.694	0.490
Products that have been developed in the organization	−0.112	0.103	−0.501	−1.08	0.281
Time that it takes to develop a new product	−0.088	0.050	−0.394	−1.77	0.050
Hiring an expert in techniques to develop products	−0.067	0.421	−0.030	−0.16	0.873
Products that have been implemented in the organization	0.045	0.091	0.216	0.500	0.619
Time that it takes to implement the product	−0.104	0.044	−0.544	−2.37	0.020
Hiring an expert in techniques to implement the product	0.139	0.430	0.061	0.323	0.747
Time that it takes to launch and introduce a new product to the Market	−0.045	0.044	−0.195	−1.02	0.311
Hiring an expert in techniques to commercialization the product	−0.021	0.364	−0.009	−0.05	0.954

Source: Own elaboration

Finally, in Table 8, we appreciate that the variables that condition innovation in commercialization are: the hiring of experts to generate ideas (38% of companies hire specialists to generate ideas), the hiring of experts for the concept of the product (37% of companies hire specialists for the concept of the product), the time to develop the product (on average companies take 5 months), the hiring of experts for the development of the product (43% of companies hire specialists to develop the product), the time to implement the new product (on average companies take 7 months) and the time to launch the new product to the market (on average, companies delay 5 months).

Table 8. Coefficients of the model for innovation in commercialization

Model	Non-standardized coefficients		Standardized coefficients	T	Sig.
	B	Standard error	Beta		
(Constant)	1.523	0.680		2.240	0.028
Good ideas that are generated within the organization	0.012	0.110	0.019	0.111	0.912
Hiring an expert in ideas generation and selection techniques	4.080	1.296	0.416	3.147	0.002
Good ideas that come from outside the organization	0.046	0.097	0.046	0.475	0.636
Time that it takes to capture external ideas	0.071	0.112	0.062	0.637	0.526
Evaluation of the viability of the product	−0.091	0.145	−0.097	−0.63	0.531
Time that it takes to make the concept of the product	−0.009	0.191	−0.006	−0.04	0.964
Hiring an expert in techniques for the concept of the product	6.309	1.723	0.639	3.661	0.000
Products that have been developed in the organization	0.249	0.299	0.262	0.833	0.408
Time that it takes to develop a new product	−1.006	0.145	−1.057	−6.96	0.000
Hiring an expert in techniques to develop products	3.781	1.218	0.393	3.101	0.003
Products that have been implemented in the organization	−0.189	0.262	−0.212	−0.72	0.473
Time that it takes to implement the product	−0.539	0.126	−0.667	−4.26	0.000
Hiring an expert in techniques to implement the product	−1.096	1.243	−0.113	−0.88	0.381
Time that it takes to launch and introduce a new product to the Market	−0.329	0.128	−0.334	−2.57	0.012
Hiring an expert in techniques to commercialization the product	−0.413	1.053	−0.041	−0.39	0.696

Source: Own elaboration

5 Discussion, Conclusion and Implications

Our findings have different implications for theory and practice on innovation management. This study contributes to the literature on innovation management by finding relationships that exist between the innovation process and the innovation results. In this way, the importance of the understanding of the innovation process is highlighted.

The variables that condition the results of innovation are: the good ideas that are generated within the organization, the hiring of experts to generate ideas, the good ideas captured from outside the organization, the time to capture ideas from the outside of the company, the amount of new products that are conceptualized in a year, the hiring of experts for the concept of the product, the hiring of experts for the development of new products, the time in developing the new product, the quantity of products that have been implemented in a year, the time to implement the new product and the time to launch the new product to the market.

However, the variables that influence and are the most significant in the results of innovation are: the good ideas that are generated within the organization for product innovation; good ideas captured from outside the organization for process innovation; the time to implement the new product for organizational innovation; the hiring of experts for the concept of the new product and the time to develop and implement the new product for the innovation in commercialization.

On the other hand, we found that the "time variable" has an inverse relationship with the innovation results. That is, the longer time spent in the innovation process, the companies can get non significates results. In this way, the time plays a key role in the innovation process. In this study, the innovation process lasts an average of 24 months from the moment an idea is generated until the product is introduced into the market. This situation in contexts of high levels competitive demands may not be adequate.

According to the "standard and basic innovation process", the phases that are most related to product innovation are the generation of ideas, the development and implementation of products. Similarly, the generation of ideas, the concept and the implementation of the product are the phases that are most related to process innovation. Furthermore, the phases of the innovation process that are more related to organizational innovation are the generation of ideas, the development and implementation of new products. Ultimately, we found that innovation in commercialization is related to all phases of the innovation process (idea generation, conceptualization, development, implementation and commercialization of products).

All these evidences show that there is a relationship between the innovation process and the innovation results, so this is why the innovation process must be structured and systematized. In short, the evidence from this study contributes to the understanding of the relationship between the innovation process and innovation results, which may be of special interest to business managers and policy makers.

Finally, this study presents some limitations that suggest future lines of research. First, due to the exploratory nature of the study, the results obtained can be generalized with some caution, because it cannot be extrapolated in a linear way to all economic sectors. Therefore, it would be advisable to carry out more causal studies with more robust statistical methods oriented to an activity sector. Secondly, complementing above, it would be advisable to carry out more descriptive studies that further specify the activities that comprise each of the phases of the innovation process, for which the use of qualitative methods could be very useful. Third, it would be convenient to carry out comparative studies at a regional and international level to validate these findings on the relationship between the innovation process and innovation results.

Acknowledgments. This article contains the results of the Research Project ID 478, which was funded by the Pontifical Catholic University of Peru. The author thanks the companies that have participated in the research. Also, to Professor Enrique Medellín at National Autonomous University of Mexico and Nils Fonstad from the Center for Information Systems Research at MIT, for their valuable contributions.

References

Acs, Z., Audretsch, D.: Knowledge spillover entrepreneurship. In: Acs, Z., Audretsch, D. (eds.) Handbook of Entrepreneurship Research: An Interdisciplinary Survey and Introduction, pp. 273–302. Springer, New York (2010)

AENOR: Gestión de la I+D+i: Requisitos del Sistema de Gestión de la I+D+i, Madrid (2002)

Bessant, J., Tidd, J.: Innovation and Entrepreneurship. Wiley, Hoboken (2007)

COTEC: Informe COTEC: Tecnología e Innovación en España. Fundación para la Innovación Tecnológica, Madrid (1998)

Cooper, R.G.: Perspective: The Stage-Gate® Idea-to-Launch Process - Update, What's New, and NexGen Systems. J. Prod. Innov. Manag. **25**(3), 213–232 (2008)

Dodgson, M., Gann, D., Phillips, N.: The Oxford Handbook of Innovation Management. Oxford University Press, Oxford (2014)

Dornberger, U., Suvelza, A., Bernal, L.: Gestión de la fase temprana de la innovación. Intelligence 4 innovation en cooperación con International SEPT Program, Leipzig (2012)

Gaubinger, K., Rabl, M., Swan, S., Werani, T.: Innovation and Product Management. A Holistic and Practical Approach to Uncertainty Reduction. Springer, Berlin (2014)

Gebauer, H.: Exploring the contribution of management innovation to the evolution of dynamic capabilities. Ind. Mark. Manag. **40**, 1238–1250 (2011)

Kotler, P., Wong, V., Saunders, J., Armstrong, G., Wood, M.B.: Principles of Marketing, Fifth European edn. Prentice Hall, London (2008)

Lam, A.: Organizaciones innovadoras: estructura, aprendizaje y adaptación. En: BBVA (ed.) Innovación perspectivas para el siglo XXI, pp. 163–177. BBVA, Madrid (2010)

OCDE: Manual de Oslo: Guía para la recogida e interpretación de datos sobre innovación. OCDE, Madrid (2005)

Sattler, M.: Excellence in Innovation Management. A Meta-analytics Review on the Predictors of Innovation Performance. Gabler Research, Wiesbaden (2011)

Seclen, J.P.: KIBS and innovation in machine tool micro enterprises: the cases of the Basque Country and Emilia-Romagna. Doctoral Thesis, University of the Basque Country (2014)

Seclen, J.P.: ¿Existe relación entre la gestión de la innovación y los resultados de innovación? un análisis exploratorio de empresas innovadoras peruanas. In: XXVI AEDEM International Conference Proceedings, pp. 991–1014 (2017)

Seclen, J.P., Ponce, F.: ¿Cómo gestionan la innovación las empresas peruanas?: Un análisis descriptivo de empresas beneficiarias de Innóvate Perú 2013 – 2015. En: V Congreso Internacional de la Red Universidad Empresa ALCUE Proceedings, Lima, Perú (2017)

Seclen, J.P., Barrutia, J.: Impacto de los SEIC sobre la innovación en las Microempresas fabricantes de máquina-herramienta del País Vasco. In: XXVII AEDEM Annual Conference Proceedings, Huelva, España (2013)

Tidd, J., Bessant, J., Pavitt, K.: Managing Innovation: Integrating Technological, Market and Organizational Change, 3rd edn. Wiley, Chichester (2005)

E-Commerce Decision-Making Factors in Peruvian Organizations of the Retail Sector

Enrique Saravia[✉]

Universidad del Pacífico, Lima, Peru
saravia_ea@up.edu.pe

Abstract. The purpose of this research is to identify, by an exploratory approach, the main factors of decision-making on adopting e-commerce technologies in Peruvian organizations of the retail sector. The study was based on the application of 80 on-line surveys and 8 interviews with experts. The main findings of the quantitative analysis, based on a structural model with a high reliability and validity index, indicate the main factors that influence decision-making on adopting e-commerce technologies, is the awareness that high investments in technologies generate profit in the long-term (motivating factor) and risks of innovation (inhibiting factors), in addition to other factors that influence at a lower level as the organizational culture, the attitudes of innovation and optimism to the use of technologies of decision-makers, people's skills, and process flexibility to adapt them to e-commerce activities, among others. The qualitative analysis allows further the study and identify specific details which can make a significant contribution to knowledge generation and practical application in the organizations, which gives us an extensive list of aspects that have to be managed appropriately, and can thus lead the investments and mitigate the risks.

1 Introduction

In recent years, the development of e-commerce is growing fundamentally because organizations can reduce their operative costs through the use of on-line services (Vieira 2010). Nevertheless, according to the report published by VISA in Latin America 2014, the development of e-commerce in Peru does not reach the levels of other countries in the region and much less the level of development of European countries who are leaders in e-commerce. Peru ranks sixth in the region, according to the development index of e-commerce named "e-Readiness" with an index of 31.5, whilst the average in Latin America is 54.8 and 82.6 in France (VISA 2014).

There are many studies in the literature which try to explain consumer behavior and identify the factors that make them select the virtual channel for on-line purchasing, instead of attending a physical establishment for their purchases. However, there are no specific studies that try to identify and explain the factors which decide on adopting e-commerce technologies by decision-makers of retail organizations, at least from the approach and depth to be explored in the current study, and far less for Peruvian organizations. Although, there is a high potential for further development of e-commerce in Peru and a vast knowledge of consumers' profile, there is a lack of

© Springer Nature Switzerland AG 2019
J. Gil-Lafuente et al. (Eds.): AEDEM 2017, SSDC 180, pp. 172–187, 2019.
https://doi.org/10.1007/978-3-030-00677-8_15

information as to the reasons which prevent Peruvian organizations from offering or promoting the digital sales channels with high intensity.

The aim of this research is to identify and explain the factors of adopting e-commerce technologies in retail organizations in Peru.

2 Conceptual Framework

Although e-commerce has many definitions, the best known refers to the process of buying and selling goods and services electronically, through transactions on the Internet, networks, or other digital technologies. More broadly, e-commerce is defined as the application of information and communication technologies to the value chain, on processes electronically produced, in a complete or partial manner (Wigand 1997).

From this broader approach, the factors of the adoption of e-commerce must be addressed from the context of innovation theories (Guerrero Cuellar and Rivas Tovar 2005). The organizations have diverse reasons to innovate, their objectives could be related to the products, markets, efficiency, quality or learning ability and introducing changes (OECD 2005). The benefits of innovation could be to improve customer service and satisfaction, increase market participation, improve the processes and reduce costs, among others. In this regard, the following hypothesis is proposed:

H_1:   The benefits of e-commerce influence decision-making on adopting technologies in the organizations of the retail sector in a positive and significant manner.

From another perspective, in a study based on relevant decision criteria, which affects decision-making on adopting technologies in Spanish organizations (E-Business Center 2008) 15 technologies adoption criteria were identified and classified in three dimensions: "Decision-making based on rationality", the "Diffusion of innovation" and the "psychology of the decision-maker". The rationality based on costs, technical features of technology, the irreversibility of the decision taken, supplier dependencies, the beliefs of decision-makers, are examples of considered variables in this study. Several of these criteria represent innovation obstacles or risks, rather than benefits. In fact, innovative activity could be obstructed by numerous factors (OECD 2005). In this way, the following hypothesis is proposed:

H_2:   The barriers, obstacles and risks of e-commerce influence decision-making on adopting technologies in the organizations in a negative and significant manner.

Many other authors also include benefits, barriers and risks, but they are classified in other dimensions. Kurnia et al. (2015) propose the "organizational context", "external context or environmental" and "technological context" dimensions. Oliveira and Martins (2011) propose the dimensions of "internal characteristics" and "external characteristics" of the organization, as well as the "individual characteristics of leaders". These approaches highlight the presence of internal factors in the organization and psychological factors of decision-makers, coinciding these last characteristics with the last dimension of the E-Business Center study.

The *"Technology Acceptance Model"* (TAM) (Davis 1986; Vankatesh and Davis 2000) defines fundamentally two dimensions to explain technologies adoption: the "perceived utility" and its "ease of use". While the "perceived utility" can be interpreted as "the benefits of innovation", "ease of use" can be associated with the internal and technological factors, which allow the organization to adapt easily to the required processes in e-commerce.

Kline's model (Kline and Rosenberg 1986) identifies basic activities of the innovation process, and these activities help identify the required people's skills and the need to adapt internal processes. Thus, the following hypothesis are proposed:

H_3:   The people's skills influence decision-making on adopting e-commerce in the organizations of the retail sector in a positive and significant manner.

H_4:   The flexibility and easy adaptation of the Processes influence decision-making on adopting e-commerce in a positive and significant manner

The *"Technology Readiness Index"* model (TRI) is based on a 16 item questionnaire in order to measure four dimensions which are related to the propensity of people to adopt and use new technologies (Parasuraman 2000; Parasuramant and Colby 2015):

OPTIMISM (Motivator):   positive view concerning technology and the belief that control is provided, flexibility and efficiency on people's lives.

INNOVATIVENESS (Motivator):   tended to become a technological and thought leader.

DISCOMFORT (Inhibitor):   perceived lack of control about technology and the sense of it being overwhelming.

INSECURITY (Inhibitor):   lack of confidence in technology, concern about its proper function and its possible consequences.

On the basis of the Technology Readiness Index (TRI) model, the following hypothesis are proposed:

H_5:   Decision-makers' attitudes of Optimism associated to the use of technology influence decision-making on adopting e-commerce in a positive and significant manner.

H_6:   Decision-makers' attitudes associated with Innovation to the use of technology influence decision-making on adopting e-commerce in a positive and significant manner.

H_7:   Decision-makers' attitudes of Discomfort associated with the use of technology influence decision-making on adopting e-commerce in the organizations of the retail sector in a negative and significant manner.

H_8:   Decision-makers' attitudes of Insecurity associated with the use of technology influence decision-making on adopting e-commerce in a negative and significant manner

Many authors also refer to the Organizational Culture as an internal factor that influences on the innovation-decisions. Specifically, Rao and Weintraub (2013) propose a model to measure the Innovative Culture in the organizations, consisting of six dimensions: Values, Leadership, Organizational Climate, Resources, Processes and Results, at the same time each one of them consists of three sub-dimensions and every sub-dimension is measured through three items. In this study, questions which are linked to the first three dimensions, were adapted. The three items of each sub-dimension were grouped together in one question. In this way, the following hypothesis are proposed:

H_9:   The Values of the Organizational Innovative Culture influence decision-making on adopting e-commerce in a positive and significant manner.

H_{10}: The Leadership of the Organizational Innovative Culture influences decision-making on adopting e-commerce in a positive and significant manner.

H_{11}: The Organizational Climate of the Organizational Innovative Culture influences decision-making on adopting e-commerce in a positive and significant manner.

3   Methodology

Based on the literature review, the Theoretical Model which was proposed, is built and illustrated in this study, and it strives to explain the adoption of e-commerce technologies in the retail organizations in Peru (Fig. 1):

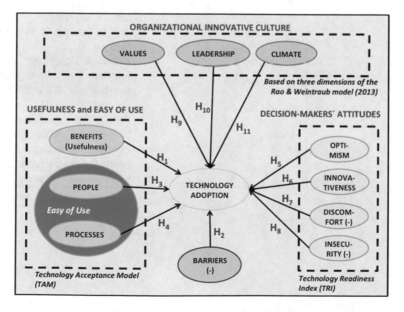

Fig. 1. Theoretical model of the study

To measure the relevant variables of this model, two instruments and measurement techniques, as well as data collection were used:

- Questionnaire to apply surveys: An on-line questionnaire was developed, based on open questions with closed answers using the 5 scales of Likert. The survey was specifically directed to executives or middle-ranking organizations of the retail sector, through a non-probabilistic sample of convenience. A sufficient number of 80 surveys was achieved, between February and June of 2017, taking into account that the *partial least squares* (PLS) only requires 80 surveys in structural models, in which the constructs are related to 7 latent variables, at most (Wong 2013). On the other hand, this technique is appropriate when the data does not fulfill the conditions of normality and/or the size of the sample is relatively small (Hair et al. 2014), and when theory is still being explored and built (Wong 2013) (Table 1).
- Guidelines to perform interviews to experts: A guideline was elaborated to interview subject matter experts, entrepreneurs who are dedicated to consultancy service, development, implementation and the maintenance of e-commerce technologies and digital marketing experts. 8 in-depth interviews were performed between February and June of 2017. The interviews were recorded and processed in Atlas TI, a software of qualitative data processing.

Table 1. Structure and variables of the questionnaire

Dimension	Source	Item of the survey/variable
BENEFITS OF INNOVATION	Based on OECD (2005), Gargallo Castel et al. (2016), Others	Permits the preservation and expansion of the market (beneco_1) Permits the reduction of costs and the increase in profitability (beneco_2) Permits to significantly improve customer service through the development of a new on-line channel (benser_1) Permits the improvement of customer satisfaction (benser_2) Permits to have more efficient internal processes and to achieve a better service (benser_3)

(continued)

Table 1. (*continued*)

Dimension	Source	Item of the survey/variable
BARRIERS OF INNOVATION	Based on E-Business Center (2008), Oliveira and Martins (2011), Kurnia et al. (2015), Others	High investment and/or the difficult access to financial sources (barinv_1) The long time required to recover the investments (barinv_2) Risk of failure (barries_1) Risk of the decision's irreversibility (barries_2) The possible high dependency of technology providers (barpro_1) The people's poor ability to adopt new technologies (barpro_2) The limited flexibility and difficulty to adapt current processes of the organization, according to the new requirements of the technological innovation (barpro_3) The low availability of technological solutions in the market (barpro_4) Contact reduction with customers and customization of the service (barpro_5)
PEOPLE	Identified variables as of the innovation process activities of Kline's model (Kline and Rosenberg 1986)	We have the talent needed to identify what the competition offers in on-line services and to get acquainted with customers' ideas (perso_1) We have the talent needed to generate technological innovation ideas (perso_2) We have the talent needed or the access to experts to develop design proposals of technological innovations (perso_3) We have the talent needed or the access to experts to develop technological innovation projects (perso_4) We have people that tend to research in technological innovation processes (perso_5)

(*continued*)

Table 1. (*continued*)

Dimension	Source	Item of the survey/variable
PROCESSES	Key characteristics according to the input of many authors	Our communication to customers is very efficient when we want to disseminate our process and service improvements (proce_1) Our processes are flexible and we have people with great capacity of adapting to new technology-based processes (proce_2) We have technological tools or we could hire technology service providers to develop technological innovation projects (proce_3) People have time and/or financial resources are available to develop technological innovation projects (proce_4)
OPTIMISM (Attitude)	*Technology Readiness Index*, TRI Model (Parasuraman and Colby, An Updated and Streamlined Technology Readiness Index: TRI 2.0, 2015)	New technologies contribute to a better quality of life (optim_1) Technology gives me more freedom of mobility (optim_2) Technology gives people more control over their daily lives (optim_3) Technology makes me more productive in my personal life (optim_4)
INNOVATIVENESS (Attitude)		Other people come to me for advice on new technologies (innov_1) In general, I am among the first in my circle of friends to acquire new technology when it appears (innov_2) I can usually figure out new high-tech products and services without help from others (innov_3) I keep up with the latest technological developments in my areas of interest (innov_4)

(*continued*)

Table 1. (*continued*)

Dimension	Source	Item of the survey/variable
DISCOMFORT (Attitude)		When I get technical support from a provider of a high-tech product or service, I sometimes feel as if I am being taken advantage of by someone who knows more than I do (disc_1) Technical support lines are not helpful because they don't explain things in terms I understand (disc_2) Sometimes, I think that technology systems are not designed for use by ordinary people (disc_3) There is no such thing as a manual for a high-tech product or service that's written in plain language (disc_4)
INSECURITY (Attitude)		People are too dependent on technology to do things for them (insec_1) Too much technology distracts people to a point that is harmful (insec_2) Technology lowers the quality of relationships by reducing personal interaction (insec_3) I do not feel confident doing business with a place that can only be reached online (insec_4)
VALUES (Innovative culture)	Adapted and based on Rao and Weintraub (2013)	We have a very strong desire to explore opportunities, we tolerate new ideas and try to implement new projects (valor_1) We encourage new ways of thinking and solutions from different perspectives, with freedom and spontaneity (valor_2) We are constantly experimenting new efforts to innovate, we are not afraid of failure and we take failure as a learning opportunity (valor_3)

(*continued*)

Table 1. (*continued*)

Dimension	Source	Item of the survey/variable
LEADERSHIP (Innovative culture)		Our leaders inspire us with a vision for the future and by searching for new opportunities (líder_1) Our leaders provide support and encouragement in innovation projects (líder_2) Our leaders use appropriate strategies to overcome organizational barriers (líder_3)
Organizational climate (innovative culture)	Adapted and based on Rao and Weintraub (2013)	We appreciate and respect the differences that exist within our community (clima_1) We are consistent with doing the things we choose to value (clima_2) We minimize the rules, policies, bureaucracy and rigidities to simplify our work (clima_3)
ADOPTION OF E-COMMERCE	–	Current development of on-line sales channel (adop_1) Development intention or future improvement of the on-line channel (adop_2)

Source: based on several authors.

4   Results

4.1   Composition of the Sample Obtained in the Surveys

A question related to the years of operation a retail company has in Peru was included in the questionnaire, since the organizations' profile with regard to this variable is relevant for the study. It is expected that mature organizations in the market have a higher probability to enter the on-line sales channel. The composition of the sample obtained in the survey was the following (Table 2):

Table 2. Composition of the sample obtained in the surveys

The company's years of operation	Distribution of the sample
Less than 1 year	4.6%
Between 1 and 3 years	21.6%
Between 3 and 10 years	13.8%
More than 10 years	60.0%

4.2   Reliability and Validity Results of the Surveys

In Table 3 the results of the reliability of data test and the convergent validity of constructs are summarized. In every dimension, reliability indexes "*Alfa de Cronbach*" superior or very close to the minimum value commonly accepted of 70% were obtained, as well as validity indexes "*Average Variance Extracted*" (AVE), much higher than the minimum value recommended of 55%. In some cases, variables were eliminated to achieve acceptable levels of reliability and validity (In Fig. 2 the number of variables, which were finally considered in each dimension can be seen).

Table 3. Reliability and validity of the surveys

DIMENSIONS		Reliability	Convergent Validity of an exploratory level (Factor Analysis)	
		Cronbach Alpha (70%)	Index of kindness Kaiser-Meyer-Olkin (0.50)	% Variance Explained AVE (55%)
ATTITUDE TOWARDS THE USE OF TECHNOLOGY	OPTIMISM	78.1%	0.694	60.6%
	INNOVATIVENESS	64.1%	0.603	58.9%
	DISCOMFORT	65.8%	0.611	59.8%
	INSECURITY	67.4%	0.615	60.8%
CULTURE	VALUES	78.4%	0.685	71.5%
	LEADERSHIP	80.7%	0.660	73.2%
	ORGANIZATIONAL CLIMATE	75.4%	0.611	71.3%
FACILITY	PEOPLE	89.9%	0.835	72.3%
	PROCESSES	78.6%	0.773	61.5%
BENEFITS	ECONOMIC	74.2%	0.500	79.9%
	SERVICES	82.6%	0.715	75.2%
BARRIERS	INVESTMENT	68.5%	0.500	76.1%
	RISK	79.1%	0.500	82.7%
	PROCESSES	65.3%	0.649	59.3%
ADOPTION		76.5%	0.500	81.7%

The factor analysis also allowed to classify the factor "Benefits" in two sub-dimensions, and the dimension "Barriers" in three.

4.3   Results of the Structural Model and Validation of Hypothesis

In order to determine significant cause-effect relationships and validate the hypothesis of the model, a structural equation model was proposed, using the SmartPLS software. In Fig. 2 the Structural Model resulting in the format of the SmartPLS software, is illustrated:

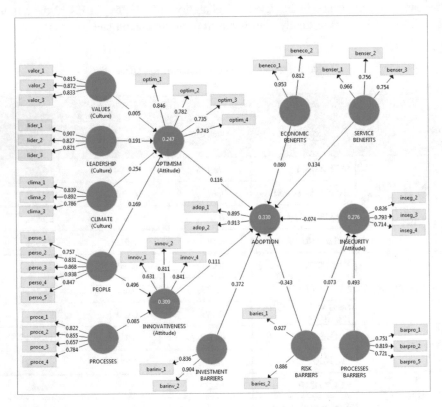

Fig. 2. Results of the Structural Model. *Source*: SmartPLS software, based on the data of the study

The structural model resulted completely consistent in cause-effect relationships between constructs, positive or negative, depending if they have been defined as motivational or inhibiting factors in the theoretical framework. For instance, according to what is expected, the "attitude of insecurity towards technology" has a negative causality (negative beta of −0.074) with the "adoption of technologies".

The "adoption of technologies", meaning, the independent variable of the model, resulted in a Pearson's coefficient r2 = 0.33, value of which exceeds the commonly minimum acceptable of 0.30. In other words, economic and service improvement benefits, an attitude of optimism and innovation towards technology and the investment's characteristics, between motivating factors, and the attitudes of insecurity towards technology as well as the barriers generated by the risks as inhibiting factors, explain 33% of the "adoption of technologies". Nevertheless, while all of this variables together explain the adoption of technologies, the investment's characteristics as a motivating factor (beta = 0.372; t = 3.081; p-value = 0.002) and the barriers generated by the risks as an inhibiting factor (beta = −0.343; t = 2.819; p-value = 0.005) influence in a significant manner.

Table 4. Structural Model Results and Hypothesis Validation

	Hypothesis	Beta	t	p-value	Hypothesis Validation*	R2
H1	(H1 a) Economic Benefits→ Adoption	0.080	0.546	0.585	Partially	
	(H1 b) Service Benefits → Adoption	0.134	0.907	0.365	Partially	
H2	(H2 a) Investment Barriers → Adoption	0.372	3.081	0.002	ACCEPTED	
	(H2 b) Risk Barriers → Adoption	-0.343	2.819	0.005	ACCEPTED	
	(H2 c) Processes Barriers → Adoption				REJECTED	
H3	People → Adoption				REJECTED	
H4	Processes → Adoption				REJECTED	
H5	Optimism (Attitude) → Adoption	0.116	0.893	0.372	Partially	0.330
H6	Innovativeness (Attitude) → Adoption	0.111	0.701	0.484	Partially	
H7	Discomfort (Attitude) → Adoption				REJECTED	
H8	Insecurity (Attitude) → Adoption	-0.074	0.496	0.620	Partially	
H9	Values (Culture) → Adoption				REJECTED	
H10	Leadership (Culture) → Adoption				REJECTED	
H11	Organizational Climate (Culture) → Adoption				REJECTED	
	Values → Optimism (Attitude)	0.005	0.024	0.981	Partially	
	Leadership → Optimism (Attitude)	0.191	1.125	0.261	Partially	0.247
	Organizational Climate → Optimism (Attitude)	0.254	1.646	0.100	Partially	
	People → Optimism (Attitude)	0.169	1.594	0.112	Partially	
	People → Innovation (Attitude)	0.496	2.677	0.008	ACCEPTED	0.309
	Processes → Innovation (Attitude)	0.085	0.479	0.632	Partially	
	Risk Barriers→ Insecurity (Attitude)	0.073	0.616	0.538	Partially	0.276
	Processes Barriers→ Insecurity (Attitude)	0.493	4.553	0.005	ACCEPTED	

* The hypothesis validated as "Partially", indicate that there is a correlation, but NOT a significant one.
Source: Study data based on software SPSS reports

Another important finding of the study is that the "innovative culture", specifically the components of values, leadership, organizational climate, besides "people's competence", achieved to explain the attitude of optimism towards technology in a 24.7%. Out of these variables, organizational climate influences decision-making on adopting technologies in retail organizations in a significant manner (beta = 0.254; t = 1.646; p-value = 0.100), according to the critical value t-students of 1.65, which corresponds to a level of significance of 10% (Wong 2013).

On the other hand, the study proved that the people's competence is very significant and has a great contribution in explaining the "attitude of innovation", with an index of r2 = 0.309. Likewise, the "process barriers" influence in a significant manner and have a great contribution in explaining the attitudes of insecurity in a 27.6% (Table 4).

4.4   In-Depth Interviews Results with Experts

The reports obtained in the in-depth interviews with experts, were systematized in the following concept map (Fig. 3):

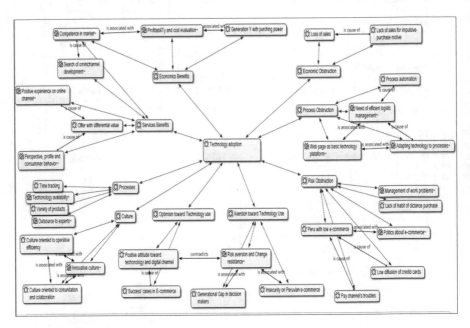

Fig. 3. Concept Map resulting from the reports of the interviews with experts. *Source*: Atlas TI software, based on the experts' reports in the interviews of the study

The concept map allows to analyze each one of the associated factors and variables, meaning, each of the branches within a network:

– In the network of "benefits", in the interviews, experts mentioned aspects associated with the two types of benefits identified in the theoretical framework. Referring to the "economic benefits", experts discussed about cost reductions and the increase in profitability, the need to develop virtual channels because the competition "demands it" and the sales potential which represents the generation of the millennial, that are in the labor market nowadays and have purchasing power. The "benefits of an improved service" were associated with the need to develop an advertising strategy and "omni-channel" sales due to the current competition and the need to offer a better service, in accordance with the consumer's profile and behavior. Today they even have more examples of positive experiences in on-line channels.

- With regard to the network of "barriers" of the e-commerce development, experts mentioned aspects associated with the three types of barriers identified in the theoretical framework. Referring to the "economic barriers", experts discussed about possible lost sales for turning to current customers that may not be familiar with on-line shopping and on the other hand, reducing impulsive buying in stores. In reference to "processes barriers", experts discussed about the need to adapt the processes to technology, the need of an efficient logistical management and the need to automate processes and develop the website as a basic technology platform, which must feed the other tools or digital channels. In the "risk barrier", the interviewees mentioned risks due to the people's uncertainty and fear before a possible people reduction and several structural problems at country level such as governmental policies that encourage e-commerce development, lack of habit or people's willingness to purchase on digital media and little credit cards penetration in the country, as well as the fear of means of payment, which generate a low e-commerce penetration.

- With regard to the network of "processes", experts quoted four key aspects in process management: the management of an extensive range of products, the availability of technology, the need to outsource processes to a technology provider and monitoring products lead times.

- In the network of attitudes to the use of technology of decision-makers, experts mentioned aspects associated with the attitudes of "optimism" and "insecurity". An escalating development of favorable attitudes to the use of digital channels and e-commerce success stories in the market generate attitudes of "optimism" to the use of technologies. The decision-makers' generation gap and the insecurity in e-commerce generate low risk tolerance levels, which flows in an attitude of "insecurity" to the use of technology of decision-makers. At the same time, experts also associated the favorable attitude to the use of digital channels (attitude of optimism) with low risk tolerance levels (attitude of insecurity) as mutually dependent variables.

- With regard to the Organizational Culture aspects, although the experts did not specifically mention the dimensions of "values", "leadership" and "organizational climate" of the theoretical framework, they mentioned three typologies of the culture needed for the development of e-commerce: (i) innovative culture, culture oriented to the operational efficiency, as well as communication and collaboration.

5 Discussion, Conclusion and Implications

The study presents a high consistency between the quantitative results obtained in the survey and the qualitative aspects identified in the analysis of interviews to experts. In that sense, the qualitative analysis not only confirms the findings obtained through surveys, moreover, and most especially, allows further the analysis to a wider understanding of the subject. In the qualitative research, an extensive list of aspects which complement the findings obtained in the quantitative, was identified, as well as valuable input that is difficult to detect in the quantitative research due to the lack of "context".

The main findings of the quantitative analysis indicate, generally, the variables of the model manage to explain decision-making on adopting e-commerce technologies in the organizations of the retail sector in Peru. Nevertheless, the awareness that high investment on technology is a long-term decision as a motivating factor and that barriers generated by risks as inhibiting factors are the factors which influence decision-making in a significant manner. Other findings of the quantitative analysis are: (1) the aspects of the organizational culture influence the attitude of optimism to the decision-makers' use of technologies in a significant manner; (2) people's skills and processes flexibility to adapt to e-commerce activities promote attitudes associated with innovation of decision-makers; and (3) inherent risks in innovation and processes adaptation affect the attitudes of insecurity of decision-markers in a significant manner.

With regard to the qualitative analysis' contribution, several details associated with the decision-making factors on adopting technologies were identified: the need to develop virtual channels because the competition "demands it" and sales potential which represents the generation of millennial, that are in the labor market nowadays and have purchasing power; the need to develop an advertising strategy and "omni-channel" sales of digital channels; the need to develop the website as a basic technology platform, which must feed the other tools or digital channels; the risk of possible lost sales for turning to current customers that may not be familiar with on-line shopping; the risk of reducing impulsive buying in stores; the need to adapt the processes to technology; the availability of technology or a reliable technology supplier, the need to outsource services of logistic distribution; the need for greater operating efficiency; a need for an efficient logistics management and automate processes; risks in aspects associated with people's uncertainty and fear before a possible people reduction; the need to promote favorable attitudes to the use of digital channels; the dissemination of e-commerce success stories in the market as means to generate attitudes of "optimism" to the use of technologies; the need to promote an innovative culture based on communication and collaboration; in addition to several structural problems at country level such as governmental policies, the lack of habit or people's willingness to purchase on digital media, little credit cards penetration in the country, as well as the fear of means of payment, among others.

This study, therefore, achieves a great contribution in knowledge generation and practical application in the organizations, since it provides an extensive list of aspects that must be managed appropriately, and can thus lead the "investments", which is the most considered factor in decision-making on adapting technologies.

References

COTEC: Innovación Tecnológica. Ideas Básicas (2001). https://www.innova.uned.es/webpages/innovaciontecnologica/mod1_tema1/InnovacionTecIdeasBasicas.pdf. Accessed Febrero de 2016

Davis, F.: A technology acceptance model for empirically testing new end-user information systems: theory and results. Doctoral dissertation, MIT Sloan School of Management, Cambridge (1986)

Davis, F.D.: Perceived usefulness, perceived easy of use, and user acceptance of information technnology. MIS Q. 13(3), 319–340 (1989)

Cuadernos delebcenter, e-business Center: Pricewaterhouse Coopers & IESE. Cuadernos de ebcenter (2008). www.iese.edu/en/files/Criterios%20de%20adopción%20de%20las%TIC_tcm4_23387.pdf. Accessed 20 de febrero de 2016

Fishbein, M., Ajzen, I.: Belief, Attitude, Intention and Behavior: An Introduction to Theory and Research. Addison-Wiley Publishing Company, Reading (1975)

Guerrero Cuellar, R., Rivas Tovar, L.A.: Comercio Electrónico en México: Propuesta de un modelo conceptual aplicado a las PyMEs. In: SOCIOTAM XV, vol. 1, pp. 79–116 (2005)

Guisado, M.: La relación entre diferentes tipos de innovación en el Sector Servicios. Un análisis desde el enfoque de Complementariedad. In: XXX AEDEM Annual Meeting. Las Palmas de Gran Canaria (2016)

Hair Jr., J., Sarstedt, M., Hopkins, L., Kuppelwieser, V.: Partial least squares structural equation modeling (PLS-SEM). An emerging tool in business research. Eur. Bus. Rev. 26(2), 106–121 (2014)

Kline, S., Rosenberg, N.: An overview of innovation. In: The Positive Sum Strategy: Hearnessing Technology for Economic Growth, 14, 640. National Academy of Sciences, Washington (1986)

Kurnia, S., Karnali, R., Rahim, M.M.: A quality study of business-to-business electronic commerce adoption within the Indonesian grocery industry: a multi-theory perspective. Inf. Manag. 52, 518–536 (2015)

Mick, D.G., Fournier, S.: Paradoxes of technology: consumer cognizance, emotions, and coping strategies. J. Consum. Res. 25, 123–143 (1998)

OECD: Oslo Manual: Guidelines for Collecting and Interpreting Innovation Data, 3rd edn. OECD/EC/Eurostat, OECD Publishing, Paris (2005)

Oliveira, T., Martins, M.F.: Literature review of information technology adoption models at firm level. Electron. J. Inf. Syst. Eval. 14(1), 110–212 (2011)

Parasuraman, A.: Technology readiness index (TRI). A multiple-item scale to measure readiness to embrace new technologies. J. Serv. Res. 2(4), 307–320 (2000)

Parasuraman, A., Colby, C.A.: An updated and streamlined technology readiness index: TRI 2.0. J. Serv. Res. 18(1), 59–74 (2015)

Rao, J., Weintraub, J.: How innovative is your company's culture? MIT Sloan Manag. Rev. 54(3), 28–39 (2013)

Vankatesh, V., Davis, F.: A theoretical extension of the technology acceptance model: four longitudinal field studies. Manag. Sci. 46(2), 186–204 (2000)

VISA: Informe sobre e-Readiness en Latinoamérica 2014 (2014). http://promociones.visa.com/lac/ecommerce/es/index.html. Accessed Marzo de 2016

Wigand, R.T.: Electronic commerce: definition, theory, and context. Inf. Soc. 13, 1–16 (1997)

Wong, K.: Partial least squares structural equation modeling (PLS-SEM) techniques using SmartPLS. Mark. Bull. 24(1), 1–32 (2013)

Emotional Legitimacy

Francisco Díez-Martín[1](✉), Camilo Prado-Román[2],
Ana Cruz-Suárez[1], and Emilio Díez-de-Castro[3]

[1] Department of Business Economics, Rey Juan Carlos University,
Paseo de los Artilleros S/N, 28032 Madrid, Spain
{francisco.diez, ana.cruz}@urjc.es
[2] European Academy of Management and Business Economics
(AEDEM) and Department of Business Economics, Rey Juan Carlos University,
Paseo de los Artilleros S/N, 28032 Madrid, Spain
camilo.prado.roman@urjc.es
[3] Department of Business Administration and Marketing,
University of Sevilla, Facultad de CC Economicas y Empresariales,
Avda. Ramon y Cajal, S/N, 41018 Seville, Spain
diez@us.es

Abstract. This paper analyzes the roll emotions play in organizational legitimacy. In contrast to literature about organizational legitimacy, we suggest emotions are crucial to provide legitimacy to some organizations where feelings usually are essential to have a bond with them. 36 interviews about 5 types of organizations were developed: religious, sport, cultural and educational organizations and NGOs. The results confirmed and extended the constructs of emotional legitimacy provided in the literature. From a managerial perspective and in order to obtain emotional legitimacy, our results confirm the need of taking long-term actions instead of being short-sighted. From a theoretical point of view, emotions arising from organizations with legitimacy are assessed and analyzed.

1 Introduction

This paper explains how some emotions described in other documents about organizational legitimacy can be considered items from constructs that are decisive to establish emotional legitimacy.

In the last 40 years, research about types of legitimacy of organizations has been focused on different factors that lead stakeholders to provide organizations with legitimacy.

Several factors have to be considered when assessing organizational legitimacy. Since the different issues covered are not common to all stakeholders, legitimacy is evaluated according to its sector, its environment and the stakeholders' perspectives. Therefore, we can confirm it is necessary to define the organization's attributes that will be examined before judging its legitimacy.

In order to synthesize these issues, we can find a large literature about it:

- *Bases of legitimacy* (Alt et al. 2015; Chun and Dyck 2015; Deephouse et al. 2017; Gray et al. 2015; Greenwood et al. 2008).

© Springer Nature Switzerland AG 2019
J. Gil-Lafuente et al. (Eds.): AEDEM 2017, SSDC 180, pp. 188–199, 2019.
https://doi.org/10.1007/978-3-030-00677-8_16

- *Categories of legitimacy* (Ahlstrom and Bruton 2001; Batchelor and Burch 2011).
- *Criteria of legitimacy* (Deephouse et al. 2017; Suddaby and Greenwood 2005).
- *Dimensions of legitimacy* (Batchelor and Burch 2011; Deephouse and Carter 2005; Deephouse and Suchman 2008).
- *Forms of legitimacy* (Batchelor and Burch 2011; Dacin et al. 2007; Greenwood et al. 2008; Patriotta et al. 2011; Richards et al. 2017; Suchman 1995; Treviño et al. 2014).
- *Levels of legitimacy* (Brønn and Vidaver-Cohen 2009; Cohen and Dean 2005; Cornelissen and Clarke 2010; Díez-Martín et al. 2013a; Fisher et al. 2016; Gardberg and Fombrun 2006; Hudson and Okhuysen 2009; Jijelava and Vanclay 2017; Pollack et al. 2012; Reast et al. 2013; Rutherford and Buller 2007; Rutherford et al. 2009; Theingi et al. 2017; Tost 2011; Zimmerman and Zeitz 2002).
- *Pillars of legitimacy* (Greenwood et al. 2008).
- *Sources of legitimacy* (Deeds et al. 2004; Gippert 2016; Kinser 2007; Lamin and Zaheer 2012; Stenholm and Hytti 2014).
- *Types of Legitimacy* (Ahlstrom and Bruton 2001; Batchelor and Burch 2011; Behram 2015; Bitektine 2011; Chung et al. 2016; Cruz-Suárez et al. 2014a; Dacin et al. 2007; Deephouse 1996; Deephouse and Carter 2005; Díez-De-Castro et al. 2018; Díez-Martín et al. 2010; Díez-Martín et al. 2013b; Freitas and Guimarães 2007; Greenwood et al. 2008; Hasbani and Breton 2016; Landau et al. 2014; Mallon 2017; McQuarrie et al 2013; Nagy et al. 2012; O'Dwyer et al. 2011; Peris-Ortiz and Díez-De-Castro 2018; Pollack et al. 2012; Rutherford et al. 2009; Schultz et al. 2014; Steverson et al. 2013; Suchman 1995; Tchokogué et al. 2017; Tost 2011; Treviño et al. 2014; Vidaver-Cohen and Simcic Brønn 2008; Zimmerman and Zeitz 2002).

Without pretending to be exhaustive, clearly many authors prefer to use one term over the other. In this regard, it is obvious the most common used terms are 'types of legitimacy'; 'levels of legitimacy' and 'forms of legitimacy'. Even though different terms are used when referring to the same concept, the research community has focused on this issue.

On the other hand and at a theoretical and empirical level, literature has defined all types of legitimacy and they have increased over time. Although Bitketine distinguished 24 legitimacy types, we identified 41 legitimacy types that have been used by researchers when we developed this paper and we assume this number must increase eventually.

2   Theoretical Background

2.1   Emotional Constructs of Organizational Legitimacy

We reviewed the literature about legitimacy in order to find how legitimacy is manifested within organizations that are considered legitimate by stakeholders. It is striking that literature has mentioned a lot of issues but without going deeper into them. However this is a key issue for us.

Emotions caused by legitimacy are a powerful expression of this legitimacy. Many of the aspects or perceptions mentioned by literature are so similar and so closely connected with each other that makes more difficult to find a difference. That leads us to consider the need of grouping these aspects that we have called: Reliability, Comprehensibility, Social Identity, Moral emotions y Belonging (Table 1).

Table 1. Types of emotions

Reliability		Social identity	
Credibility	Hunter and Bansal (2007), Suchman (1995)	Competent	Nagy et al. (2012)
Compliance	Treviño et al. (2014)	Effective	Nagy et al. (2012)
Trust	Kumar and Das (2007)	Rational	Deephouse (1996)
Trustworthy	Bansal and Clelland (2004), O'Dwyer et al. (2011)	Stability	Suddaby (2010)
Comprehensibility		Wise	Suchman (1995)
Accountability	Suchman (1995), Park et al. (2012)	Security	Marin and Ruiz (2007)
Interpretability	Suchman (1995)	Persistence	Suchman (1995)
Predictability	Vidaver-Cohen and Brønn (2008, Das and Ten (1998)	Accomplish	Suchman (1995), Pawlowski and Wiley-patton (2006)
Visibility	Chiu and Sharfman (2011)	**Moral emotions**	
Belonging		Decent	Suchman (1995)
Have our best interests at heart	Suchman (1995)	Honest	Suchman (1995)
Share our values	Suchman (1995)	Worthy	Nagy et al. (2012), Zimmerman and Zeitz (2002)
Sense of belonging	Moisander et al. (2016)	Congruency of values	Zimmerman and Zeit (2002)

Reliability. Hope, confidence and encouragement come from the relationship between the organization and the stakeholders. These feelings are associated to credibility, compliance, trust and trustworthy. Credibility is an affective emotion that arises from the belief in the organization and in its legitimacy in itself. Suchman (1995) explains how credibility together with other feelings, reinforce each other. Compliance is set up to ensure the organization meets the regulations. It is impossible to conceive of legitimate organizations acting without complying with the standards (Treviño et al. 2014). Trust refers to the confidence that one will find what is desired from the partner (Das and Teng 1998b). As Kumar and Das (2007) asserts, trust exists when there is legitimacy but this is not the only condition a legitimate organization must fulfill.

Comprehensibility. Because understanding organizations must precede its evaluation, different elements have been created in order to get this comprehensibility: accountability; interpretability; predictability; visibility. Suchman (1995) highlights how important it is to report and explain all areas of the organization using both accountability and legitimacy together. Obviously, these concepts are connected. Accounting allows the organization to be much more clearly understood and this encourages legitimacy (Park et al. 2012). "Legitimacy enhances both the stability and the comprehensibility of organizational activities, and stability and comprehensibility of ten enhance each other" (Suchman 1995: 574). When an organization is economically, socially and culturally understood (interpretability) it becomes much more visible (Chiu and Sharfman 2011) and predictable (Das and Teng 1998a; Suchman 1995; Vidaver-Cohen and Simcic Brønn 2008). Since we are focusing on emotions and not data or facts, members of the organization will give it emotional legitimacy not because data supports it but because they consider it transparent, understandable, etc.

Social identity. A social identity "tends to focus on those attributes that are perceived to be central, distinctive, and more or less enduring to the target refers to attributes that reflect group membership" (Ashforth et al. 2016: 32). Social identity is not a binary variable (yes/no) but there is a wide scope from low or poorly defined identity up to a strongly defined identity. Stakeholders' attitude can provide organizations with legitimacy when these achieve a recognizable identity. Emotions arising from social identity of the organization are associated to positive evaluations: competent, effective, rational, stability, security, persistence, accomplish. Organizations with a clear and recognizable social identity firstly transmit stability (Suddaby 2010) and security in relationships (Marin and Ruiz 2007) because stakeholders know who they are dealing with avoiding doubts and hesitations. The second issue is that social identity is pursued and established by the organization in order to convey the appropriate image. Generally the image arising from social identity is associated to a qualified and effective organization (Nagy et al. 2012) that will act rationally. It is difficult to change the social identity. Actually, it should be sustained over time so that stakeholders' emotions will not be disturbed and that is why perseverance is one of the dimensions of legitimacy highlighted by Suchman (1995). Accomplish implies that organizations will be able to carry out their activity. It dispels doubts because stakeholders keep a relationship with the organization that provides them security (Pawlowski and Wiley-Patton 2006) and this will help them to achieve their goals.

Moral emotions. Moral emotions "deal with felt obligations and rights as well as feelings of approval and disapproval based on moral intuitions and principles" (Moisander et al. 2016: 966). The deep sense of business ethics and its relevance in our society contributed to create moral emotions deriving from terms like decent, honest, worthy and congruency of values. Society at large, particularly relevant stakeholders, values an entity as: decent (Suchman 1995), honest (Suchman 1995), and worthy (Nagy et al. 2012; Vidaver-Cohen and Simcic Brønn 2008) because the institution's principles are solid and similar to those of society (Cruz-Suárez et al. 2014b). "Congruency of values means that one partner accepts the decisions and behaviors of the other party as one's own. There is, as if, a suspension of judgment derived from the assumption that it is the right thing to do" (Kumar and Das 2007: 1432).

Belonging. Research about the sense of belonging has been focused on several aspects: the faith in religious institutions (Martinez 2017), the feelings of the citizens (Beugelsdijk et al. 2017), the rejection to tourism by the local citizens (Rasoolimanesh et al. 2016), the emotional engagement with the university (Hadžiahmetović and Dinç 2017) and also the study of how and why people belong to a group (Smith et al. 1998). However we found greater interest for this issue regarding educational institutions (Freeman et al. 2007; Johnson et al. 2007). How stakeholders identify the organization's principles, which we assume are in line with the same values stakeholders have in mind and heart, is a key element in the sense of belonging.

Even though the review of the literature has given us a broad spectrum of emotions and feelings linked to organizational legitimacy, we are concerned whether all issues about emotions are included. That is primarily due to the fact that all emotions we described come mainly from theoretical studies and, but for very few exceptions, emerge from empirical reviews. As a result, we decided to extend the review to prove if it is possible to consider some other emotions associated to organizational legitimacy.

3 Data and Measurements

We target different groups of people who could pass some filters. Firstly they should admit they provide the organization with legitimacy based mainly on their beliefs, emotions or affectivity. Some of their feelings involved were based on strong religious beliefs and others on its social commitment and its willingness to work, help and find solutions to social problems. Further feelings stemmed from sports, tradition or personal and family bonds. Some interviewees argued that their organizations were legitimate due to their environmental awareness or because they were able to channel their efforts and work for the common good. As a second selection filter, interviewees should have maintained a relationship with their organizations at least for 10 years. The selection of interviewees was not chosen at random. After the first 20 respondents, no new inputs were basically obtained.

This study used semi-structured interviews as an instrument to gather all the information. It was necessary to lead the interviews in the same direction for all interviewees but also allowing them to express their views freely. The interviews were recorded and literally transcribed. Three respondents did not agree to be recorded so they were interviewed by two researchers. One of them was taking notes throughout the interview allowing the other researcher to focus on the questions ensuring a dynamic meeting.

4 Results

The results have supported the enhancement of the identified constructs in order to group emotions. We were not able to find indicators of 'affective emotions' or 'relational identification' in other studies about legitimacy, but we could distinguish them as a couple of new constructs through the contributions of the groups interviewed. As a result, we could add two more constructs in addition to the current five. These new constructs are described below:

Affective emotions. "are positive or negative bonds and commitments that actors have" (Moisander et al. 2016: 4). The construct of 'affective emotions' emerges from terms that enhance its content. Firstly, affective emotions like appreciation, admiration, gratitude and love are conceptually very similar and they represent emotions raised by organizations in stakeholders. Secondly, other terms close to affective emotions are associated to what we call "quality of emotions". In this case, interviewees use terms like joy, satisfaction, happiness and empathy.

Relational identification suggests the bond between stakeholders and the organization generates emotions. Besides being very relevant to stakeholders this bond makes them use positive terms and place value on the organization as a result. These emotions group into two concepts: (a) the sense of belonging to a group and (b) the personal and professional growth derived from being part of the group.

Being part of the group lead to different emotions like: personal fulfilment, companionship, personal attention and bonds, fraternity, assistance and mutual support and friendship. The second concept mentioned before was related to: respect, personal recognition, positivity, education and culture, greater knowledge, grow and mature, learning, obtaining rewarding experience, capacity and overcome. The first consequence of the relationship between the stakeholders and organizations was the reduction of uncertainty (Nagy et al. 2012). The second one is how relational identification increases invulnerability to questioning (Baum and Oliver 1991) and reduces or helps us cope with negative stigmas (Nagy et al. 2012).

Both *affective emotions* and *relational identification* are constructs described in theory about emotions but they were not covered by current literature about organizational legitimacy. Conversely, comprehensibility (accountability, interpretability, predictability, visibility) and *reliability* (credibility, compliance, trust, trustworthy) are perfectly defined in current literature about legitimacy. However no contributions about these constructs have arisen in the interviews. Similarly, the *sense of belonging* shows a special bond in the interviews but it is impossible to be clearly defined by the interviewees. We have been able to find some contributions to other constructs. Regarding *Moral emotions*, besides other items we have identified in literature like decent, honest, worthy and congruency of values, new items emerged from the results of the interviews: aligned with my beliefs, humility, modesty, altruism, solidarity and loyalty.

On the other hand, the concept *Social identity* defines the organization with a differentiating and unique identity (competent, effective, rational, stability, wise, security, persistence, accomplish) has been enriched by new items such as awareness, achievement of objectives, proper functioning, hope, bond, overcoming, consistency, cooperation and to feel important.

5 Discussion of Results and Conclusions

5.1 Discussion of Results

Our study has focused on emotions when these are linked to organizational legitimacy. From a managerial point of view, where rationality prevails, emotions may have a limited relevance. Because legitimacy is provided by stakeholders based on their emotions and knowledge, we assume legitimacy is partially rational. This paper supports the fact that emotions are crucial to provide an organization with legitimacy.

Focusing on emotional legitimacy involves choosing organizations to be evaluated according to their typology and attributes. Some relevant stakeholders provide organizations with legitimacy based on their emotions. Emotions can be the first criterion or even the only one to be considered by stakeholders. These organizations are easy to recognize in sectors like competitive sports, religious organizations, non-profit organizations, political parties, public organizations and educational institutions. Emotional legitimacy could also be related to other sectors, but these may be isolated cases of particular organizations. This study exposes not only a broader theoretical framework of legitimacy but also a research agenda to be investigated. According to literature about legitimacy, we gathered the more significant items linked to emotional variables and five of them were identified: reliability, comprehensibility, belonging, social identity and moral emotions. We assumed not all the issues were addressed in prior studies about organizational legitimacy.

Since we found certain gaps regarding social psychology, the possibility of including them in emotional legitimacy was verified. That is the reason why we decided to interview people who should have a relationship with organizations susceptible to achieve legitimacy through their stakeholders' emotions. Results proved our initial suspicion is confirmed.

Many of those interviewed have clearly stated that some Affective and Relational aspects should be included in emotional legitimacy as significant constructs. These two constructs, in addition to the five items mentioned above, were named Affective emotions and Relational identification. Our study represents a contribution to the growing literature on legitimacy by suggesting a type of comprehensible legitimacy that supplements the most widely studied types.

The critical question underlying this study is: Are emotions generated by legitimacy or is legitimacy generated by emotions? Based on the literature, emotions exist when there is legitimacy. The rational approach, based on a cause and effect relationship, asserts emotions emerge from organizations once they have achieved legitimacy. Rationality lead us to confirm when organizations have legitimacy, emotions like confidence, sense of belonging etc. arise in stakeholders.

This rational approach relies on casual relationships and no empirical evidence has been provided. In fact, these connections could contradict the ones mentioned above. When emotional legitimacy is provided by stakeholders with powerful emotions, legitimacy represents the effect and emotions the cause in this relationship. Sports organizations or political parties achieve legitimacy when their supporters have a sense of belonging or affectivity. In this case, emotions represent the cause and not the result. Thus, emotional legitimacy is not subordinated to other types of legitimacy. Legitimacy does not represent the effect once the organization has already been provided with legitimacy derived from further criteria.

5.2 Concluding Remarks and Theoretical and Practical Implications

Our study has provided new contributions to literature on organizational legitimacy. In some cases it is possible to see coincidences between emotional legitimacy and some other types. For instance, moral emotions and moral legitimacy are apparently connected as well as comprehensibility and cognitive legitimacy. That is why we aimed to define the differences between them.

Referring to legitimacy on a traditional way means to consider legitimacy arises from perceptions. Inevitably, perceptions are based on results and knowledge. As a result, sometimes perceptions that derive from the valuations of stakeholders cannot be substantiated by indicators or judgements but in other cases the indicators are solid and gather essential data of what has been assessed. Perception can be built on a weak, wide or excellent basis but judgements need to be sustained by indicators and data.

Emotional legitimacy concerns feelings and beliefs but not judgements or perceptions. For example, stakeholders establish a link with a religious organization not because they agree with everything but because they have faith. Thus, stakeholders acquire comprehensibility due to their beliefs excluding perceptions based on indicators. Emotional legitimacy stem from feelings that are independent from both empirical facts and more or less accurate results from indicators. Concerning experts, organizations should pursue different strategies to keep or develop its legitimacy depending on where the legitimacy comes from perceptions or exclusively emotions.

Regarding future research, substantial efforts are required in order to progress with the assessment of emotional legitimacy in organizations. It is crucial to gain experience and consider all relevant items mentioned above as elements to define the constructs. Sectors and organizations where emotional legitimacy is essential and belongs to its basic criterion should be identified accurately.

References

Ahlstrom, D., Bruton, G.D.: Learning from successful local private firms in China: establishing legitimacy. Acad. Manag. Exec. **15**(4), 72–83 (2001)

Alt, E., Díez-de-Castro, E.P., Lloréns-Montes, F.J.: Linking employee stakeholders to environmental performance: the role of proactive environmental strategies and shared vision. J. Bus. Ethics **128**(1), 167–181 (2015)

Ashforth, B.E., Schinoff, B.S., Rogers, K.M.: "I Identify with her"; "I identify with him": unpacking the dynamics of personal identification in organizations. Acad. Manag. Rev. **41**(1), 28–60 (2016)

Bansal, P., Clelland, I.: Talking trash: legitimacy, impression management, and unsystematic risk in the context of the natural environment. Acad. Manag. J. **47**(1), 93–103 (2004)

Batchelor, J.H., Burch, G.F.: Predicting entrepreneurial performance: can legitimacy help? Small Bus. Inst. J **7**(2), 30–45 (2011)

Baum, J.A.C., Oliver, C.: Institutional linkages and organizational mortality. Adm. Sci. Q. **36**(2), 187–218 (1991)

Behram, N.K.: A cross-sectoral analysis of environmental disclosures in a legitimacy theory context. J. Manag. Sustain. **5**(1), 20–37 (2015)

Beugelsdijk, S., Kostova, T., Roth, K.: An overview of Hofstede-inspired country-level culture research in international business since 2006. J. Int. Bus. Stud. **48**(1), 30–47 (2017)

Bitektine, A.: Toward a theory of social judgments of organizations: the case of legitimacy, reputation, and status. Acad. Manag. Rev. **36**(1), 151–179 (2011)

Brønn, P.S., Vidaver-Cohen, D.: Corporate motives for social initiative: legitimacy, sustainability, or the bottom line? J. Bus. Ethics **87**(Suppl. 1), 91–109 (2009)

Chiu, S.-C., Sharfman, M.: Legitimacy, visibility, and the antecedents of corporate social performance: an investigation of the instrumental perspective. J. Manag. **37**(6), 1558–1585 (2011)

Chun, D., Dyck, L.R.: Firm strategy, societal norms and cognition: the American Auto Industry from 1968 to 2008. Am. J. Manag **15**(4), 86–101 (2015)

Chung, J.Y., Berger, B.K., DeCoster, J.: Developing measurement scales of organizational and issue legitimacy: a case of direct-to-consumer advertising in the pharmaceutical industry. J. Bus. Ethics **137**(2), 405–413 (2016)

Cohen, B., Dean, T.: Information asymmetry and investor valuation of IPOs: top management team legitimacy as a capital market signal. Strateg. Manag. J. **26**(7), 683–690 (2005)

Cornelissen, J.P., Clarke, J.S.: Imagining and rationalizing opportunities: inductive reasoning and the creation and justification of new ventures. Acad. Manag. Rev. **35**(4), 539–557 (2010)

Cruz-Suárez, A., Prado-Román, A., Prado-Román, M.: Cognitive legitimacy, resource access, and organizational outcomes. RAE-Revista de Administração de Empresas **54**(5), 575–584 (2014a)

Cruz-Suárez, A., Prado-Román, C., Díez-Martín, F.: Por qué se institucionalizan las organizaciones. Revista Europea de Dirección y Economía de la Empresa **23**(1), 22–30 (2014b)

Dacin, M.T., Oliver, C., Roy, J.-P.: The legitimacy of strategic alliances: an institutional perspective. Strateg. Manag. J. **28**(2), 169–187 (2007)

Das, T.K., Teng, B.-S.: Between trust and control: developing confidence in partner cooperation in alliances. Acad. Manag. Rev. **23**(3), 491–512 (1998a)

Das, T.K., Teng, B.-S.: Resource and risk management in the strategic alliance making process. J. Manag. **24**(1), 21–42 (1998b)

Deeds, D.L., Mang, P.Y., Frandsen, M.L.: The influence of firms' and industries' legitimacy on the flow of capital into high-technology ventures. Strateg. Organ. **2**(1), 9–34 (2004)

Deephouse, D.L.: Does isomorphism legitimate? Acad. Manag. J. **39**(4), 1024–1039 (1996)

Deephouse, D.L., Carter, S.M.: An examination of differences between organizational legitimacy and organizational reputation. J. Manag. Stud. **42**(2), 329–360 (2005)

Deephouse, D.L., Suchman, M.: Legitimacy in organizational institutionalism. In: Greenwood, R., Oliver, C., Suddaby, R., Sahlin, K. (eds.) The Sage Handbook of Organizational Institutionalism, vol. 49, pp. 50–77. Sage Publications, Thousand Oaks (2008)

Deephouse, D.L., Bundy, J., Tost, L.P., Suchman, M.C.: Organizational legitimacy: six key questions. In: Renate M., Greenwood, R., Oliver, C., Lawrence T. (eds.) The SAGE Handbook of Organizational Institutionalism, 2nd edn. Sage, Thousand Oaks (2017)

Díez-de-Castro, E., Peris-Ortiz, M., Díez-Martín, F.: Criteria for evaluating the organizational legitimacy. A typology for legitimacy jungle. In: Peris-Ortiz, M., Diez de Castro, E. (eds.) Organizational Legitimacy - Challenges and Opportunities for Businesses and Institutions. Springer, London (2018)

Díez-Martín, F., Blanco González, A., Prado Román, C.: Legitimidad como factor clave del éxito organizativo. Investigaciones Europeas de Dirección Y Economía de La Empresa **16**(3), 127–143 (2010)

Díez-Martín, F., Prado-Roman, C., Blanco-González, A.: Beyond legitimacy: legitimacy types and organizational success. Manag. Decis. **51**(10), 1954–1969 (2013a)

Díez-Martín, F., Prado-Román, C., Blanco-González, A.: Efecto del plazo de ejecución estratégica sobre la obtención de legitimidad organizativa. Investigaciones Europeas de Dirección Y Economía de La Empresa **19**(2), 120–125 (2013b)

Fisher, G., Kotha, S., Lahiri, A.: Changing with the times: an integrated view of identity, legitimacy, and new venture life cycles. Acad. Manag. Rev. **41**(3), 383–409 (2016)

Freeman, T.M., Anderman, L.H., Jensen, J.M.: Sense of belonging in college freshmen at the classroom and campus levels. J. Exp. Educ. **75**(3), 203–220 (2007)

De Freitas, C.A.S., de Guimarães, T.A.: Isomorphism, institutionalization and legitimacy: operational auditing at the court of auditors. *Revista de Administração Contemporânea*, 11 (spe1), 153–175 (2007)

Gardberg, N.A., Fombrun, C.J.: Corporate citizenship: creating intangible assets across institutional environments. Acad. Manag. Rev. **31**(2), 329–346 (2006)

Gippert, B.J.: The sum of its parts? Sources of local legitimacy. Cooper. Conflict **51**(4), 522–538 (2016)

Gray, B., Purdy, J.M., Ansari, S.: From interactions to institutions: microprocesses of framing and mechanisms for the structuring of institutional fields. Acad. Manag. Rev. **40**(1), 115–143 (2015)

Greenwood, R., Oliver, C., Sahlin, K., Suddaby, R.: The SAGE Handbook of Organizational Institutionalism. Sage, London (2008)

Hadžiahmetović, N., Dinç, M.S.: The mediating role of affective commitment in the organizational rewards– organizational performance relationship. Int. J. Hum. Resour Stud. **7**(3) (2017)

Hasbani, M., Breton, G.: Discursive strategies and the maintenance of legitimacy. Engl. Lang. Lit. Stud. **6**(3), 1–15 (2016)

Hudson, B.A., Okhuysen, G.A.: Not with a ten-foot pole: core stigma, stigma transfer, and improbable persistence of men's bathhouses. Organ. Sci. **20**(1), 134–153 (2009)

Hunter, T., Bansal, P.: How standard is standardized MNC global environmental communication? J. Bus. Ethics **71**(2), 135–147 (2007)

Jijelava, D., Vanclay, F.: Legitimacy, credibility and trust as the key components of a social licence to operate: an analysis of BP's projects in Georgia. J. Clean. Prod. **140** (2017)

Johnson, D.R., Soldner, M., Leonard, J.B., Alvarez, P., Inkelas, K.K., Rowan-Kenyon, H.T., Longerbeam, S.D.: Examining sense of belonging among first-year undergraduates from different racial/ethnic groups. J. Coll. Stud. Dev. **48**(5), 525–542 (2007)

Kinser, K.: Sources of legitimacy in U.S. for-profit higher education. In: Private Higher Education in Post-communist Europe, pp. 257–276. Palgrave Macmillan US, New York (2007)

Kumar, R., Das, T.K.: Interpartner legitimacy in the alliance development process. J. Manage. Stud. **44**(8), 1425–1453 (2007)

Lamin, A., Zaheer, S.: Wall street vs. main street: firm strategies for defending legitimacy and their impact on different stakeholders. Organ. Sci. **23**(1), 47–66 (2012)

Landau, D., Drori, I., Terjesen, S.: Multiple legitimacy narratives and planned organizational change. Hum. Relat. **67**(11), 1321–1345 (2014)

Mallon, M. R.: Getting buy-in: financial stakeholders' commitment to strategic transformation. Manag. Res: J. Iberoam. Acad. Manag. **15**(2) (2017). MRJIAM-06-2016-0667

Marin, L., Ruiz, S.: "I need you too!" Corporate identity attractiveness for consumers and the role of social responsibility. J. Bus. Ethics **71**(3), 245–260 (2007)

Martinez, F.: On the role of faith in sustainability management: a conceptual model and research agenda. J. Bus. Ethics 1–21 (2017)

McQuarrie, F.A.E., Kondra, A.Z., Lamertz, K.: Government, coercive power and the perceived legitimacy of Canadian post-secondary institutions. Canadian Journal of Higher Education Revue Canadienne D'enseignement Supérieur **43**(2), 149–165 (2013)

Moisander, J.K., Hirsto, H., Fahy, K.M.: Emotions in institutional work: a discursive perspective. Organ. Stud. **37**(7), 963–990 (2016)

Nagy, B.G., Pollack, J.M., Rutherford, M.W., Lohrke, F.T.: The influence of entrepreneurs' credentials and impression management behaviors on perceptions of new venture legitimacy. Entrepreneurship Theor. Pract. **36**(5), 941–965 (2012)

O'Dwyer, B., Owen, D., Unerman, J.: Seeking legitimacy for new assurance forms: the case of assurance on sustainability reporting. Acc. Organ. Soc. **36**(1), 31–52 (2011)

Park, H.-S., Auh, S., Maher, A., Singhapakdi, A.: Marketing's accountability and internal legitimacy: implications for firm performance. J. Bus. Res. **65**(11), 1576–1582 (2012)

Patriotta, G., Gond, J.-P., Schultz, F.: Maintaining legitimacy: controversies, orders of worth, and public justifications. J. Manag. Stud. **48**(8), 1804–1836 (2011)

Pawlowski, S.D., Wiley-patton, S.: Building legitimacy for it innovations: organizing visions and discursive strategies of legitimacy. In: Academy of Management Best Conference Paper (2006)

Peris-Ortiz, M., Díez-De-Castro, E. (eds.): Organizational Legitimacy - Challenges and Opportunities for Businesses and Institutions. Springer, London (2018)

Pollack, J.M., Rutherford, M.W., Nagy, B.G.: Preparedness and cognitive legitimacy as antecedents of new venture funding in televised business pitches. Entrepreneurship Theor. Pract. **36**(5), 915–939 (2012)

Rasoolimanesh, S.M., Roldan, J.L., Jaafar, M., Ramayah, T.: Factors influencing residents perceptions toward tourism development: differences across rural and urban world heritage sites. J. Travel Res. **56**, 1–16 (2016)

Reast, J., Maon, F., Lindgreen, A., Vanhamme, J.: Legitimacy-seeking organizational strategies in controversial industries: a case study analysis and a bidimensional model. J. Bus. Ethics **118**(1), 139–153 (2013)

Richards, M., Zellweger, T., Gond, J.-P.: Maintaining moral legitimacy through worlds and words: an explanation of firms' investment in sustainability certification. J. Manag. Stud. **54** (5), 676–710 (2017)

Rutherford, M.W., Buller, P.F.: Searching for the legitimacy threshold. J. Manag. Inq. **16**(1), 78–92 (2007)

Rutherford, M.W., Buller, P.F., Stebbins, J.M.: Ethical considerations of the legitimacy lie. Entrepreneurship Theor. Pract. **33**(4), 949–964 (2009)

Schultz, P.L., Marin, A., Boal, K.B.: The impact of media on the legitimacy of new market categories: the case of broadband internet. J. Bus. Ventur. **29**(1), 34–54 (2014)

Smith, H.J., Tyler, T.R., Huo, Y.J., Ortiz, D.J., Lind, E.A.: The self-relevant implications of the group-value model: group membership, self-worth, and treatment quality. J. Exp. Soc. Psychol. **34**, 470–493 (1998)

Stenholm, P., Hytti, U.: In search of legitimacy under institutional pressures: a case study of producer and entrepreneur farmer identities. J. Rural Stud. **35**, 133–142 (2014)

Steverson, B.K., Rutherford, M.W., Buller, P.F.: New venture legitimacy lies and ethics: an application of social contract theory. J. Ethics Entrepreneurship **3**(1), 73–92 (2013)

Suchman, M.C.: Managing legitimacy: strategic and institutional approaches. Acad. Manag. Rev. **20**(3), 571–610 (1995)

Suddaby, R.: Challenges for institutional theory. J. Manag. Inq. **19**(1), 14–20 (2010)

Suddaby, R., Greenwood, R.: Rhetorical strategies of legitimacy. Adm. Sci. Q. **50**(1), 35–67 (2005)

Tchokogué, A., Paché, G., Nollet, J., Stoleru, R.-M.: Intra-organizational legitimization strategies used by purchasing managers. J. Purchasing Supply Manag. **23**(3), 163–175 (2017)

Theingi, T., Theingi, H., Purchase, S.: Cross-border remittance between emerging economies: an institutional perspective. J. Bus. Ind. Mark. **32**(6), 786–800 (2017)

Tost, L.: An integrative model of legitimacy judgments. Acad. Manag. Rev. **36**(4), 686–710 (2011)

Treviño, L.K., den Nieuwenboer, N., Kreiner, G.E., Bishop, D.G.: Legitimating the legitimate: a grounded theory study of legitimacy work among ethics and compliance officers. Organ. Behav. Hum. Decis. Process. **123**(2), 186–205 (2014)

Vidaver-Cohen, D., Simcic Brønn, P.: Citizenship and managerial motivation: implications for business legitimacy. Bus. Soc. Rev. **113**(4), 441–475 (2008)

Zimmerman, M. a., & Zeitz, G.J.: Beyond Survival: achieving New Venture Growth by Building Legitimacy. Acad Manag. Rev. **27**(3), 414–431 (2002)

Hierarchization of Factors Involved in the Failure of Startups

Gerardo Gabriel Alfaro Calderón[1]([⊠]), Víctor Gerardo Alfaro García[2],
and Hugo Alejandro Rivera Betancourt[1]

[1] Facultad de Contaduría y Cs. Administrativas,
Universidad Michoacana de San Nicolás de Hidalgo, Morelia, Mexico
ggalfaroc@gmail.com, hugoalex.riverab@gmail.com
[2] REDCID - CYTED, Universitat de Barcelona, Barcelona, Spain
valfaro06@gmail.com

Abstract. In this research we determine the main factors by which start-ups or startups fail in a period of no more than five years in the city of Morelia Michoacán, using the methodology AHP (Analytical Hierarchical Process), which allows us to establish the level of priority of each of the factors identified in the business failure, a phenomenon that we live on a daily basis in the city, which can be seen worldwide.

1 Introduction

Currently, entrepreneurship is a resource that allows countries to achieve a high degree of growth, in economic aspects with a greater purchasing power in the different sectors, and in turn, the social part is highly benefited because the growth in employability in a country allows a greater part of the population to attain increasing levels of schooling high generating a better quality of life.

For a long time it was believed that large companies, the most hierarchical, the most complex, were responsible for the generation of employment, welfare and satisfaction, in recent years, various investigations have raised something very different, which should lead us to change the concept of reverence for big business (Varela 2008).

It is now widely recognized that entrepreneurship plays an important role in economic development, including regional development, and entrepreneurs are key agents of change in market economies (OCDE 2012).

Entrepreneurs and business manager are terms that are generally taken as synonyms, when in fact the differences between one and the other are duly marked. An entrepreneur is not necessarily a good business manager, nor an business manager is necessarily an entrepreneur (Bucardo et al. 2015).

Entrepreneurship goes beyond going out and getting a job; it is about seeking independence, self-realization, innovation and earning the respect of others. Entrepreneurship is necessary for society, fosters growth and development, and is based on a philosophy of individual initiative (Dávila et al. 2014).

The ventures are also known as the creation of startups for their term in English. It is important to define or determine what is a startup or a company because it has

© Springer Nature Switzerland AG 2019
J. Gil-Lafuente et al. (Eds.): AEDEM 2017, SSDC 180, pp. 200–213, 2019.
https://doi.org/10.1007/978-3-030-00677-8_17

characteristics and circumstances very different from those faced by a company and therefore it is necessary to define them.

According to Ries (2011), startups "are a human institution designed to create a new product or service under conditions of extreme uncertainty." On the other hand (Blank 2013), it defines it as: "A temporary organization in search of a scalable and replicable business model." In terms of functions or circumstances, we understand that an entrepreneur must be capable, solve problems and manage risks and uncertainties.

Considering that a startup is a newly created company it is necessary to define what an entrepreneur is and for this part of the theories that Schumpeter worked on are analyzed, for him an entrepreneur is the founder of a new company, an innovator that breaks with the form traditional way of doing things, with the established routines. Must be a person with leadership skills, and with a special talent to identify the best way to act. The entrepreneur has the ability to see things as nobody else sees them.

The entrepreneur according to Varela (2008), "It is the person or group of people who are capable of perceiving an opportunity of production or service, and formulate freely and independently a decision of achievement and allocation of natural, financial, technological and human resources needed to start the business, which in addition to creating additional value for the economy, generates work for him and often for others. In this process of creative leadership, the entrepreneur invests money, time and knowledge and participates in the assembly and operation of the business, risking their resources and their personal prestige but seeking monetary and personal rewards."

Entrepreneurship contributes to the creation of new and small companies that are the best providers of jobs, can stimulate economic activity and integrate unemployed or disadvantaged people into working life, OECD (Potter et al. 2013).

Entrepreneurship is necessary for society, fosters growth and development, and is based on a philosophy of individual initiative. It is very common that there are new ventures everywhere, unfortunately, there are failures everywhere (Dávila et al. 2014). On the other hand Blank (2013), mentions that in the last years it has been discovered that the startups are not small versions of the big companies, the skills that the entrepreneurs require are not seen in the traditional books of administration and management skills.

From the above, the failures of entrepreneurship that we have as a country and as a state can be seen in the following tables (Graphs 1 and 2):

From the graph it is observed that in Mexico a large percentage of companies do not manage to pass of the first year of life, which is an alarming fact because with the statistics of mortality of companies that appear in the country, it will be difficult to find in entrepreneurship, an opportunity for economic growth and employability.

On the other hand, in the State, 30% of the undertakings do not exceed the first year of life and 83% fail to reach the five years of life, and compared with the national average it can be seen that Michoacán is below this. INEGI (2016), due to the foregoing, it is a priority to determine, what are the main hierarchical causes for which startups in Morelia Michoacán fail?

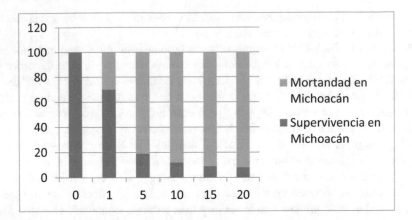

Graph 1. Rate of mortality and survival of companies in Michoacán. Source: Own elaboration based on INEGI statistics (2016)

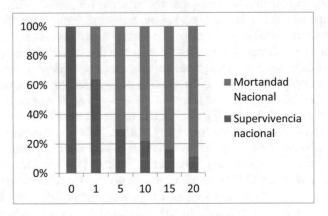

Graph 2. Mortality rate and survival of companies at the national level.   Source: Own elaboration based on INEGI statistics (2016)

1.1   Literature Review

In the development of this research we focus on that information that reveals important facts about the success and failure factors of startups or companies, which allowed the identification of the variables and their respective dimensions, through the works of the following authors (Table 1):

The total of variables and dimensions from the literature review are shown below (Table 2).

From the above, the need to use the hierarchy of variables and their respective dimensions through the Analityc Hierarchic Process (AHP) is derived.

Table 1. Authors who mention each of the variables of the investigation.

Variables	Autores
Factores Personales	Dávila et al. (2014), Cacciotti et al. (2016), Valencia (2016), (Deeb 2014), Feinleib (2011), Cisneros et al. (2011), Carretero (2008), TOLAMA (2016), Pérez and Avilés (2016), BBVA (2016), Fuentes et al. (2016), Aguilasocho et al. (2014), Farias et al. (2016), Alom et al. (2016), de Oliveira et al. (2015), Shahid et al. (2016), Poornima and Mathew (2014), Rodríguez (2011), Elizundia (2011), Devece et al. (2016), Simón (2017), Alfaro and González (2011), Valencia et al. (2016), Maqueda (2012), García-Gasulla (2015), Miller (2015), Fernández and de la Rivaa (2014)
External environment	Salamzadeh and Kawamorita (2015), Dávila et al. (2014), Valencia (2016), Tovak (2014), Deeb (2014), Hirai (2015), López J. (2016), TOLAMA (2016), Taylor (2008), Pérez and Avilés (2016), Flores and González (2009), Némethné (2010), Farias et al. (2016), Rodríguez (2011), Devece et al. (2016), Valencia et al. (2016)
Organizational factors	Salamzadeh and Kawamorita (2015), Dávila et al. (2014), Valencia (2016), Tovak (2014), Deeb (2014), Skok (2016), Hirai (2015), Feinleib (2011), Cisneros et al. (2011), Carretero (2008), López J. (2016), Nobel (2011), López A. (2016), Fuentes et al. (2016), Aguilasocho et al. (2014), Flores and González (2009), Cabrera et al. (2011), Farias et al. (2016), de Oliveira et al. (2015), Shahid et al. (2016), Rodríguez (2011), Devece et al. (2016), March and Mora (2007), Simón (2017), Ávalos (2009), Alfaro and González (2011), Valencia et al. (2016), García-Gasulla (2015)
Financial factors	Salamzadeh and Kawamorita (2015), Dávila et al. (2014), Cacciotti et al. (2016), Tovak (2014), Deeb (2014), Skok (2016), Hirai (2015), Feinleib (2011), Cisneros et al. (2011), Carretero (2008), López J. (2016), TOLAMA (2016), Taylor (2008), López A. (2016), BBVA (2016), Cabrera et al. (2011), Némethné (2010), Farias et al. (2016), Alom et al. (2016), Shahid et al. (2016), Poornima and Mathew (2014), Mures and García (2011), Rodríguez (2011), Elizundia (2011), Consuelo (2014), Romero et al. (2015), Vargas (2015), Tascón and Castaño (2012), March and Mora (2007), Simón (2017), Gutiérrez (2013), Alfaro and González (2011), Valencia et al. (2016), Maqueda (2012)
Market factors	Dávila et al. (2014), Valencia (2016), Tovak (2014), Deeb (2014), Skok (2016), Hirai (2015), Feinleib (2011), López J. (2016), Taylor (2008), Nobel (2011), BBVA (2016), Fuentes et al. (2016), Aguilasocho et al. (2014), Flores and González (2009), Cabrera et al. (2011), Farias et al. (2016), Alom et al. (2016), de Oliveira et al. (2015), Poornima and Mathew (2014), Rodríguez (2011), March and Mora (2007), Alfaro and González (2011), Valencia et al. (2016), García-Gasulla (2015)
Human capital	Salamzadeh and Kawamorita (2015), Dávila et al. (2014), Valencia (2016), Tovak (2014), Deeb (2014), Skok (2016), Hirai (2015), Carretero (2008), López J. (2016), TOLAMA (2016), Nobel (2011), López A. (2016), BBVA (2016), Fuentes et al. (2016), Flores and González (2009), Cabrera et al. (2011), Simón (2017), Alfaro and González (2011), Valencia et al. (2016)

Table 2. Dimensions found.

Dimensiones encontradas en la revisión de literatura			
1	Lack of proper selection	33	Lack of indicators
2	Disarm on the computer	34	Planning
3	Lack of training	35	Strategy
4	Poor team management	36	Management practices
5	A demotivated team	37	Problems in the execution
6	The team does not have what it takes	38	Conflict with shareholders
7	Compensation problems	39	Accept disadvantageous contracts
8	Individualism	40	Inadequate organizational structure
9	Staff turnover	41	Changes in the market
10	Theft by staff	42	Adaptability
11	Lack of market study	43	Economic crisis and/or politics
12	Problems with the product	44	Problems with criminal groups
13	Ignore consumers	45	Problems with suppliers
14	Bad location	46	Competition
15	Lack of marketing strategy	47	Culture (Mexican context)
16	Price problems	48	Discontent of interest groups (nearby communities, media or authorities)
17	Target market poorly selected	49	Appearance of new technologies
18	Market problems (customers)	50	Legislative reforms
19	Idea or business model	51	Incorrect or non-existent leadership
20	Inadequate promotion/advertising	52	Personality and attitudes
21	Bad money management	53	Lack of skills to undertake
22	Lack of economic resources	54	Lack of skills to sell
23	Excess of operating expenses	55	Lack of social skills
24	Access to resources	56	Risk tolerance
25	Lack of investors	57	Bad time management
26	Lack of credit with suppliers	58	Reasons to open the business
27	Delay of payments from your customers	59	Inexperience
28	Quality and efficiency	60	Schooling of the entrepreneur
29	Innovation	61	Lack of social relationships
30	Operation	62	Fear to fail
31	Knowledge management	63	Age of the entrepreneur
32	Lack of focus		

Source: Own elaboration based on the selected articles.

2 Analityc Hierarchic Process (AHP)

The AHP is a tool that allows ordering factors by level of importance, to be able to develop this process it is necessary the opinion of experts in the subject that is being developed with the objective of categorizing the levels of importance of each one of the factors that are they go hierarchize.

To assign the level of importance, the Saaty scale is used, in which they are categorized from 1 to 9 with odd numbers, in this way the one to one factors are confronted, where the rating of 1 represents the same degree of importance, 3 would be three times more important one over another, 5 would be five times more important and so on until the nine that would represent a broad superiority of one factor over another and at the end the results of the variables are added to obtain the value of 100% of each of the facts.

To start from the normalized paired matrix with which the preferences of the factors are identified, the following vector is normalized, making a division between the degree of preference and the total sum of preferences of the column, of so that the sum of the following vectors is equal to 1 reflecting the unit and finally we obtain the average of the vectors by line to obtain in this way the percentage of importance of each of the lines or in this case the factors.

In the following section of methodology the example of how the hierarchy of variables and dimensions is carried out is shown.

2.1 Methodology

For the analysis, a review was made of articles, theses and books that referred to entrepreneurship, startups and entrepreneurs. From this review, a selection of 50 articles was made that mentioned the factors that influence the success and failure of startups. Once the articles were selected, a list was prepared with the factors mentioned in each of them, in order to concentrate them in 6 major areas, personal factors, external environment factors, organizational factors, financial factors, market factors and human capital factors., which allowed us to identify, according to the literature, the failure variables of startups and the possible dimensions of each of them.

After determining possible variables and dimensions, the frequency with which the different authors made mention of the failure factors was identified, from which a comparison was made, in order to obtain the variables and dimensions that were repeated with greater frequency of according to the articles reviewed. The most representative variables and dimensions for the failure of startups according to the literature were identified from the frequency counting.

2.2 Interviews with Experts

(a) Preparation of measurement instrument.
(b) Identification of the candidates for application of the instrument.
(c) Interview with 10 experts with experience in failure.
(d) Interview with 10 experts with successful experience.
(e) Interview with 10 academic experts.
(f) Preparation of results table.
(g) Analysis of the results.

Once the variables and their dimensions were identified, an interview was conducted with 30 experts on the subject divided into three categories: experts from startups with experience of failure, experts from startups with successful experience and finally academic experts from the area.

To determine which prospects met these characteristics, the following bases were taken into account:

Experts with experience of failure	Entrepreneurs with experience of closing a business in a period of less than 5 years
Experts with successful experience	Entrepreneurs with business survival experience in a period of more than 5 years
Academic experts	Academics who currently work in management positions related to entrepreneurship with at least 2 years of experience

This base was determined from the mortality statistics of companies of the National Institute of Geography and Information Statistics (INEGI), where it is identified that in Mexico 70% of the companies disappear after 5 years of being created.

It is important to mention that according to the factors identified in the literature, the interviews were aimed at determining which variables and dimensions the experts give greater importance to avoid the failure of startups.

The interviews with experts allowed to ponder the importance of each one of the factors to proceed with the AHP methodology.

2.3 AHP Analityc Hierarchic Process

1. Identification of priorities developed by the experts.
2. Development of hierarchy tables for variables according to the Saaty scale.
3. Development of dimension hierarchy tables according to the Saaty scale.
4. Elaboration of tables of results of the hierarchy of variables and dimensions (Table 3).

In order to carry out this methodology, the interviews with experts allowed us to know the degree of importance that each one of the variables and dimensions has for the failure of the companies, once this information was obtained, hierarchy tables were developed through which the following results were reached.

Finally, the importance levels of each of the factors are obtained, where the last column shows in percentage the hierarchy of the factors worked, as shown in Table 4.

For the purposes of our research, we take all the hierarchical variables and dimensions and perform the corresponding analysis that will be observed next in the results section.

Table 3. Shows the development of the AHP tool with the dimensions of the variable of financial factors.

	1	2	3	4	5	6	7
Bad money management	1.00	1.00	3.00	3.00	5.00	5.00	7.00
Excess of operating expenses	1.00	1.00	3.00	3.00	5.00	5.00	7.00
Lack of economic resources	0.33	0.33	1.00	1.00	3.00	5.00	5.00
Lack of credit with suppliers	0.33	0.33	1.00	1.00	3.00	3.00	5.00
Delay of payments from your customers	0.20	0.20	0.33	0.33	1.00	3.00	5.00
Access to resources	0.20	0.20	0.20	0.33	0.33	1.00	3.00
Lack of investors	0.14	0.14	0.20	0.20	0.20	0.33	1.00
	3.209	3.209	8.733	8.866	17.533	22.333	33

Source: Own elaboration based on interviews with experts.

Table 4. Percentage of hierarchy of the dimensions of financial factors.

0.51	0.63	0.53	0.32	0.26	0.263	0.191	0.38520118
0.17	0.21	0.32	0.32	0.26	0.263	0.191	0.24699432
0.10	0.07	0.11	0.32	0.26	0.263	0.191	0.18717785
0.06	0.02	0.01	0.04	0.15	0.146	0.191	0.08716591
0.06	0.02	0.01	0.01	0.03	0.029	0.106	0.03758252
0.06	0.02	0.01	0.01	0.03	0.029	0.106	0.03758252
0.06	0.02	0.01	0.00	0.01	0.005	0.021	0.0182957

Source: Own elaboration based on interviews with experts.

2.4 Results

2.4.1 Hierarchy of Variables

Table 5. Variables with percentages of importance.

Variables that affect the failure of startups	% of importance in the failure of startups
Human capital factors	0.339
Organizational factors	0.253
Market factors	0.214
Personal factors	0.104
Financial factors	0.068
External factors	0.022

Source: Own elaboration based on the results of the AHP methodology.

Table 6. Dimensions of human capital with percentages of importance.

Dimensions of human capital	% Incidence in failure
Lack of proper selection	0.194
Lack of training	0.149
Disarm in the team	0.146
A demotivated team	0.120
The team does not have what it takes	0.099
Poor team management	0.087
Compensation problems	0.070
Individualism	0.058
Staff turnover	0.041
Theft by staff	0.036

Source: Own elaboration based on the results of the AHP methodology.

Table 7. Dimensions of organizational factors with percentages of importance.

Dimensions of organizational factors	% Incidence in failure
Quality and efficiency	0.230
Innovation	0.226
Lack of focus	0.095
Strategy	0.093
Knowledge management	0.085
Planning	0.069
Operation	0.051
Lack of indicators	0.051
Management practices	0.036
Problems in the execution	0.025
Accept disadvantageous contracts	0.019
Inadequate organizational structure	0.010
Conflict with shareholders	0.010

Source: Own elaboration based on the results of the AHP methodology.

Table 8. Dimensions of market factors with percentages of importance.

Dimensions of market factors	% Incidence in failure
Lack of market study	0.228
Ignore consumers	0.155
Problems with the product	0.128
Lack of marketing strategy	0.128
Bad location	0.110
Target market poorly selected	0.078
Market problems (customers)	0.066
Price problems	0.044
Inadequate promotion/advertising	0.033
Idea or business model	0.030

Source: Own elaboration based on the results of the AHP methodology.

Table 9. Dimensions of personal factors with percentages of importance.

Dimensions of personal factors	% Incidence in failure
Personality and attitudes	0.147319815
Lack of skills to sell	0.147319815
Incorrect or non-existent leadership	0.109153104
Risk tolerance	0.099733952
Lack of skills to undertake	0.084797432
Bad time management	0.084797432
Lack of social skills	0.08479743
Reasons to open the business	0.07325717
Lack of social relationships	0.040779972
Inexperience	0.033244651
Fear to fail	0.033244651
Schooling of the entrepreneur	0.033244651
Age of the entrepreneur	0.028309927

Source: Prepared by the authors based on the results of the AHP methodology.

Table 10. Dimensions of financial factors with percentages of importance.

Dimensions of financial factors	% Incidence in failure
Bad money management	0.385
Excess of operating expenses	0.247
Lack of economic resources	0.187
Lack of credit with suppliers	0.087
Delay of payments from your customers	0.038
Access to resources	0.038
Lack of investors	0.018

Source: Prepared by the authors based on the results of the AHP methodology.

Table 11. Dimensions of external factors with percentages of importance.

Dimensions of external factors	% Incidence in failure
Adaptability	0.287
Economic crisis and/or politics	0.185
Changes in the market	0.141
Problems with criminal groups	0.126
Competition	0.074
Problems with suppliers	0.062
Culture (Mexican context)	0.045
Discontent of interest groups (nearby communities, media or authorities)	0.033
Appearance of new technologies	0.028
Legislative reforms	0.020

Source: Prepared by the authors based on the results of the AHP methodology.

2.5   Conclusions

This research shows us the variables that directly affect newly created companies and the main factors of failure hierarchized in such a way that we observe that the factors of human capital, organizational and market are the main factors of failure in startups and with less importance personal, financial and external factors (Tables 5, 6, 7, 8, 9, 10 and 11).

By performing the analysis of each of the variables and their respective dimensions we can determine which aspects of the variables according to the hierarchy are relevant, from which future research is proposed to eliminate dimensions that are not relevant for the study of the failure of startups.

Taking into account the human factor variable and its hierarchical dimensions, we can see that the most important variables affecting the newly created companies are the lack of personnel selection, lack of training, disharmonies in the team and a team unmotivated, as you can see these four variables represent more than 60% of the degree of importance of the total dimensions.

By performing a brief analysis we can infer that some of the dimensions that do not show a higher hierarchy depend or are derived from those that are better ranked, a subsequent study is proposed to determine cause-effect relationships that allow to shorten the number of dimensions and in turn a better study of the variable.

From our second variable according to the hierarchy we can see that the most representative dimensions were quality and efficiency, innovation, lack of focus and strategy, which in turn represent a degree of importance according to the hierarchical analytical process above 60% dividing the remaining percentage among 6 more dimensions.

According to the results, we find that the most representative dimensions of the variable of market factors and with a percentage higher than 60% are lack of market research, ignoring consumers, problems with the product and lack of marketing strategy. With which we can make the same inference as in the factors of human capital, some of the dimensions with lower hierarchy are consequence or are derived from those that have a higher hierarchy therefore it is proposed to carry out a subsequent investigation that allows to do a detailed analysis with the aim of eliminating dimensions that are not significant and in turn grouping those with greater affinity for a precise study of the behavior of this variable with respect to the failure of the newly created companies.

On the other hand, the variable of personal factors details that the dimensions of personality and attitudes, lack of skills to sell, incorrect or nonexistent leadership and tolerance to risk are the main individual factors ranked for the failure of a newly created company. Of which it is possible to understand that the attitudes and abilities of the people are factors that can cause the business failure in spite of having knowledge or relevant academic experience.

Regarding the financial factor that although we could imagine would be one of the variables with a higher hierarchical level and which in turn could be the main cause of business failure, we find that is the fifth variable of six according to the hierarchy with the methodology described above and we can perceive that the first three dimensions according to their hierarchy represent more than 80% of importance are the mismanagement of money, excess of operating expenses, lack of economic resources.

Ultimately we find hierarchical to external factors that although they are shown with the lowest percentage of importance, we cannot leave them aside taking into account that any of its dimensions can be a factor of failure of new companies and according to the results of its dimensions we find that the first three according to the hierarchy represent more than 60% and are the following: Adaptability, economic and/or political crisis and changes in the market, which although are dimensions that do not depend on the own company, are important characteristics that must be taken into account to prevent and take measures to prevent the failure of start-ups.

Acknowledgements. Project supported by "Red Iberoamericana para la Competitividad, Innovación y Desarrollo" (REDCID) project number 616RT0515 in "Programa Iberoamericano de Ciencia y Tecnología para el Desarrollo" (CYTED).

References

Aguilasocho, R.D., Montoya, G.E., Guerra, R.: Factores que afectan la competitividad de las pymes agrocítricas manufactureras en Michoacán. Mercados y negocios **15**(2), 45–69 (2014)

Alom, F., Asri, A.M., Moten, A.R., Ferdous, A.S.: Success factors of overall improvement of microenterprises in Malaysia: an empirical study. J. Glob. Entrepreneurship Res. **6**(1), 1–13 (2016)

Alfaro, G., González, F.: Gestión del conocimiento en pequeñas y medianas empresas. GAC FEGOSA, Morelia, México (2011)

BBVA: 6 errores de startups fintech que llevan al fracaso. Noticias BBVA (2016)

Blank, S.: The Four Steps to the Epiphany: Successful Strategies for Products That Win. K S Ranch (2013)

Bucardo, C., Saavedra, G.M., Camarena, A.M.: Hacia una comprensión de los conceptos de emprendedores y empresarios. SUMA DE NEGOCIOS, pp. 98–107 (2015)

212   G. G. A. Calderón et al.

Cabrera, M.M., López, L.P., Ramírez, M.: La competitividad empresarial: un marco conceptual para su estudio. Universidad Central, Bogotá (2011)

Cacciotti, G., Hayton, C., Mitchell, J.R., Giazitzoglu, A.: Conceptualización del miedo al fracaso en el emprendimiento. J. Bus. Ventur. 302–325 (2016)

Carretero, P.A.: La guía del Emprendedor: De la idea a la empresa. Asociación de jovenes empresarios de Valencia, Valencia (2008)

Cisneros, M.E., Martín, G.E., Castelán, G.B., Puga, M.C.: FACTORES DE EMPRENDI-MIENTO QUE INFLUYEN EN EL DESEMPEÑO DE LA MICRO Y PEQUEÑA EMPRESA. Universidad Nacional Autónoma de México, México (2011)

Consuelo, M.: Declive organizativo, fracaso y reestructuración organizacional en las empresas colombianas. Contaduría y Administración 59(3), 235–260 (2014)

Dávila, M., Layrisse, F., Lozano F., G., Riojas, E., Urbina, H. : El libro del fracaso. Ciudad de México (2014)

de Oliveira, J., Escrivão, F.E., Seido Nagano, M., Ferraudo, A.S., Daniela, R.: What do small business owner-managers do? A managerial work perspective. J. Glob. Entrepreneurship Res. 5(1), 1–21 (2015)

Deeb, G.: 13 errores de las startups. Entrepreneur (2014)

Elizundia, C.M.: Factores de emprendimiento que influyen en el desempeño de la micro y pequeña empresa. UNAM, Ciudad de México, México (2011)

Devece, C., Peris, O.M., Rueda, A.C.: Entrepreneurship during economic crisis: Success factors and paths. J. Bus. Res. 69(11), 5366–5370 (2016)

Mizumoto, F.M., Artes, R., Lazzarini, S.G., Hashimoto, M., Bedê, M.A.: A sobrevivência de empresas nascentes no estado de São Paulo: um estudo sobre capital humano, capital social e práticas gerenciais. R. Adm. 45(4), 343–355 (2010)

Farias, A., Escrivão, F., Seido, N., Philippsen, J.L.: A change in the importance of mortality factors throughout the life cycle stages of small businesses. J. Glob. Entrepreneurship Res. 6 (1), 1–18 (2016)

Feinleib, D.: Why Startups Fail: And How Yours Can Succeed. Apress (2011)

Fernández, S.C., de la Rivaa, B.: Entrepreneurial mentality and culture of entrepreneurship. Procedia Soc. Behav. Sci. 139(22), 137–143 (2014)

Flores, R., González, S.F.: LA COMPETITIVIDAD DE LAS PYMES MORELIANAS. Cuadernos del CIMBAGE, pp. 85–104 (2009)

Fuentes, N., Osorio, G., Mungaray, A.: Capacidades intangibles para la competitividad microempresarial en México. Problemas del Desarroll, pp. 83–106 (2016)

García-Gasulla, S.M.: La Orientación al Mercado y el Emprendimiento en el Marco de los Modelos de Excelencia. Análisis del Impacto en el Rendimiento Empresarial Sector Automatismos Comunidad Valenciana. Doctoral Thesis. Universidad Politécnica de Valencia, Valencia, Valencia, España (2015)

Gutiérrez, F.J.: Variables y modelos para la evaluación del fracaso empresarial. Propuesta de una metodología de fronteras basada en percentiles. Doctoral Thesis. Universidad de León, León, España (2013)

Hirai, A.: Gestión de riesgo en los emprendimientos. Recuperado el 6 de Diciembre de 2016, de (2015). http://egesoftware.blogspot.mx/2015/04/gestion-del-riesgo-en-los.html

INEGI: Esperanza de vida de los negocios a nivel nacional y por entidad federativa. Recuperado el 5 de Diciembre de 2016, de (2016). http://www.inegi.org.mx/inegi/contenidos/investigacion/Experimentales/Esperanza/doc/evn_ent_fed.pdf

López, A.: Cinco factores que alejan a los emprendedores del éxito. Tec Rev. (2016)

López, J.: Fracasan en México 75% de emprendimientos. El Financiero, 18 de Enero de 2016

March, C.I., Mora, E.R.: Creación de empresas de base tecnológica: factores de éxito y fracaso, vol. 5, pp. 97–120 (2007)

Maqueda, R.G.: Determinación y análisis de factores de éxito en proyectos de base tecnológica. Protocolo de investigación. Guadalajara, Jalisco, México: Centro de Investigación y de Estudios Avanzados del Instituto Politécnico Nacional (2012)

Miller, D.J.: High Growth Student Startups at US Colleges and Universities. Doctoral Thesis. George Mason University, Arlington, USA (2015)

Mures, Q.M., García, G.A.: Factores determinantes del fracaso empresarial en Castilla León. Economía y empresa **21**(51), 95–115 (2011)

Némethné, G.A.: Competitiveness of small and medium sized enterprises - a possible analytical framework. HEJ, 1–14 (2010)

Nobel, C.: Why companies fail—and how their founders can bounce back. Harvard Bus. Rev. (2011)

OCDE: Evaluación de la OCDE del sector de las nuevas empresas basadas en el conocimiento. OCDE Publishing, México (2012)

Pérez, P.C., Avilés, H.M.: Explanatory factors of female entrepreneurship. SUMA DE NEGOCIOS **7**(15), 25–31 (2016)

Poornima, W.K., Mathew, M.: Potential for opportunity recognition along the stages of entrepreneurship. J. Glob. Entrepreneurship Res. **4**(1), 1–24 (2014)

Potter, J., Marchese, M., Feldman, M., Kemeny, T., Lawton-Smith, H., Pike, A.: The Local Dimension of SME and Entrepreneurship Issues and Policies in Mexico. OECD Publishing, Paris (2013)

Ries, E.: El método Lean start-up. Crown Publishing Group, Sillicon Valley (2011)

Rodríguez, R.A.: El emprendedor de éxito. Mcgrawhill, Ciudad de México, México (2011)

Romero, E.F., Melgarejo, M.Z., Vera, C.M.: Fracaso empresarial de las pequeñas y medianas empresas (pymes) en Colombia. Suma de Negocios **6**(13), 29–41 (2015)

Salamzadeh, A., Kawamorita, K.H.: Startup companies: life cycle and challenges. In: Proceedings of the 4th InternationalConference on Employment, Education and Entrepreneurship (EEE), Belgrade, Serbia (2015)

Shahid, Q.M., Saeed, S., Mehmood, W.S.: The impact of various entrepreneurial interventions during the business plan competition on the entrepreneur identity aspirations of participants. J. Glob. Entrepreneurship Res. **6**(1), 1–18 (2016)

Simón, M.V.: Análisis de los factores de éxito en el proceso de creación. Doctoral Thesis. Universidad de Valencia, Valencia, España (2017)

Skok, D.: For Entrepreneurs. Recuperado el 7 de Diciembre de 2016, de (2016). http://www.forentrepreneurs.com/es/5-razones-por-las-que-las-startups-y-los-nuevos-negocios-fracasan/

Tascón, F.M., Castaño, G.F.: Variables and models for the identification and prediction of business failure: revision of recent empirical research advances. Revista de Contabilidad **15**(1), 7–58 (2012)

Taylor, B.: Four Reasons Most Startups Fail (And How Yours Can Succeed). Harvard Bus. Rev. (2008)

TOLAMA, J.: FRACASO, EL MIEDO QUE IMPIDE A LOS MEXICANOS EMPRENDER. Expansión (2016)

Tovak, S.: 9 razones por las que las startups fracasan. Entrepreneur (2014)

Valencia, R.: El Fracaso de la Micro, Pequeña y Mediana Empresa de Base Tecnológica en México. Monterrey (2016)

Valencia, R., Olavarría, D., Vargas, M., Stapley, E.: Tech Startup Failures in Mexico. Recuperado el 8 de Diciembre 2016 de (2016). https://thefailureinstitute.com/wp-content/uploads/2016/09/Tech-Startup-Failure-Research-final.pdf

Varela, R.: Innovación Empresarial. Pearson Educación de Colombia, Santa Fe de Bogotá (2008)

Vargas, C.J.: Modelos de Beaver, Ohlson y Altman: Son realmente capaces de predecir la bancarrota en el sector empresarial costarricense. Tec. Empresarial **8**(3), 29–40 (2015)

Public's Behaviour in Front of Sports: Case of Spanish Football

Miguel Prado-Román[✉], Alberto Prado-Román, Paola Plaza-Casado, and Iria Paz-Gil

European Academy of Management and Business Economics (AEDEM), Universidad Rey Juan Carlos, Madrid, Spain
miguel.prado@urjc.es

Abstract. Football is more than a sport, it's a business that offers an outstanding return for the participants in this business model. However, in recent years it appears that the economic situation of the sport is not as prominent as before and they have begun to detect declines in audiences, reduction of the number of fans in stadiums, etc. In this way, we will formulate a model that will allow us to identify the aspects that influence the loyalty of consumers in sports area. Thus, carry out an analysis of structural equations that will allow us to test the degree of loyalty in sport will be defined both by the quality of the service, as for the level of consumer satisfaction. Thus, we will proceed to an analysis of structural equations that allow us to test that relevant factors in the behavior of the followers will be the reputation of the soccer teams and the degree of satisfaction of the followers.

1 Introduction

Sports market is, nowadays, an appealing tool for business' incomes generation. Thus, sponsorship of sports entities is identified to provide higher rentability than more traditional media, such as advertising (Dominguez and Labrador 2011). In sports field, one sport stands out in the whole world: football. Football is the most popular sport in the world (Conlin 2015) and one of most played in the world (Vacas 2014).

Despite football's economic and social potential, the current model presents a problem that can create serious consequences. The problem is that everything is based in big football teams. This leads us to think about the absence of a solid conceptual basis that allows football managers to develop a quality sports offer adapted to actual fan's interests and able to generate an increase of the audience and of the economic potential of this sport.

Thus, it's noted an important lack of attention by sports entities towards business management and marketing disciplines (Luna-Arocas and Mundana 1998). This lack of interest may generate problems in the management of a sport entity's offer. In this way, Sanz et al. (2005) address that the interest in sports marketing is still recent.

Therefore, appears the need of delve into the study of sport entities' fans. So, this research aims to reach the next goals: (1) Analyse the sports entities' image and reputation implications on consumers' degree of satisfaction in sports market. (2) Analyse

© Springer Nature Switzerland AG 2019
J. Gil-Lafuente et al. (Eds.): AEDEM 2017, SSDC 180, pp. 214–224, 2019.
https://doi.org/10.1007/978-3-030-00677-8_18

sports entities' image, sports entities' reputation and fans' degree of satisfaction implications on loyalty feeling of fans in sports market.

2 Theoretical Framework

Sports entities should be able to offer services and offers to their clients for them to keep acquiring match tickets, seeing matches on TV, buying merchandising, etc. This is based on the reason than the development of the economic activity of any business will depend, to a large extent, on the ability that each business shows for keeping their clients (Castañeda and Luque 2008).

Nevertheless, appears in sports entities an important shortage of attention towards marketing and business management disciplines (Luna Arocas and Mundana 1998). This lack of interest on the analysis of this business fields may affect the proper development of the activity of the sports entity, influencing, among others, the management of plenty stadiums for maximizing the fan experience and with that the degree of satisfaction of consumers. In this way, Sanz et al. (2005) address that the interest in sports marketing is very newly, and identify three key aspects (Manaserro-Mas et al. 1998). The first one is related with the generalization of the use of brands and sports equipment both in sports and no sports practices. The second one is own to the transformation of sports into a mass phenomena, which is the reason why clubs try to maximize the attendance to sports events. The third reason focuses on interests and behaviours of people who play sports nowadays, that caused a huge proliferation of sports facilities, public and private, that have to be properly managed to make them profitable and to satisfy clients.

According to the previous exposition, it may be fathomed that sports marketing will be compounded by activities detonated to analyse needs and wishes of sports consumers through exchange process (Mullin et al. 1995). This way, sports marketing will have two main objectives: the development of sports products and services and the development of industrial services through sports promotions (Mullin et al. 1995).

According to the previous exposition, it's identified the need to develop an analyse of the factors that may affect one fan on his decision of enjoying the different activities that sports entities offer. In this way, it will be proceeded to formulate a model that allows to analyse the aspects related with sports activities by fan's level of satisfaction with sports offer, the sports entities image and their reputation, in order to affect the future fan behaviour (Fig. 1).

Hereunder we make a detailed analyse of each of the constructs that define our study model and identify the effect that may have on consumer's behaviour through sports entities.

2.1 Sports Satisfaction

There are in literature several approaches to consumer's satisfaction. One stream had analysed satisfaction focusing on cognitive aspects, evaluative process of goods offered by the entity to the consumers (Howard and Sheth 1969; Churchill and Suprenant 1982; Bearden and Teel 1983). Other stream had considered satisfaction as a result of

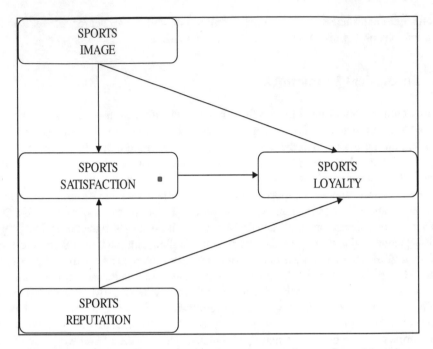

Fig. 1. Analyse of public's behaviour beyond sports Source: Own elaboration

commercial relationship between consumers and business from an affective perspective (Hunt 1977; Woodruff et al. 1983; Halstead et al. 1994; Price et al. 1995). Lastly, satisfaction had been analysed integrating the previous perspectives, both cognitive and affective. Thus, it's understood that satisfaction must be analysed from a more complete point of view that includes consumer's emotional evaluations (subjectives) as much as service or product consumption technical evaluations (objectives) (Westbrook and Reilly 1983; Mano and Oliver 1993; Oliver 1993).

Based on the different study approaches to satisfaction, it's identified a huge heterogeneity on the conceptualization of satisfaction. On one side, satisfaction is defined from an emotional perspective. Thus, Price et al. (1995) consider that satisfaction makes reference to affective aspects related to use experience. In this way, Hasltead et al. (1994) define it as an affective response associated to an specific transaction as a result of the comparison between the result of the product with some standard previous to purchase. On other side, satisfaction is defined from a cognitive perspective. Churchill and Suprenant (1982) define it as a response to the purchase of products or the use of services that appears from the comparison between costs and benefits related to anticipated consequences. In addition, Bearden and Teel (1983) define it as a function of expectatives and opinions about the attributes of the product and the confirmation or not of that expectatives.

Despite the difficulty for obtaining a consensus both in satisfaction analyse and conceptualization, in our research we define a consumer's global satisfaction with an sports entity as "the global satisfaction or dissatisfaction with the organization, based on every encounter and experiences with that particular organization on a period of time" (Bitner and Hubbert 1994).

Regarding the identification of the relevance of satisfaction in our research, it was been identified as a key aspect in sports marketing (Redondo and Cuadrado 2002). Thus, consumer's satisfaction will be an essential condition on economic development of sports entities, since, among other aspects, consumers would accept in a more tolerant way possible price increases on their sports season passes or sports event tickets (Reichneld and Sasser 1990). Therefore, we should consider consumer's degree of satisfaction if we want to understand his needs and inquisitiveness (Luna-Arocas and López 2007).

If a sports entity wants to get success in its activities, it must generate a certain level of satisfaction in its consumers, that is essential (Díaz et al. 2007). This way, sports entities must consider consumers' satisfaction as a differentiation tool, and not only as the result of a good service. Thus, sports entities will be able to cover consumers' needs and wishes, involving future repurchase behaviour.

Based on this, we formulate the following hypothesis:

Hypothesis H1: The level of satisfaction affects positively the fan's loyalty towards his football team.

2.2   Sports Image

The image that a fan has about one football team is not only related to it is perceived ability, based on past experiences (Lewis and Sourelli 2006), but also includes subjective aspects, based on fan's assessment. This assessment is own to the inclusion of emotional or symbolic values, together with the assessment of objective and easily measurable aspects (as sports results) (Gardner and Levy 1995). In this way, we define a football team projected image as the addition of fans' beliefs, sensations and impressions about it (Barich and Kotler 1991).

The importance of the image that a football team projects to market is own to its ability to create an emotional connection with its fans and with their satisfaction with that image. For that, a football team must be able to thoroughly take care of its image, since a positive image will provoke a confidence, interest and preference feeling among fans (De la Fuente and Rey 2008). Therefore, we can state that the image projected by a football team will be decisive in fans' degree of satisfaction (Loureiro and González 2010).

Based on previously exposed, we formulate the next study hypothesis:

Hypothesis H2: The image of a football team affects positively the fan's degree of satisfaction.

We also must keep in mind that the football team image will affect the purchase behaviour and the future intention of rebuying/revisiting (De la Fuente and Rey 2008). Thereby, fans' loyalty will be influenced by the football team projected positive image (Calvo et al. 2015).

Thus, we formulate the following study hypothesis:

Hypothesis H3: The image of a football team affects positively the fan's loyalty to his football team.

2.3 Sports Reputation

The entities reputation in sports field will be defined as the reflect that an organization has in time of how it's seen by its influencers and how they express it through their thoughts and words (Arbelo and Pérez 2001). Thereby, the fans' feeling about sports entities, both positive and negative, will affect its global perception. This obey to the identification of the relationship between reputation and quality of offered products (Shapiro 1983). Thus, we can understand that sports entities reputation will affect consumers' perception and their subsequent degree of satisfaction.

Based on previously exposed, we formulate the next study hypothesis:

Hypothesis H4: The reputation of a football team affects positively the fan's degree of satisfaction.

Besides, it's identified the existence of a direct influence of reputation on loyalty (Andreassen and Lindestad 1998; Robertson 1993; Yoon et al. 1993). This way, the consumer would increase his sales and his ability of coming again to sports events organized by the sports entity.

Thereby, we formulate the following study hypothesis:

Hypothesis H5: The reputation of a football team affects positively the fan's loyalty towards his football team.

3 Methodology

The research scene is sports sector, more specifically, football market. This is because football is the most popular sport in the world, ahead other sports such as basketball, golf or American football (Conlin 2015). Thereby, it's identified as a valid scene for the analysis of our model. Thus, and in order to corroborate the previously formulated model, we select football fans as our target population, collecting data through an electronic survey.

The questionnaire was designed based on three questions. The first block has the aim to filter among the ones polled, for that who appreciate the quality of sports entities' services were football fans. The second block is compounded by questions destined to measure the different aspects defining each construct included in the proposed model, using an ascendant Likert scale from 1 to 5. The third and last block is compounded by questions aimed to segment the fans group that had been studied in the survey.

Thereby, the sample chosen for our research is formed by men (56% of the sample) and women (44% of the sample), predominating survey respondents with an age below 30 years old (57% of the sample) and singles as predominating marital status (65% of the sample). The total sample is compounded by 413 surveys. Nevertheless, valid surveys are 241. Lastly, data were collected during the months of January and February in 2016, and the survey was distributed to the sample by Internet. The election of these months for the distribution of the survey was because in these dates European football is in the middle of the season and, thus, we avoid the influence of fans spirit or despondency in their responses, motivated for the results of their teams at the end of the season or the hopeful anticipation beyond a new season plenty of possible success.

To conclude, we must point that we'd used structural equations model as data treatment model. In this way, we proceeded to analyse reliability and discriminant validity of the scales of measurement of the different latent variables that form the model. Following, we proceeded to analyse the structural relationships among latent variables of the models and the verification of the previously formulated hypothesis.

4 Results

Once the collecting data process finished, we analysed the scales of measurement of the latent variables compounding the model of study.

We made a reliability analysis of the scales of measurement of the latent variables identified on the model, and we had to eliminate some items since those ones didn't reached the recommendable values, showing values higher than 0.7 (Cronbach Alpha and composite reliability) and than 0.5 (extracted variance). Thus, we proceeded to confirm the indicators of latent variables that appeared as reliable measures (Table 1).

Table 1. Reliability analysis of the scales of Quality of service, Satisfaction, Reputation and Loyalty

Dimension	Item	Standardized weights	T value	R^2	α Cronbach	Composite reliability	Extracted variable
Satisfaction	SAT1	0.828	13.387	0.685	0.879	0.881	0.649
	SAT3	0.764	11.900	0.584			
	SAT5	0.824	13.295	0.679			
	SAT6	0.804	12.819	0.647			
Image	IM1	0.793	12.727	0.629	0.867	0.870	0.693
	IM2	0.758	11.931	0.575			
	IM3	0.936	16.490	0.877			
Reputation	R1	0.862	14.443	0.743	0.877	0.875	0.638
	R2	0.787	12.577	0.620			
	R3	0.823	13.443	0.677			
	R4	0.715	10.975	0.511			
Loyalty	LEA1	0.870	14.768	0.757	0.906	0.915	0.729
	LEA2	0.871	14.795	0.759			
	LEA3	0.918	16.135	0.842			
	LEA4	0.747	11.737	0.558			

X^2 (g.l. = 84) = 350.907 (p = 0.000); NFI = 0.851; NNFI = 0.852; CFI = 0.882; IFI = 0.883; GFI = 0.791; AGFI = 0.702; RMSEA = 0.130

Source: Own elaboration

Following, discriminant validity is analysed among latent variables using the average extracted variance test method and the confidence interval test method (Table 2). The results of the analysis of discriminant validity shows it existence, so the independence of the studied variables is preserved. Despite it, there is one only case which discriminant validity analysis indicates that the variables of image and reputation may be measuring similar aspects. Nevertheless, the theoretical development of mentioned variables shows the existence of a solid basis that oblige us to consider them independently.

Table 2. Average extracted variance test and confidence interval test results

	Satisfaction	Image	Reputation	Loyalty
Satisfaction	**0.661**	0.388	0.506	0.501
Image	(0.52; 0.73)	**0.696**	0.819	0.417
Reputation	(0.62; 0.80)	(0.86; 0.95)	**0.638**	0.445
Loyalty	(0.62; 0.80)	(0.55; 0.74)	(0.57; 0.76)	**0.737**

Source: Own elaboration

Lastly, the causal relationships among latent variables compounding the model of study were analysed, to verify the hypothesis previously formulated. The results of the model's estimation process are shown on Table 3.

Table 3. Estimated structural equations for the study model

Hypothesis	Relationship between variables	Standardized charge	T-value
H1	The level of satisfaction affects positively the fan's loyalty towards his football team	0.485**	4.792
H2	The image of a football team affects positively the fan's degree of satisfaction	−0.112	−0.506
H3	The image of a football team affects positively the fan's loyalty to his football team	0.289	1.504
H4	The reputation of a football team affects positively the fan's degree of satisfaction	0.812**	3.565
H5	The reputation of a football team affects positively the fan's loyalty towards his football team	0.061	0.277

$^{*}P < 0.05$; $^{**}P < 0.01$; n = 214
R^2 (Satisfaction) = 0.507; R^2 (Loyalty) = 0.571 X^2 (g.l. = 82) = 350.901 (p = 0.000);
NFI = 0.851; NNFI = 0.847; CFI = 0.881; IFI = 0.882; GFI = 0.791; AGFI = 0.694;
RMSEA = 0.132
Source: Own elaboration

The results obtained in this research offer important information for understanding football fans' behaviour. About the process of creating a certain level of satisfaction in fans, it's noted that the football team reputation can explain the 50.7% ($R^2 = 0.507$). So it's proven that the fan's satisfaction is positive and significantly affected by football team reputation ($\beta = 0.812$; $p < 0.01$). Nevertheless, in spite of the a priori identification of the projected image in market of a football team as relevant for its fans, the research shows that this relevance doesn't exist. Thereby, hypothesis 4 is accepted and hypothesis 2 is rejected.

Thus, the reputation of a football team is fundamental for fans to be satisfied. This could be because fans aren't unaware to public opinion and would wish their opinion agreeing on team's external assessment. Fans would understand that their football team should be managed in a way that doesn't ignore external determinants, and not only maximizing their internal resources (such as a quality use of their players to achieve their goals), but also considering the whole society, which should help in consideration to their support. Therefore, the football team must be able to analyse constantly its own reputation on market and influence it through internal and external actions (communication campaigns, social actions, etc.) to improve it. In this way, the level of satisfaction of their fans would increase to see improved the acknowledgment of their football team and feeling proud of it. Besides, it was identified that the imaged projected by a football team to the market is not really important for the follower. In spite of it, it's a factor that must be taken into account because it could affect the global assessment of the team and its reputation.

Regarding the process of generating loyalty of the follower towards his team, it's observed that the level of satisfaction showed by the follower explains 57.1% ($R^2 = 0.571$). Thus, is demonstrated that the satisfaction of the follower affects positive and significantly his degree of loyalty towards his football team ($\beta = 0.496$; $p < 0.01$), so the hypothesis 1 is accepted. Notwithstanding, the research reveals that both the image projected by the football team and its reputation aren't relevant factors for that their fans could reduce their level of loyalty towards their football team. For that, hypothesis 3 and 5 are rejected.

Thus, it had been detected that satisfied followers will be those that can feel loyal to it. This doesn't mean that if a follower isn't always satisfied, he will turn his back permanently to his team, since it can be observed a very intense loyalty towards his team. Nevertheless, if the fan doesn't feel satisfied he could stop temporary following his team or watch less matches. In this way, interests of followers should be cover and not only think about reaching sports results needed for the entity, no mind how. Thus, they must take care of details as relevant as fluent communication between sports entity members and mass media or close contact between players and fans, among others.

In addition, it had been identified that both reputation and image of a football team won't affect its fans loyalty. Nevertheless, they are relevant factors for the rest of followers to respect the values that the football team would transmit. So the football team should not only satisfy its own followers needs, but take care of every aspect that can influence its global perception by external agents. This way, it should take care of aspects such as social responsibility, social and economic activities, etc., that could affect the good reputation of the sports entity and the coherence between the image to be projected and the actions that are carried out.

5 Conclusions

In the research carried out, we proceeded to analyse the football team's fans behaviour. The election of this field of study was justified for the consideration of being the most popular sport in the world (Conlin 2015) and one of the most played ones (Vacas 2014). Therefore, the aim was determining those factors really important in present and future behaviour of football team's followers and with that providing a basis for football teams to design offers that really suit their followers.

Thereby, the research revealed that fans identify as relevant factors to them the reputation of their football team. Thus, football teams should be able to understand that their management should focus also externally and contribute with its resources to improve the conditions of the society they inhabit. That not only will cause a huge improvement of public opinion towards the team, and with that, its reputation, but also will increase the feeling of proudness of the fan towards his team. Therefore, the follower will feel very satisfied of belonging to his team, being able to downplay the sports results in his present or future mood. Besides, if the fan is satisfied, he will feel his loyalty feeling towards his team, avoiding negative behaviours, such as stop attending the stadium, watching his team's matches on TV or purchasing its merchandising, among others.

During the elaboration of this research, we had found several limitations. The principal one is that the research is centred on specific geographical field and commercial sector, like football sports field in Spain, that, despite allowing the generalization of the results of the research to other sports fields, it doesn't ensure its adaptation.

Based on exposed above, there had been identified future lines of research. Thereby, it's proposed as a future line of research the identification of the different groups of consumers that play on sports field, to identify the variables that will result more relevant on the definition of their level of loyalty or the inclusion of new study variables in the model. In addition, the possibility of the application of the model for different sports fields arises and, with that, the possibility of offering a more complete view of the process of generating the consumer's level of loyalty.

References

Andreassen, T.W., Lindestad, B.: The effect of corporate image in the formation of customer loyalty. J. Serv. Res. **1**(1), 82–92 (1998)

Arbelo, A.A., Pérez, P.G.: La Reputación Empresarial como Recursos Estratégico: Un Enfoque de Recursos y Capacidades. Ponencia presentada al XI Congreso Nacional de Acede, Zaragoza (2001)

Barich, H., Kotler, P.: A framework for marketing image management. Sloan Manag. Rev. **32**(2), 94–104 (1991)

Bearden, W., Teel, E.: Selected determinant of consumer satisfaction and complaint reports. J. Mark. Res. **20**, 21–28 (1983)

Bitner, M.J., Hubbert, A.R.: Encounter satisfaction versus overall satisfaction versus quality. In: Rust, R.T., Oliver, R.L. (eds.) Service Quality: New Directions in Theory and Practice, pp. 72–94. Sage Publications, Thousand Oaks (1994)

Calvo, C.P., Martínez, V.A.F., Juanatey, O.B., Lévy, J.P.M.: Medición de la influencia en la intención de compra del valor basado en el consumidor de las marcas del distribuidor. Cuadernos de Gestión 15(1), 93–118 (2015)

Castañeda, J.A., Luque, T.: Estudio de la Lealtad del Cliente a Sitios Web de Contenido Gratuito. Revista Europea de Dirección y Economía de la Empresa 17(4), 115–138 (2008)

Churchill, G.A.J.R., Surprenant, C.: An investigation into the determinants of customer satisfaction. J. Mark. Res. 19, 491–504 (1982)

Conlin, L.: Los 10 deportes más populares en el mundo en el 2015, Febrero 2015. http://es. express.live/

De la Fuente, M.H., Rey, G.F.: Análisis de la calidad percibida, imagen corporativa, satisfacción y lealtad de los clientes de los supermercados: Una aplicación para un supermercado de Chile. Proyecto Social: Revista de relaciones laborales 12(12), 77–102 (2008)

Díaz, A.M., Del Río, A.B., Suárez, L., Vázquez, R.: Satisfacción, Barreras de Cambio y Lealtad hacia el Servicio. XIX Encuentro de Profesores Universitarios de Marketing, Vigo, p. 165. ESIC, Madrid (2007)

Domínguez, M., Labrador, I.: El patrocinio deportivo: una opción más rentable que la publicidad… pero no exenta de riesgos, Febrero 2011. www.eleconomista.es

Gardner, B.E., Levy, S.: The product and the brand. Harvard Bus. Rev. 33, 33–39 (1995)

Halstead, D., Hartman, D., Schmidt, S.L.: Multisource effects on the satisfaction formation process. J. Acad. Mark. Sci. 22(2), 114–129 (1994)

Howard, J.A., Sheth, J.N.: The Theory of Buyer Behavior. Wiley, New York (1969)

Hunt, H.K.: CS/D: overview and future research directions. In: Hunt, H.K (ed). Conceptualization and Measurement of Consumer Satisfaction and Dissatisfaction. Marketing Science Institute, Cambridge (1977)

Lewis, B., Soureli, M.: The antecedents of customer loyalty in retail banking. J. Consum. Behav. 5(1), 15–31 (2006)

Loureiro, C.S.M., González, M.F.J.: Calidad y satisfacción en el servicio de urgencias hospitalarias: Análisis de un hospital de la zona centro de Portugal. Investigaciones Europeas de Dirección y Economía de la Empresa 16(2), 27–41 (2010)

Luna-Aroca, R., López, A.: Necesidades Formativas en Marketing y Gestión Pública del Deporte. Portaldeportivo La Revista Año 1, Núm. 1, Julio-Agosto 2007

Luna-Arocas, R., Mundina, J.J.: La Satisfacción del Consumidor en el Marketing del Deporte. Revista de Psicología del Deporte 13, 147–155 (1998)

Manassero-Mas, M.A., García-Buades, E., Ferrer-Pérez, V.A.: El Papel de Marketing en el Deporte. Revista de Psicología del Deporte 13, 115–120 (1998)

Mano, H., Oliver, R.L.: Assessing the dimensionality and structure of the consumption experience: evaluation, feeling and satisfaction. J. Consum. Res. 20, 451–466 (1993)

Mullin, B.J., Hardy, S., Sutton, W.A.: Marketing Deportivo. Paidotribo. Barcelona (1995)

Oliver, R.L.: Cognitive, affective, and attribute bases of the satisfaction response. J. Consum. Res. 20, 418–430 (1993)

Price, L.L., Arnould, E.J., Deibler, S.L.: Consumers' emotional responses to service encounters: the influence of the service provider. J. Serv. Ind. Manag. 6(3), 34–63 (1995)

Redondo, J.C., Cuadrado, G.: El Marketing Deportivo como Herramienta de Gestión. In: II Congreso de Ciencias del Deporte. Asociación Española de Ciencias del Deporte, INEF Madrid, pp. 295–301 (2002)

Reichheld, F.F., Sasser, W.E.: Zero defections: quality comes to services. Harvard Bus. Rev. 68, 105–111 (1990)

Robertson, T.S.: How to reduce market penetration cycle times. Sloan Manag. Rev. **35**(1), 87–96 (1993)

Sanz, I., Redondo, J.C., Gutiérrez, P., Cuadrado, G.: La Satisfacción en los Practicantes de Spinning: Elaboración de una Escala para su Medición, Motricidad. Eur. J. Hum. Mov. **13**, 17–36 (2005)

Shapiro, C.: Premiums for high quality products as returns to reputations. Q. J. Econ. **98**, 659–679 (1983)

Vacas, J.E.: Cuáles son los deportes más practicados del mundo, Mayo 2014. http://sportadictos.com/

Westbrook, R.A., Reilly, M.D.: Value–percept disparity: an alternative to the disconfirmation of expectations theory of consumer satisfaction. Adv. Consum. Res. **10**, 256–261 (1983)

Woodruff, R., Cadotte, E., Jenkins, R.: Modeling consumer satisfaction using experience-based norms. J. Mark. Res. **20**, 296–304 (1983)

Yoon, E., Guffey, H.G., Kijewski, V.: The effects of information and company reputation on intentions to buy a business service. J. Bus. Res. **27**, 215–228 (1993)

The LOCAL WORK PLANS (LWP)
and Territorial Economic System (TES):
Assessment and Evaluation

Cosimo Cuomo[1(✉)] and Domenico Marino[2] ⓘ

[1] Regione Calabria, Catanzaro, Italy
c.cuomo@regcal.it
[2] Mediterranea University of Reggio Calabria, Calabria, Italy
dmarino@unirc.it

Abstract. Territorial Economic System (TES) is a multidimensional concept that encompasses economic and social dimensions. Whereas the production system has a mainly material connotation, technical knowledge and social capability have a mainly immaterial nature (Latella and Marino 1996). It is important for a description of the TES to define two dimensions: the proximity and the resiliency. Each territory shows first of all a different degree of proximity which does not necessarily mean contiguity, but can have a functional meaning (Veltz 1991; Dupuy and Gilly 1993). There is, in fact, an industrial organization, cultural and temporal proximity.

The LOCAL WORK PLANS (LWP) are a tool integrating active policies for employment with policies for local development.

Incentivizing innovative investments is another policy pillar that should be implemented within Calabria's TES, and once again the LWPs are the natural and privileged setting for this implementation.

1 Introduction

Each territory shows first of all a different degree of proximity which does not necessarily mean contiguity, but can have a functional meaning (Veltz 1991; Dupuy and Gilly 1993). There is, in fact, an industrial organization, cultural and temporal proximity.

The resiliency shows the problem of the spatial evolution in the forms of the production, which leads to the question of the historic dynamics and the evolutive trajectories (Gillespie 1991). It is the capability of the system in the self-organization and in the metabolizing of the change in the external environment.

Proximity and resiliency are a way to express the concepts of local interaction and self-organization.

The territory assess economic dynamic as a *self-reinforcing mechanism*: a positive (or negative) feedback that characterizes the evolution of a dynamic system. The concept of self-reinforcing mechanism can be expressed as a dynamic system, with path dependence and a positive feedback, which tends to a large variety of asymptotic states. Every evolutionary step of the system influences the next one and then the evolution of the

J. Gil-Lafuente et al. (Eds.): AEDEM 2017, SSDC 180, pp. 225–237, 2019.
https://doi.org/10.1007/978-3-030-00677-8_19

entire system, thus generating *path dependence*. Such a system has a high number of asymptotic states, and the initial state (Time zero), unpredicted shocks, or other kind of fluctuations, can all conduct the system in any of the different domains of the asymptotic states (Arthur 1988). Furthermore, the system selects the state in which placing itself. Such dynamics are well known in physics, in chemistry as well as in biology and the final asymptotic state it is called the *emergent structure*. The concept of positive feedback in fact is relatively new for the economic science. The latter generally deals with problems of optimal allocation of scarce/insufficient resources, thus the feedback is usually considered to be negative (decreasing utility and decreasing productivity). Self reinforcing mechanism dynamic can be used to assess many different economic problems with different origins: from those related to the international dimension, to those typical of the industrial economy, and, last but not least, problems related to regional economics. Many scholars have assessed multiple equilibria and their inefficiency (Marshall 1890; Arrow and Hahn 1971; Brown Heal 1979; Scarf 1981). Multiple equilibria depend on the existence of increasing returns to scale. If the mechanism of self-reinforcing is not counterbalanced by any opposite force, the output is a local positive feedback. The latter, in turn, amplifies the deviation from some states. Since these states derive from a local positive feedback, they are unstable by definition, so multiple equilibria exist and are efficient. If the *vector field* related to a given dynamic system is regular and its critical points follow some particular rules, then the existence of other critical points or of stable cycles (also called *attractors*) turns out (Marino 1998). The multi-attractors systems have some particular properties that are very useful for our research (Marino 1998). Strict path dependence is therefore manifested, and the final state of the system will depend on the particular trail it has been covering during its dynamic evolution from an (instable) equilibrium towards another (instable) equilibrium, and so on. Accordingly, the system's dynamic is a non-ergodic one.

2   The Territorial Economic System

The TES is the physical space in which economic agents interact; the equilibrium properties of this system depend on its structure and, if the space is complex, on the particular attraction basin in which the system stays. The increasing returns, the multiple equilibria, the history dependence can found a meaning in the complex space (Krugman 1994).

By introducing the notion of Territorial Economic System (TES) (Fig. 1) as unit of analysis, it is possible to move towards the increase of interpretative capability when a synthesis among production system, technological knowledge at territorial level and local institution is searched. A TES then consist of interconnection among production system, technological knowledge and *social capabilities*. Each of these dimensions encompasses some factors which determine the performance of the TES (see Table 1).

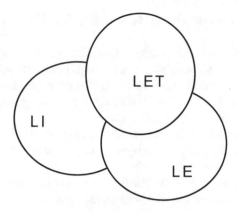

Fig. 1. The elements of the TES

Table 1. .

Firm Level (FL)	Extra Territorial Level (ETL)	Territorial Level (TL)
Access to: - Contextual and codified knowledge - Local and regional infrastructure networks	Codified knowledge: technological, organisational and communication codes	*Intangible elements:* - Available knowledge and social capabilities
Receptiveness: - Size - Organisational structure - Constitutive structure - Innovative experiences - Business networks		*Physical elements:* - Available infrastructure - Production system - Material resources

Latella and Marino (1996)

TES is a multidimensional concept that encompasses economic and social dimensions. Whereas the production system has a mainly material connotation, technical knowledge and social capability have a mainly immaterial nature (Latella and Marino 1996). It is important for a description of the TES to define two dimensions: the proximity and the resiliency. Each territory shows first of all a different degree of proximity which does not necessarily mean contiguity, but can have a functional meaning (Veltz 1991; Dupuy and Gilly 1993). There is, in fact, an industrial organization, cultural and temporal proximity.

The resiliency shows the problem of the spatial evolution in the forms of the production, which leads to the question of the historic dynamics and the evolutive trajectories of each TES (Gillespie 1991). It is the capability of the system in the self-organization and in the metabolizing of the change in the external environment. Proximity and resiliency are a way to express the concepts of local interaction and self-organization.

Within TESs economic dynamic takes the form of a *self-reinforcing mechanism*: a positive (or negative) feedback that characterizes the evolution of a dynamic system.

3   TES: Reflections on the LOCAL WORK PLANS Tool (LWP)

What we have outlined can be a schedule to orient the evolutionary path of a Local Work Plan that exhibits all the characteristics of a Regional economic system, and can thus be characterized as a very modern and efficient development planning unit.

The **LOCAL WORK PLANS (LWP)** are a tool integrating active policies for employment with policies for local development.

The *strategic objective* consists of supporting processes for the growth of skills in a specific regional network, and of self-generating strategies for change in order to strengthen the local economy.

The *general objective* is to experiment with a model of local regional cohesion in order to favour a *regional approach to employment policies.*

The *operational objective* consists of strengthening network economies and enhancing production supply chains, by combining local production businesses/local networks and young professionals possessing innovative skills and innovation abilities, in order to contribute to raising the levels of competitiveness in local networks.

The idea was born from the realization that generalist-type active policies (employment increase, self-employment incentives, occupational reintegration support, etc.), in light of the impacts produced in the territory especially from 2008 until now, have almost never tapped into the potential of local development, in essence the production strengths found in the regional territory; in order to overcome this critical issue, the LWPs reverse the approach towards work policies by starting with identifying the potential for local development through the involvement of young new graduates possessing professional skills, innovation and an enthusiasm for change.

The model is inspired by the following *principles*:

- *focus on people*, as a *starting point* for the active policies cycle, consisting of the *needs of the final recipients* of intervention measures, in this case the youth;
- *focus on local places-territories-networks*, as areas of cohesion and social growth, oriented towards enhancing the potential for development, directly linked to territorial and business capital.

The LWPs are inspired by the following *values*:

- *relational value,* corresponding to the *territorial system's* ability to tap into the real potential for local development and to create a community between *businesses, institutions and people*, able to generate efficiency, desire for change, competitiveness and good employment;
- *reputational value,* corresponding to the quality of the *skills* level reached in a specific *territorial network*, able to integrate *innovation processes*, in the hands of new generations, with the heritage of knowledge and local cultural tradition.

The *key players* of the LWP operational model are:

- *14 Project Partnerships* composed of Local Entities, individuals and/or corporations, Public Administrations and Bodies, bodies governed by public law, trades and employers' associations, unions, environmental and cultural associations, representatives of the private social sector, universities, public research centres, organizations of manufacturers and other development players possessing various interests, able to concretely contribute to the establishment and implementation of the relative LWP;

- **Young Men and Women** under the age of 35, with a Bachelor's Degree and residency for at least 6 months in the Calabria Region; The 150 young men and women currently employed in the LWP activities have been selected based on the professional profiles matching those indicated in the LWP offers, and considered strategic for strengthening the local economy;
- **Businesses** corresponding to *production supply chains and/or micro supply chains, development polarity*, and therefore having an approach oriented towards consolidating regional networks for local development.

There are 3 LWP *implementation phases*:

*(Phase 1) establishing **institutional partnerships** (phase which is already completed, 14 LWPs have been established)* with the responsibility of drafting the LWP proposal, through a participatory process of sharing and with the purpose of identifying the *potentials for local development*, recognizable in:

- *productive supply chains and/or micro supply chains*, corresponding to the production concentrations characterizing the regional context, or the places of reference (for example: Production community, typical quality productions, quality agricultural/food supply chains, tourism networks, hospitality networks in a rural setting, widespread hospitality, etc…);
- *development polarity,* cultural attractions (for example: archeological areas and parks, natural parks, museums, cultural assets), research centres (for example: incubation and spin-offs of companies at UNICAL, research and analysis centre in the agricultural/food chain, etc…);
- *social economy supply chain*, production of services for wellness and social integration, innovative systems for sustainability, etc…

Furthermore, the partnerships established for the presentation of the LWPs have identified the professional profiles required to strengthen, in terms of skills, the supply chain economies ascribable to the potentiality of local networks.

(Phase 2) the *youth* (beneficiaries Action 5- voucher) at the centre of regional development, (phase being implemented with a Regional Notice, through the selection of 150 young men and women, based on the professional profiles chosen by the partnerships).

The young men and women apply for candidacy by drafting a *PIAL- Work Integration and Start-up Project*, describing their path towards professional growth, related to the local development project.

The PIAL is divided into several operational phases:

- specializing one's skills;
- internships in companies and/or facilities and/or research centres, leaders in the production industries concerning the respective LWP;

- working and learning in the field, in collaboration with one of the supply chains/ development polarities identified by the local networks.

Moreover, the young men and women will receive a monthly voucher of € 800.00 for 9 months, in order to acquire the services functional to their work integration/self-integration (a newly graduated young man or woman or in the professional start-up phase).

The specialization of one's skills may lead to an additional € 4,800, which in turn can lead to participating in training courses, study trips, and internships in Italy or abroad, within the leading settings or environments of the developing industries related to the respective LWP.

At the same time, in addition to being a time for the young man or woman's growth, this phase may be an opportunity for companies to strengthen their relations with markets and/or an opportunity to increase competitiveness.

The skills acquired by the young men and women during their field-work phase, will become an additional asset to the productive-economic fabric of the LWP, while simultaneously contributing to the development of synergies for future opportunities, even with international cooperation networks.

(Phase 3) integration into the labour market

At the end of the program, lasting 9 months, the young men and women will be able to benefit from the following incentives, consistent with the financial resources provided by the individual LWPs:

- favoured recruitment by one of the supply chains/development polarities and/or business networks. The hiring incentive will occur based on the occupational skills, under a permanent contract with the companies/conglomerates having production branches within the LWP area selected by the Calabria Region, that have expressed their willingness to hire.
- self-employment incentives, stimulating corporate partnerships up to 3 shareholders. The new business should be encouraged by the local production networks, welcoming this new company as a new party of the network, bringing innovative processes for the local economy.

Piani Locali per il Lavoro (PLL)	Soggetto capofila	Provincia	Settore/ filiera
PLL - "Un'opportunità di crescita per il territorio: Lo Stretto sostenibile"	Comunità Montana Versante dello Stretto	RC	Turismo, enogastromia, artigianato
PLL - Ionio-Tirreno	Comunità Montana dell'Area Grecanica	Reggio Calabria	Agroalimentare, agroforestale
PLL - "SILAvoro"	Unione dei comuni della Presila	CS e CZ	Agro- silvo-pastorale, artigianato, turismo
PLL - "Terra fra i due mari tra accoglienza diffusa e Distretto della diversità"	Unione dei comuni Monte Contessa	CZ e RC	Turismo rurale ed accoglienza
PLL - "SiAMO IL LAVORO. Dai percorsi produttivi ai percorsi lavorativi, nuove opportunità per le PMI"	Associazione dei comuni Crotonesi	KR	Agricoltura, turismo
PLL - "Terre Jonico-silane"	Comune di Rossano	CS	Turismo culturale e sostenibile
PLL - "Cantieri per l'occupazione"	Comune di Bisignano	CS	Industria
PLL - Area Basso Tirreno Cosentino "ORASILAVORA"	Comune di Amantea	CS	Turismo, artigianato
PLL - Cosenza area Crati-Savuto. Innovazione ed ecosostenibilità	Associazione dei comuni KRATOS 2020	CS	Edilizia e innovazione
PLL - "Vibo Vale"	Associazione dei comuni "Vibo Vale"	VV	Turismo
PLL - Marco Polo	Comune di Catanzaro	CZ	Manifatturiero
PLL - "Goethe: professionalità e formazione nel turismo sociale e accessibile"	Unione comuni Versante Ionico	CZ	Turismo
PLL - " NEOS POL.J.S"	Comune di Castrovillari	CS	Agroalimentare , ambiente, turismo
PLL - Del Lametino e del Reventino	Associazione dei comuni del PIT 14 –Lamezia Terme	CZ	Economia sociale, innovazione sociale

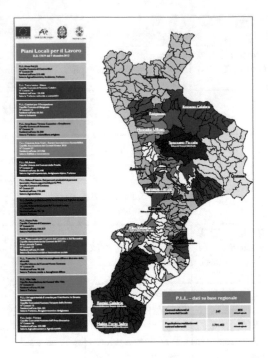

4   LWP and TES: Analysis and Policy Highlights

LWPs exhibit great versatility because they are able to operate efficiently on certain causes that have led to a deficit in the development of Calabria's regional production system. This production system is very interesting, complex and full of contradictions.

These seeming contradictions are a constant found in underdeveloped areas, and perhaps these LWPs efficiently operating on these contradictions may become a useful tool for initiating development processes.

It's easy to find consensus when identifying weaknesses in Calabria's manufacturing sector: it's below-average size, under-capitalization, the lack of openness to the markets, both in the sense of low competition as well as outlet markets limitations and organizational limitations. These are some of the elements explaining a sort of structural delay in the sector. The decline overtime of the external industry contributed to the weakening of an economic sector where local resources, both financial and human, appear substantially asphyxiated. As previously mentioned there are exceptions, however, they are certainly not frequent.

The basic belief is that solid economic systems are not born unless all the "typical" paths of economic growth trend are taken. This is where LWPs can succeed. The belief that we can jump from a rural economy to a society of services belongs in the past. The contribution that the manufacturing sector provides to local economies is crucial: the relative size may even be limited, however, the regional economic networks have a strong need for a stable manufacturing department. They especially need those external effects which the local industrial networks produce, from simple clusters of companies to real districts, in terms of organizational skills, networking, business and market culture.

The characteristics of Calabria's production system have not prohibited us, during the last decade, from attempting to trace a path towards the development of specialized production within the manufacturing sector. We find ourselves in front of a production system that is rather cohesive and rooted in the social context. Being a local entrepreneur is rooted in one's territory and is learned in the field, usually in a family-type business that is carried on by the children; skilled labour is selected with mechanisms from outside of the market, always within the entrepreneurs' close circle of social relations. This very strong local aspect could definitely be a strength. And in many ways it is. It's not difficult to witness production colonisations that overturn the production and social context of a territory, especially in underdeveloped areas. In the case of Calabria, instead, the regional roots are strong but the other side of the coin is that this system is very cohesive from a regional point of view, and risks becoming impermeable to change, closed-off to external markets, failing to invest in human capital and innovation, in essence, poorly competitive. The internal cohesion pays a high price in terms of being open to external markets. The small and often micro-dimension of businesses is another weakness in this framework. This micro-dimension is definitely a feeble attempt at a short-sighted reduction of labour costs, in order to regain the competitiveness that should instead be searched for using other tools, in particular by investing in knowledge, human capital and innovation. In light of these considerations, the production system appears to be a system afraid of growth, afraid to confront the market suffering by a perverse "Peter Pan" syndrome that prohibits it to take advantage of the strengths that do exist.

The LWPs are important because they are able to reassess the cohesion and rooting, while nevertheless projecting the production system to a larger scale and pushing it towards innovation.

Therefore, instead of being concerned with regional size, urgent measures are undertaken aimed at creating an environment that favours business development. The image of a business afraid of growth is certainly linked to the external environment, to a territory not as favourable if not at times hostile towards businesses. In this context, excessive localism and dwarfism of businesses would only be an endogenous form of defense against an environment that fails to help the development of businesses and entrepreneurs, almost like looking for camouflage. The presence of a strong criminal economic component is definitely at the basis of this tendency, as well as the relations with the often inefficient public administration, the lack of infrastructures, and the lack of cooperation and at times loan-shark behaviour of credit institutions, which all contribute greatly to the businesses' choice to refuse to grow and merely survive in the local area where it feels strong, protected and even competitive.

We therefore need territorial policies to support the TES. Firstly we need safety policies, in order for the criminal economy to avoid suffocating the legal one, and for the entrepreneur to avoid losing faith in the institutions and above all the market. But we also need infrastructural policies that lower the costs of businesses to reach outlet markets, credit policies that rebuild a virtuous relationship between the banking system and businesses, where the bank and especially the local bank becomes a player in local development, instead of becoming the undertaker of businesses as it often happens. Finally, we need policies that lower "transaction costs" between the citizen and the PA, imagining administrative simplifications, creating one-stop shops, supporting participation.

These are the system measures favouring the LWPs that must be designed in synergy with the evolution of Projects.

Finally, with regards to businesses, investing in human capital is important and fundamental. This is one of the biggest bets that the LWP program must win in order to reach the objective of greater competitiveness. The basic level of training of entrepreneurs is rather high, especially if compared with that of other Italian areas, and this could be an interesting point of strength to be able to implement policies that raise the level of human capital. These policies however, also need to affect the workforce, which registers certain weaknesses given that the casualness and irregularity characterizing a part of the workforce is definitely less suited for these types of policies.

Incentivizing innovative investments is another policy pillar that should be implemented within Calabria's TES, and once again the LWPs are the natural and privileged setting for this implementation.

5 Conclusions

LWPs are created to tap into the drive for change in local networks, through the active and responsible participation of those who wish to witness the renewal of society (bodies, companies, young men and women, associations), in order to guarantee increasingly inclusive employment levels that are also aligned with the needs of local communities. By applying the interpretative TES pattern to the LWPs we notice how the entire extraterritorial area should be developed and the policies that must be made to achieve this objective. It is a matter of incentivizing:

- the promotion of exportation and relations with external markets
- *joint venture* businesses
- the promotion of investments in external commercial structures
- the promotion of external production investments
- the promotion of production cooperations, sub-supply relations from and to external businesses
- the promotion of technological cooperation with external businesses
- the promotion of regional marketing policies
- the promotion of inter-regional cooperative relations

With these procedures we can create a systemic environment that favours development.

References

Aoki, M.: Analysis of an open model of share markets with several types of participats. UCLA, Working Paper (1999)

Aoki, M.: A new model of labour dynamics: ultrametrics, Okun's law, and transient dynamics. UCLA, Working Paper (2003)

Aoki, M.: New frameworks for macroeconomic modelling: some illustrative examples. UCLA, Working Paper (2004)

Arthur, W.B.: Self-reinforcing mechanisms in economics. In: Anderson, P.W., Arrow, K.J., Pines, D. (eds.) The Economy as an Evolving Complex System. Addison-Wesley, Reading (1988)

Arthur, W.B.: Competing technologies, increasing returns and lock in by historical events. Econ. J. **99**, 116–131 (1989a)

Arthur, W.B.: Complexity and the economy. Science **284**, 107–109 (1989b)

Arthur, W.B.: Out-of-equilibrium and agent-based modelling. In: Hand-Book for Computational Economics, vol. 2. North-Holland, Amsterdam (2005)

Barkley Rosser, J.: A reconsideration of the role of discontinuity in regional economic models. Chaos, Solitons Fractals. **18**(2003), 451–462 (2003)

Bailey, M.N., Lawrence, R.Z.: Do we have a new economy. National Bureau of Economic Research. Working Paper No. 8243, April 2001

Barewald, F.: History and Structure of Economic Development. India Book House, Mumbai (1969)

Barro, R.J., Sala-i-Martin, X.: Economic Growth. McGraw-Hill, New York (1995)

Baumol, W.J., Oates, W.E.: The Theory of Environmental Policy. Cambridge University Press, Cambridge (1988)

Becattini, G.: "Dal "settore" al "distretto" industriale. Alcune considerazioni sull'unità di indagine nell'economia industriale". Rivista di Economia e Politica Industriale 5(1), 7–21 (1979)

Becattini, G., Rullani, E.: Sistema locale e mercato globale. Economia e Politica industriale **80**, 24–48 (1993)

Behrens, R., Wilkinson, P.: Housing and urban passenger transport policy and planning in South African cities: a problematic relation? In: Harrison, P., Huchzermeyer, M., Mayekiso, M. (eds.) Confronting Fragmentation: Housing and Urban Development in a Democratising Society, pp. 154–172. University of Cape Town Press, Cape Town (2003)

Bentollila, S., Dolado, J.J., Franz, W., Pissarides, C.: Labour flexibility and wages: lessons from Spain. Econ. Policy **9**(18), 53–99 (1994)

van den Berg, L., Braun, E., van der Meer, J.: National Urban Policies in the European Union. Euricur, Rotterdam (2004)

Bianchi, P.: Le politiche industriali dell'Unione Europea. Il Mulino, Bologna (1995)

Boix Domènech, R.: Economía del conocimiento, tecnología y territorio en España. CDTI and UAB, mimeo (2006)

Borjas, G.J.: The economic analysis of immigration. In: Ashenfelter, O., Card, D. (eds.) Handbook of Labour Economics, vol. 3A. North Holland, New York (1999)

Borjas, G.J., Hilton, L.: Immigration and the welfare state: immigrant participation in means-tested entitlement programmes. Q. J. Econ. **111**(2), 575–604 (1996)

Borjas, G.J.: Self-Selection and the earnings of immigrants. Am. Econ. Rev. **77**(4), 531–553 (1987)

Breschi, S., Lissoni, F.: Knowledge spillovers and local innovation systems: a critical survey. Ind. Corp. Chang. **10**(4), 975–1005 (2001)

Briggs, X.: Democracy as Problem Solving. Civic Capacity in Communities Across the Globe. MIT Press, Cambridge (2007, Forthcoming)

Brusco, S., Sabel, C.: Artisan production and economic growth. In: Wilkinson, F. (ed.) The Dynamics of Labour Markets Segmentation. Academic Press, London (1981)

Buesa, M.: El sistema regional de innovación de la Comunidad de Madrid, informe sobre Situación económica y social de la Comunidad de Madrid 2001, Consejo Económico y Social de la Comunidad de Madrid (2002)

Castells, M.: The City and the Grassroots. University of California Press, Berkeley (1983)

Coase, R.H.: The nature of the firm. Economica 4(386), 386–405 (1937)

Conway, P., Nicoletti, G.: Product market regulation in the non-manufacturing sectors of OECD countries: measurement and highlights. OECD Economics Department Working Papers, No. 530 (2006)

Conway, P., de Rosa, D., Nicoletti, G., Steiner, F.: Regulation, competition and productivity convergence. OECD Economics Department Working Papers, No. 509 (2006)

Danson, M., Halkier, H., Cameron, G.: Regional governance, institutional change and regional development. In: Danson, M., Halkier, H., Cameron, G. (eds.) Governance, Institutional Change and Regional Development. Ashgate, Aldershot (2000)

David, P.: Clio and the economics of QWERTY. Am. Econ. Rev. In: Papers and Proceedings, vol. 75, pp. 332–337 (1985)

Foster, G.: Traditional Societies and Technological Change. Harper & Row, New York (1973)

Frenken, K.: Technological innovation and complexity. Urban and Regional Research Centre Utrecht (URU), Working Paper (2005)

Friedmann, J.: The world cities hypothesis. Dev. Chang. 17(1), 69–84 (1986)

Gordon, R.J.: The jobless recovery: does it signal a era of productivity growth? Brook. Pap. Econ. Act. 1, 271–306 (1993)

Hall, P.: Creative industries and economic development. Urban Stud. 37(4), 639–649 (2000)

Hirschman, A.O.: The Strategy of Economic Development. Yale University Press, New Haven (1959)

Jacobs, J.: Cities and the Wealth of Nations. Vintage Books, New York (1984)

Kauffman, S.A.: The Origin of Order. Self-organisation and Selection in Evolution. Oxford University Press, Oxford (1993)

Krugman, P.: History versus expectations. Q. J. Econ. 106, 651–667 (1991a)

Krugman, P.: Geography and Trade. MIT Press, Cambridge (1991b)

Krugman, P, Venables, A.J.: The seamless world: a spatial model of international specialization, 1230. C.E.P.R. Discussion Papers (1995)

Krugman, P.R., Venables, A.J., Fujita, M.: The Spatial Economy. The MIT Press, Cambridge (1999)

Lafuente, E., Vaillant, Y., Vendrell-Herrero, F.: Territorial servitization: exploring the virtuous circle connecting knowledge-intensive services and new manufacturing businesses. Int. J. Prod. Econ. 192, 19–28 (2016)

Latella, F., Marino, D.: Diffusione della conoscenza ed innovazione territoriale: verso la costruzione di un modello. Quaderni di Ricerca di Base dell' Università Bocconi, n. 2 (1996)

Lester, R., Piore, M.: Innovation: The Missing Dimension. Harvard University Press, Cambridge (2004)

Lublinski, A.E.: Does geographic proximity matter? Evidence from clustered and non-clustered aeronautic firms in Germany. Reg. Stud. 37(5), 453–467 (2003)

Marino, D.: Territorial economic systems and artificial interacting agents: models based on neural networks. Int. J. Chaos Theory Appl. 3(1/2), 23–29 (1998)

Marino, D., Trapasso, R.: The new approach to regional economics dynamics: path dependence and spatial self-reinforcing mechanisms. In: Fratesi, U., Senn, L. (eds.) Growth and Innovation of Competitive Regions, pp. 329–367. Springer, Berlin (2009)

Markusen, A.: Fuzzy concepts, scanty evidence, policy distance: the case for rigor and policy relevance in critical regional studies. Reg. Stud. **33**(9), 869–884 (1999)

Myrdal, G.: Economic Theory and Underdeveloped Regions. Duckworth, London (1957)

Nelson, R.R., Winter, S.G.: In search a useful theory of innovation. Res. Policy **6**, 36–76 (1977)

Nelson, R.R., Winter, S.G.: An Evolutionary Theory of Economic Change. The Belknap Press of Harvard University Press, Cambridge (1982)

Niosi, J., Zhegu, M.: Aerospace clusters: local or glocal knowledge spillovers? Ind. Innov. **12**(1), 1–25 (2005)

Porter, M.: The Competitive Advantage of Nations. The Free Press, New York (1990)

Rodríguez Pose, A., Bwire, A.: The economic (in)efficiency of devolution. Environ. Plan. A **36**, 1907–1928 (2003)

Scott, A., Storper, M.: Production, Work, Territory: The Geographical Anatomy of Industrial Capitalism. Allen & Unwin, Boston (1986)

Scott, A., Storper, M.: Regions, globalisation, development. Reg. Stud. **37**(6/7), 579–593 (2003)

Storper, M.: Regional economies as relational assets. In: Lee, R., Wills, J. (eds.) Geographies of Economies. Arnold, London (1997)

Storper, M.: Civil society: three ways into a problem. In: Douglas, M., Friedman, J. (eds.) Cities for Citizens, pp. 239–246. Wiley, Chichester (1998)

Storper, M., Walker, R.: The Capitalist Imperative: Territory, Technology and Industrial Growth. Basil Blackwell, New York (1999)

Storper, M., Venables, A.J.: Buzz: face-to-face contact and the urban economy. J. Econ. Geogr. **4**(4), 351–370 (2004)

Taylor, P.J.: World City Network: A Global Urban Analysis. Routledge, London (2004)

Taylor, P.J.: Leading world cities: empirical evaluations of urban nodes in multiple networks. Urban Stud. **42**(9), 1593–1608 (2005)

Thrift, N., Olds, K.: Refiguring the economic in economic geography. Prog. Hum. Geogr. **20**, 311–337 (1996)

Tirmarche, O., Le Galès, P.: Life after industrial decline in St. Etienne: robust SMEs, deterritorialization, and the making of a local mode of governance. In: Crouch, C., Le Galès, P., Trigilia, C., Voelzkow, H. (eds.) Changing Governance of Local Economies: Responses of European Local Production Systems, pp. 160–180. Oxford University Press, Oxford (2004)

Venables, A.J.: Equilibrium locations of vertically linked industries. Int. Econ. Rev. **37**(2), 341–359 (1996)

Vendrell-Herrero, F., Wilson, J.R.: Servitization for territorial competitiveness: taxonomy and research agenda. Compet. Rev. **27**(1), 2–11 (2017)

Ward, S.V.: Planning the Twentieth Century City: The Advanced Capitalist World. Wiley, Chichester (2002)

Wheeler, S.: The new regionalism: key characteristics of an emerging movement. J. Am. Plan. Assoc. **68**(3), 267–278 (2002)

Williamson, O.E.: The economics of organization: The transaction cost approach. Am. J. Sociol. **87**(2), 233 (1981)

Using Method of Expertons in Bidder Selection on the Spanish Public Procurement Process

Jaime Gil Lafuente[(✉)] and José Humberto González Rodríguez[(✉)]

Universidad de Barcelona & European Academy of Management and Business Economics, Barcelona, Spain
{j.gil, jhumberto.gonzalez}@ub.edu

Abstract. The method of expertons is a methodology that has been developed in the field of fuzzy logic. This methodology allows qualitative and subjective assessments to be made by a group of experts on one or several criteria that are desired to be evaluated for decision making. In this sense, this chapter presents a proposal for the application of the method of expertons in response to article 150.2 of the PSCL, for which a practical assumption is presented, the results of which allow the application of this methodology to be positively evaluated, without forgetting the origin of the result, and that opens the possibility for future research.

1 Introduction

Public procurement represents an important activity to states and their governments, for both central and local governments, including regional governments, and it affects the citizens (the taxpayers) because government procurement represents between approximately 10 and 15% of the gross domestic product (GDP) on average, according to the World Trade Organization (WTO)[1]. In the case of the European Union, public authorities spend around 14% of GDP on the purchases of services, works and supplies[2].

Public procurement is defined as the process by which public governments or authorities, such as government departments or local authorities, purchase work, goods or services from companies, where these authorities are the buyer in many sectors such as energy, transport, waste management, social protection and the provision of healthcare or education services[3].

On the other hand, public procurement processes are regulated by different directives and rules in the case of Europe, and by royal decree in the case of Spain, which define it, delimit it and, above all, regulate what its procedure should be in terms of publicity and criteria for awarding the public procurement contracts by the public entities that issue them.

---

[1] See WTO web: https://www.wto.org/english/tratop_e/gproc_e/gproc_e.htm.
[2] See European Commission website: https://ec.europa.eu/growth/single-market/public-procurement_es.
[3] See European Commission website: https://ec.europa.eu/growth/single-market/public-procurement_es.

© Springer Nature Switzerland AG 2019
J. Gil-Lafuente et al. (Eds.): AEDEM 2017, SSDC 180, pp. 238–249, 2019.
https://doi.org/10.1007/978-3-030-00677-8_20

In the specific case of Spain, without entering into the publicity aspect of the public procurement offers, since it is not subject of this chapter, special mention must be made of the contract awarding criteria, where articles 150.1 and 150.2 of the Royal Legislative Decree 3/2011, November 14, which approves the consolidated text of the Public Sector Contracts Law (PSCL) indicate that the public entity issuing the contract may apply a single award criteria (in this case, the low price criterion) like it can apply several award criteria (in this case, qualitative criteria).

In the case that the public entity decides to apply the single award criterion or low-price criterion, according to article 150.1 of the PSCL, this process is very simple, because the contract will be awarded to the company that submits the lowest bid.

But in the case that the public authority decides to apply several award criteria or qualitative criteria, according to article 150.2 of the PSCL, this process becomes slightly more complex for two reasons: (1) For the public decision maker (that public authority that offers the public tender) it becomes a situation of multi-criteria decision-making (MCDM) and (2) they have to choose between applying objective criteria by figures or percentages obtained through the mere application of the formulas established in the information sheets, or applying subjective criteria (value judgments). See Fig. 1.

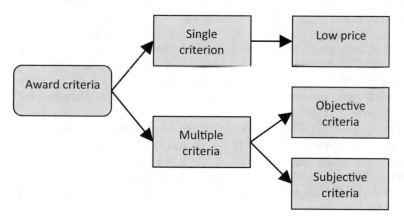

Fig. 1. Possible award criteria according to article 150.1 and 150.2 of the PSCL. Source: PSCL and authors' elaboration.

This last point, where public authorities decide to apply subjective criteria in the public procurement process, is relevant to the objective of this chapter for two reasons: (1) the award criteria definition, award criteria application and, finally, the awarding of the public contract to the bidder represents an MCDM scenario for public administrations; (2) the application of subjective criteria in bidder selection in the public procurement process implies the need to apply other evaluation methods, because, on the one hand, public entities use subjective evaluations and on the other hand, those evaluations imply a certain level of uncertainty and discretion.

Finally, in relation to what has already been explained in the previous paragraph, the objective of this chapter is to make a proposal to be able to use the method of expertons, a methodology related to fuzzy logic, as a methodology to respond to MCDM and scenarios of uncertainty.

2 Conceptual Framework

Taking into account the ideas presented in the introduction, the development of a conceptual framework will be based on two pillars. The first pillar will be the Spanish legal framework, as has previously been stated in the introduction, and the second pillar will be the academic contributions made by other authors, evidently, contributions related to fuzzy logic (and its several models such as fuzzy AHP and fuzzy TOPSIS, for example).

Spanish Legal Framework
In the case of Spain, public procurement is regulated by the Royal Legislative Decree 3/2011, November 14, which approves the consolidated text of the Public Sector Contracts Law, a regulation that defines public contracts, how they must be publicized, how they are regulated, how bidders are selected and how those contracts are awarded, among other aspects, without forgetting that there are more considerations that are not mentioned in this chapter, which focuses on bidder selection and contract awarding.

According to PSCL, public sector contracts are delimited or classified into construction contracts, public works concessions, public services management, supply, services, collaboration between the public and private sectors, and mixed contracts.

Evidently, the above classification (Table 1) is not exhaustive, but it does represent a general classification of the contracts contemplated in the PSCL. For more detailed information about the contract types and their variants, it is recommended to read Chap. 2, Sects. 1, 2 and 3 of the PSCL.

Leaving aside the different contract types contemplated in the PSCL, the most important element for the purpose of this chapter is the one that defines article 150.2, which has previously been referred to in the introduction, since it refers to the use of multiple criteria for the process of bidder selection. No mention will be made of article 150.1 of the PSCL (single criterion for the process of bidder selection), since it is not relevant for the purpose of this chapter, because it makes mention of the use of the lowest price criterion as the single criterion.

Why is article 150.2 of the PSCL important for the purpose of this chapter? Because this article opens the possibility of evaluating the bidders submitted to a public procurement offer qualitatively. In other words, the public entities that wish to carry out public procurement offers may have other requirements that go beyond the single lowest price criterion, such as (Nieto-Morote and Ruz-Vila 2012) technical capacity, experience, management capability, financial stability, past performance, past relationships, reputation and occupational health and safety (see Table 1).

But very exceptionally, article 150.2 of the PSCL offers the possibility that the award criteria can be valuated (quantified) through value judgment. In order to be able to give this type of valuation, beyond exposing it in the information sheet of a public procurement offer, a committee of experts must be established. This committee must consist of at least three members who are not integrated in the entity that is proposing

Table 1. Aspects to take into account for multiple criteria. Source: (Nieto-Morote and Ruz-Vila 2012)

Concept	Description
Technical capacity	The contractor must demonstrate that it has the technical capacity to perform the activities of the specific project for which it is seeking the prequalification
Experience	The contractor must demonstrate its participation in other previous projects, especially if they are similar to the project that will be executed
Management capability	The contractor must demonstrate that it is capable of planning, organizing and controlling a project
Financial stability	The client must reach an informed opinion regarding the overall financial position and capability of the contractor
Past performance	Considering the past performance of each contractor, the project manager will have a higher or lower degree of confidence in the possible contractors regarding the quality, time and cost control requirements
Past relationship	The client must collect and analyze information about the contractor's past relationship with other entities which participate in the construction activities
Reputation	The project manager must have an overall estimation or opinion about how good the contractor is
Occupational health and safety	To encourage contactors to establish and maintain effective systems to manage the risks to the health and safety of their employees, arising from the nature of the work

the public procurement offer and must have the appropriate expertise to make such valuations. Another option is for that subjective valuation to be carried out by a specialized technical body.

Academic Contributions Made by Other Authors

Related to the above, it is pertinent to take into account the academic contributions made by other authors regarding the application of a single criterion or multiple criteria, since depending on the needs of the public entity it may be advisable to develop a public procurement offer based on either a single criterion or multiple criteria (Bergman and Lundberg 2013; Lundberg and Bergman 2017).

In the case that the public administration wants to make a public procurement offer where the work or service is more or less standard, the quality is similar among the different bidders and there are a large number of bidders, it is more convenient to apply a single criterion of lower price, because other more qualitative factors lose meaning and the public administration can opt for more advantageous prices (Lundberg and Bergman 2017).

On the other hand, if the public administration wants to make a public procurement offer where the work or service is not standard, there are small number of bidders and quality is the most relevant aspect, it is more convenient to apply multiple criteria, because in this case the price may be irrelevant (Lundberg and Bergman 2017).

If the public entity decides to make a public procurement offer through multiple criteria, it must develop an MCDM scenario, which implies that decision making can be simple or complex depending on the number and types of criteria applied (Jaskowski et al. 2010; Nieto-Morote and Ruz-Vila 2012).

According to the aforementioned academic contributions, the MCDM scenarios developed on multi-attribute decision making (MADM), as in the case of public procurement bidder selection using multiple subjective criteria, imply that bidder selection can be subjective and it requires decision makers to reach a consensus (Jaskowski et al. 2010).

The use of subjective criteria implies that the decision-making process goes beyond the use of quantitative assessments through using statistical tools, since these are usually imprecise and there is a certain level of inherent uncertainty. In addition, in the case of qualitative criteria, linguistic variables could be used for their description (Machacha and Bhattacharya 2000). In this sense, the combination of MCDM techniques with fuzzy logic techniques is revealed as a useful tool to respond to what has been stated in this paragraph and in the previous one (Machacha and Bhattacharya 2000; Pamucar et al. 2017; Zadeh 1965).

Apart from the previous combination, there are other fuzzy logic techniques that can give an adequate response to decision making with qualitative and subjective criteria. In this case, and in relation to the objective of this chapter, one technique would be the application of the methodology of expertons in the processes of bidder selection when qualitative and subjective criteria are applied, since it is a methodology that allows grouping or tabulating the subjective valuations from different experts. At the same time, these subjective valuations are recorded on a scale of 0 to 1, where 0 is the lowest valuation and 1 is the highest valuation. Furthermore, this methodology allows for the collection of valuations from non-homogeneous groups of experts, but thanks to the development of "Fuzzy Random Subsets", in the end, a homogenous result is obtained with which different conclusions can be obtained based on the valuations from the group or groups of experts (Kaufmann and Gil Aluja 1993).

3   Sample and Methodology

As already mentioned in the introduction, the objective of this chapter is to make a proposal to assess the bidders that are presented to a certain public procurement offer, where this offer is made under the assumption of multiple qualitative criteria, criteria that at the same time are also subjective. To carry out the assessment of the bidders, taking into account article 150.2 of the PSCL, a fictional group of ten experts is created that will value the bidders in question with four qualitative characteristics, which are defined for use in this fictional public procurement offer. The expert evaluation of the group of experts will be developed by applying the methodology of expertons. Also, it must be taken into account that the public procurement offer in this example is about a public work and only three bidders are presented.

The four criteria contained in the fictional public procurement offer are defined below:

C_{1k}: Adequacy of the execution price of public work, with k = 1, 2, 3 for each bidder.
C_{2k}: Quality (adequacy of the quality management system developed by the bidder), with k = 1, 2, 3 for each bidder.
C_{3k}: Work execution period (compliance capacity of each bidder), with k = 1, 2, 3 for each bidder.
C_{4k}: Maintenance of the infrastructure once the public work is completed (capacity of each bidder), with k = 1, 2, 3 for each bidder.

Therefore, taking the above into account, there are four characteristics that each group of experts must assess independently and for each bidder, giving a total of twelve independent valuations, initially. For this public procurement offer some weightings (as a value between 0 and 1) and ideal values (also values between 0 and 1) have also been defined for each of the four characteristics (see Table 2), this will allow subsequent assessment of the suitability of each bidder.

Table 2. Ideal value of each criterion and its weight. Source: Authors' elaboration

	C_1	C_2	C_3	C_4
Ideal value	.900	.700	.800	.600
Weightings	350	.200	.350	.100

According to the methodology of expertons, explained on the book titled "Técnicas especiales para la gestión de expertos" by Kaufmann and Gil Aluja (1993), the different valuations by each group of ten experts, with respect to the different bidders and the different characteristics, are recorded as follows (Tables 3, 4 and 5):
First bidder:

Table 3. Regarding the first bidder, valuations from groups of experts taking into account the four characteristics previously defined. Source: Authors' elaboration

C11	MIN	MAX	C12	MIN	MAX	C13	MIN	MAX	C14	MIN	MAX
EXP. 01	.600	.800	EXP. 01	.400	.700	EXP. 01	.400	.600	EXP. 01	.800	1.000
EXP. 02	.500	.700	EXP. 02	.600	.900	EXP. 02	.900	1.000	EXP. 02	.600	.800
EXP. 03	.800	.800	EXP. 03	.900	1.000	EXP. 03	1.000	1.000	EXP. 03	.300	.500
EXP. 04	.100	.300	EXP. 04	.600	.600	EXP. 04	.900	.900	EXP. 04	1.000	1.000
EXP. 05	.800	1.000	EXP. 05	.200	.300	EXP. 05	.800	.800	EXP. 05	.500	.700
EXP. 06	.700	.700	EXP. 06	.500	.500	EXP. 06	.700	.800	EXP. 06	.900	1.000
EXP. 07	.500	.700	EXP. 07	.300	.300	EXP. 07	.100	.300	EXP. 07	.600	.700
EXP. 08	.400	.600	EXP. 08	.800	.900	EXP. 08	.500	.600	EXP. 08	.400	.500
EXP. 09	.600	.600	EXP. 09	.800	1.000	EXP. 09	1.000	1.000	EXP. 09	.700	.700
EXP. 10	.400	.400	EXP. 10	1.000	1.000	EXP. 10	.800	.900	EXP. 10	.600	.800

Second bidder:

Table 4. Regarding the second bidder, valuations from groups of experts taking into account the four characteristics previously defined. Source: Authors' elaboration

C21	MIN	MAX	C22	MIN	MAX	C23	MIN	MAX	C24	MIN	MAX
EXP. 01	.400	.500	EXP. 01	.600	.800	EXP. 01	1.000	1.000	EXP. 01	.400	.600
EXP. 02	.800	.900	EXP. 02	.700	.700	EXP. 02	.800	1.000	EXP. 02	.400	.400
EXP. 03	.600	1.000	EXP. 03	.500	.600	EXP. 03	.600	.700	EXP. 03	.500	.700
EXP. 04	.400	.400	EXP. 04	.700	.800	EXP. 04	.600	.800	EXP. 04	.800	.900
EXP. 05	.300	.500	EXP. 05	.300	.500	EXP. 05	.200	.400	EXP. 05	.100	.200
EXP. 06	.500	.700	EXP. 06	.500	.500	EXP. 06	.700	.700	EXP. 06	.400	.500
EXP. 07	1.000	1.000	EXP. 07	.200	.300	EXP. 07	1.000	1.000	EXP. 07	.500	.500
EXP. 08	.700	.900	EXP. 08	.700	.900	EXP. 08	.100	.300	EXP. 08	.200	.300
EXP. 09	.400	.600	EXP. 09	1.000	1.000	EXP. 09	.400	.500	EXP. 09	.300	.600
EXP. 10	.700	.800	EXP. 10	.900	1.000	EXP. 10	.900	.900	EXP. 10	.500	.700

Third bidder:

Table 5. Regarding the first bidder, valuations from groups of experts taking into account the four characteristics previously defined. Source: Authors' elaboration

C31	MIN	MAX	C32	MIN	MAX	C33	MIN	MAX	C34	MIN	MAX
EXP. 01	.700	.900	EXP. 01	.700	.800	EXP. 01	.600	.700	EXP. 01	.500	.800
EXP. 02	.500	.600	EXP. 02	.400	.700	EXP. 02	.500	.500	EXP. 02	1.000	1.000
EXP. 03	.600	.600	EXP. 03	.500	.600	EXP. 03	.800	.900	EXP. 03	.700	.900
EXP. 04	.700	.800	EXP. 04	.400	.400	EXP. 04	.100	.200	EXP. 04	1.000	1.000
EXP. 05	.400	.700	EXP. 05	.100	.300	EXP. 05	.400	.400	EXP. 05	.100	.100
EXP. 06	.700	.700	EXP. 06	1.000	1.000	EXP. 06	.500	.600	EXP. 06	.700	1.000
EXP. 07	.900	1.000	EXP. 07	.600	.900	EXP. 07	.800	.800	EXP. 07	.800	.900
EXP. 08	.200	.400	EXP. 08	.100	.100	EXP. 08	.800	.900	EXP. 08	.500	.600
EXP. 09	1.000	1.000	EXP. 09	.500	.600	EXP. 09	.100	.400	EXP. 09	.100	.400
EXP. 10	.500	.600	EXP. 10	.900	1.000	EXP. 10	1.000	1.000	EXP. 10	.900	.900

Once the valuations have been obtained for each group of experts and for each bidder, the next step is to process this information to obtain standardized frequency tables, as many tables as groups of experts, which will lead to what is known as "Fuzzy Random Subsets".

The second step consists in comparing these "Fuzzy Random Subsets" with ideal values of different characteristics (Table 2), which at the same time are considered other "Fuzzy Random Subsets". This comparison is made using differences, row by row, where the value obtained is subtracted from the ideal value. This process is also called Hamming distances calculation (the calculation formula is shown below) where the idea is to evaluate the distance between the value obtained and the ideal value, with a greater distance implying a worse result and a smaller distance a better result[4]. At the same time, the obtained values from Hamming distances calculation are collected in tables too.

Table 6. Example of distance calculation. "Reordered distances" refers to the fact that the maximum value cannot be less than the minimum value. Source: Authors' elaboration

Ideal values			Values obtained from group of experts			Distances			Reordered distances		
C1	MIN	MAX	C1	MIN	MAX		MIN	MAX		MIN	MAX
.000	1.000	1.000	.000	1.000	1.000	.000	.000	.000	.000	.000	.000
.100	1.000	1.000	.100	1.000	1.000	.100	.000	.000	.100	.000	.000
.200	1.000	1.000	.200	.900	1.000	.200	.000	.100	.200	.000	.100
.300	1.000	1.000	.300	.900	1.000	.300	.000	.100	.300	.000	.100
.400	1.000	1.000	.400	.900	.900	.400	.100	.100	.400	.100	.100
.500	1.000	1.000	.500	.700	.800	.500	.200	.300	.500	.200	.300
.600	1.000	1.000	.600	.500	.800	.600	.200	.500	.600	.200	.500
.700	1.000	1.000	.700	.300	.600	.700	.400	.700	.700	.400	.700
.800	1.000	1.000	.800	.200	.300	.800	.700	.800	.800	.700	.800
.900	1.000	1.000	.900	.000	.100	.900	.900	1.000	.900	.900	1.000
1.000	.000	.000	1.000	.000	.100	1.000	.100	.000	1.000	.000	.100

The third step consists in multiplying the obtained values from Hamming calculations applied to each characteristic and bidder by their weightings. These last values are summed, each bidder's values, and finally we obtain the final distance of each bidder (see results section).

To calculate the distances between the valuations of each group of experts and the ideal values for each characteristic, the "Hamming Distance" calculation is used as shown below:

$$\delta(I, C_k) = \sum \left\{ w_1 \left(\left| \underline{c_{1i}} - \overline{c_{1e}} \right| + \left| \overline{C_{1i}} - \underline{C_{1e}} \right| \right) + w_2 \left(\left| \underline{c_{2i}} - \overline{c_{2e}} \right| + \left| \overline{C_{2i}} - \underline{C_{2e}} \right| \right) \right. $$
$$\left. + w_3 \left(\left| \underline{c_{3i}} - \overline{c_{3e}} \right| + \left| \overline{C_{3i}} - \underline{C_{3e}} \right| \right) + w_4 \left(\left| \underline{c_{4i}} - \overline{c_{4e}} \right| + \left| \overline{C_{4i}} - \underline{C_{4e}} \right| \right) \right\}$$

Where:

I: represents the ideal values of each characteristic.
C_k: represents the valuation of each characteristic and of each bidder.
W_k: represents the value of each weighting.

---

[4] In this case we have used a variant of Hamming distance calculations, for more details on "Hamming distances calculation" it is recommended to consult the book written by Kaufmann and Gil Aluja (1993).

4 Results

Considering the previous point, the results obtained from applying the "Hamming Distance" are shown below (Table 7, 8 and 9):

First bidder results:

Table 7. Weighted distances between the ideal values of each characteristic and each valuation. Source: Authors' elaboration

DIST. C1	MIN	MAX	DIST. C2	MIN	MAX	DIST. C3	MIN	MAX	DIST. C4	MIN	MAX
.000	.000	.000	.000	.000	.000	.000	.000	.000	.000	.000	.000
.100	.000	.000	.100	.000	.000	.100	.000	.000	.100	.000	.000
.200	.000	.035	.200	.000	.000	.200	.000	.035	.200	.000	.000
.300	.000	.035	.300	.000	.020	.300	.000	.035	.300	.000	.000
.400	.035	.035	.400	.040	.040	.400	.035	.035	.400	.000	.010
.500	.070	.105	.500	.040	.060	.500	.035	.070	.500	.000	.020
.600	.070	.175	.600	.060	.080	.600	.035	.105	.600	.020	.030
.700	.140	.245	.700	.080	.120	.700	.105	.105	.700	.040	.080
.800	.245	.280	.800	.080	.100	.800	.105	.140	.800	.030	.050
.900	.315	.350	.900	.040	.100	.900	.140	.175	.900	.020	.030
1.000	.000	.035	1.000	.020	.060	1.000	.070	.105	1.000	.010	.030
$E(\overline{C_1})$	.080	.118	$E(\overline{C_2})$	.033	.053	$E(\overline{C_3})$	.048	.073	$E(\overline{C_4})$	.011	.023

Second bidder results:

Table 8. Weighted distances between the ideal values of each characteristic and each valuation. Source: Authors' elaboration

DIST. C1	MIN	MAX	DIST. C2	MIN	MAX	DIST. C3	MIN	MAX	DIST. C4	MIN	MAX
.000	.000	.000	.000	.000	.000	.000	.000	.000	.000	.000	.000
.100	.000	.000	.100	.000	.000	.100	.000	.000	.100	.000	.000
.200	.000	.000	.200	.000	.000	.200	.000	.035	.200	.000	.010
.300	.000	.000	.300	.000	.020	.300	.000	.070	.300	.010	.020
.400	.000	.035	.400	.020	.040	.400	.035	.070	.400	.020	.030
.500	.035	.140	.500	.020	.040	.500	.070	.105	.500	.030	.060
.600	.105	.175	.600	.060	.080	.600	.105	.105	.600	.050	.090
.700	.140	.210	.700	.080	.100	.700	.105	.175	.700	.010	.030
.800	.175	.280	.800	.040	.100	.800	.175	.210	.800	.010	.010
.900	.210	.315	.900	.040	.060	.900	.105	.140	.900	.000	.010
1.000	.035	.070	1.000	.020	.040	1.000	.070	.105	1.000	.000	.000
$E(\overline{C_1})$	.064	.111	$E(\overline{C_2})$	.025	.044	$E(\overline{C_3})$	.060	.092	$E(\overline{C_4})$	.012	.024

Third bidder results:

Table 9. Weighted distances between the ideal values of each characteristic and each valuation. Source: Authors' elaboration

DIST. C1	MIN	MAX	DIST. C2	MIN	MAX	DIST. C3	MIN	MAX	DIST. C4	MIN	MAX
0	0	0	0	0	0	0	0	0	0	0	0
0.1	0	0	0.1	0	0,000	0.1	0	0	0.1	0	0
0.2	0	0	0.2	0.02	0.04	0.2	0	0.07	0.2	0.01	0.02
0.3	0	0.035	0.3	0.02	0.04	0.3	0.035	0.07	0.3	0.01	0.02
0.4	0	0.035	0.4	0.04	0.04	0.4	0.035	0.07	0.4	0.01	0.02
0.5	0.035	0.07	0.5	0.06	0.08	0.5	0.105	0.105	0.5	0.02	0.02
0.6	0.035	0.14	0.6	0.06	0.12	0.6	0.14	0.175	0.6	0.02	0.04
0.7	0.14	0.175	0.7	0.1	0.14	0.7	0.175	0.21	0.7	0.06	0.07
0.8	0.21	0.28	0.8	0.04	0.08	0.8	0.21	0.21	0.8	0.04	0.07
0.9	0.245	0.28	0.9	0.04	0.06	0.9	0.035	0.105	0.9	0.03	0.06
1	0.035	0.07	1	0.02	0.04	1	0.035	0.035	1	0.02	0.03
$E(\overline{C_1})$	.064	.099	$E(\overline{C_2})$	036	.058	$E(\overline{C_3})$	.070	.095	$E(\overline{C_4})$	.020	.032

The average values from previous tables represent the average value or final value, in this case, of Hamming distances of each characteristic of each bidder, where the maximum value represents a maximum global valuation of that characteristic and the minimum value represents a minimum global valuation of that characteristic. The following table (Table 10) summarizes the different averages and offers the total for each bidder or the Hamming distances total:

Table 10. Summary of different averages and totals for each bidder. Source: Authors' elaboration

BIDDER 01	MIN	MAX	BIDDER 02	MIN	MAX	BIDDER 03	MIN	MAX
$E(\overline{C_1})$	.080	.118	$E(\overline{C_1})$	.064	.111	$E(\overline{C_1})$	.064	.099
$E(\overline{C_2})$	.033	.053	$E(\overline{C_2})$	.025	.044	$E(\overline{C_2})$	.036	.058
$E(\overline{C_3})$	.048	.073	$E(\overline{C_3})$	.060	.092	$E(\overline{C_3})$	.070	.095
$E(\overline{C_4})$	.011	.023	$E(\overline{C_4})$	.012	.024	$E(\overline{C_4})$	.020	.032
Sum	.171	.266	Sum	.161	.271	Sum	.190	.284

Finally, related to the total value (sum of averages) of Hamming distances for each bidder it is necessary to select the best bidder or best option according to the information obtained. In this case, the third bidder is rejected because it shows the greatest distances (min = .190 and max = .284). Therefore, it is necessary to choose between the first and second bidder. The first bidder is the best choice because its maximum value shows the lowest result (.266). Although for its minimum value it does not show the lowest result (.171), the distance between its maximum and minimum value is lower than the second bidder (see Fig. 2).

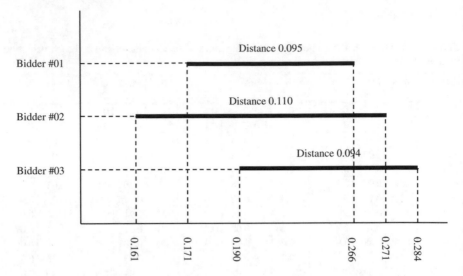

Fig. 2. Sum of Hamming distances of each bidder. Source: Authors' elaboration.

5 Discussion, Conclusion and Implications

According to the objective of this chapter, through a methodological proposal and a practical example, this chapter has presented an "ex ante" alternative to assess the suitability of bidders that may be submitted to a public procurement offer. In this methodological proposal, the use of fuzzy logic in multiple decision-making environments has been laid out, but above all it has focused on the methodology of expertons with the intention of responding to article 150.2 of the PSCL.

The results show that the methodology of expertons is feasible in its application. It is also true that in the end, a process of weighted averages has been used to obtain the final result. Although the use of weighted averages could clash with the concept of the methodology of expertons, this process is valid for three reasons: (1) the methodology of expertons allows facing a multiple decision-making environment with qualitative and subjective factors; (2) it allows analysis of the set of valuations made by a group of experts and (3) it helps them reach a consensus.

As already mentioned above, the method of expertons could not only make it possible to respond to article 150.2 of the PSCL, but it could also reduce discretion in decision making, since this methodology facilitates the selection of a bidder, which must be done by consensus.

Finally, a methodological proposal has been presented here that should not be left in the exposition in this chapter, since it could be applied to other issues that could be investigated, relating to multiple decision making.

References

Bergman, M.A., Lundberg, S.: Tender evaluation and supplier selection methods in public procurement. J. Purch. Supply Manag. **19**, 73–83 (2013)

España. Real Decreto Legislativo 3/2011, de 14 de noviembre, por el que se aprueba el texto refundido de la Ley de Contratos del Sector Público. Boletín Oficial del Estado **276**, 117729–117914 (2011)

Jaskowski, P., Biruk, S., Bucon, R.: Assessing contractor selection criteria weights with fuzzy AHP method application in group decision environment. Autom. Constr. **19**, 120–126 (2010)

Kaufmann, A., Gil Aluja, J.: Técnicas especiales para la gestión de expertos. Editorial Milladoiro (publisher), Vigo (1993)

Lundberg, S., Bergman, M.A.: Tendering design when price and quality is uncertain. Int. J. Public Sect. Manag. **30**(4), 310–327 (2017)

Machacha, L., Bhattacharya, P.: A fuzzy-logic-based approach to project selection. IEEE Trans. Eng. Manag. **47**(1), 65–73 (2000)

Nieto-Morote, A., Ruz-Vila, F.: A fuzzy multi-criteria decision-making model for construction contractor prequalification. Autom. Constr. **25**, 8–19 (2012)

Pamučar, D., Mihajlović, M., Obradović, R., Atanasković, P.: Novel approach to group multi-criteria decision making based on interval rough numbers: hybrid DEMATEL-ANP-MAIRCA model. Expert Syst. Appl. **88**, 58–80 (2017)

Zadeh, L.: Fuzzy sets. Inf. Control **8**(3), 338–353 (1965)

The Approach of the Entrepreneur Microecosystem for University Entrepreneurial Education: Model M2E EMFITUR

Ricardo Hernández Mogollón[✉], Antonio Fernández Portillo,
Mari Cruz Sánchez Escobedo, and José Luis Coca Pérez

University of Extremadura, Cáceres, Spain
ricardoh@arrakis.es, {antoniofp,maricruzse,jlcoca}@unex.es

Abstract. Sometimes one doubts whether entrepreneurial talent can and should be taught or a person is simply born with enterprising talent. However, entrepreneurial Education often lacks a rigorous, long-term, systematic and well-connected approach to the agents of entrepreneurship. This work addresses this issue and its context, and proposes solutions and bases for action. Starting from the Ecosystem Model, a real model of University Entrepreneurial Education (EEU) is provided which is based on a microcosm or open system, called EMFITUR Micro Ecosystem Model of Entrepreneurship, M2E, developed in an unfavorable context and with a double Bottom Up dynamic: on the one hand, Model for the Contribution of Value-Objectives-Content-Action and, on the other hand, Subject/Degree-Faculty-Campus-University and Educational System in general, not necessarily in sequential order.

Keywords: University Entrepreneurial Education · Micro ecosystem approach

1 Introduction

1.1 The Economic Problem

The growing interest that has been awakened by the creation of companies as a solution to unemployment problems (Birch 1979, 1987; Birley 1987; Kirchhoff and Phillips 1988; Storey 1982, 1994; White and Reynolds 1996), to problems regarding economic and regional development (Dubini 1989; Kent 1982; Sexton 1986; Storey 1994; Reynolds et al. 1999, 2000, 2001, 2002) and as a way to promote innovation (Acs and Audretsch 1988) (Hisrich and Peters 1989; Drucker 1964, 1986; Schumpeter 1912, 1942) has led Public Administrations, both in the USA and in Europe, to arbitrate measures and programs to support the creation of new companies and the promotion of entrepreneurship.

At the same time, the study of the creation of companies as a scientific field has been on the uprise since the 1980's. (Brockhaus 1987; Sexton and Bowman-Upton 1988; Hisrich 1988; Stevenson and Harmelin 1990; Bygrave and Hofer 1991; Blenker 1992; Hornaday 1992; Bowmen and Steyaert 1992; Johannison 1992; Nueno 1994, 2005; Veciana 1999; Genescà et al. 2003; etc.). One of the priorities of the current strategy for

© Springer Nature Switzerland AG 2019
J. Gil-Lafuente et al. (Eds.): AEDEM 2017, SSDC 180, pp. 250–275, 2019.
https://doi.org/10.1007/978-3-030-00677-8_21

local and regional development is the promotion of certain territories with competitive capacity where innovation is to be stimulated along with entrepreneurial talent and flexibility in the production system in order to achieve a competitive advantage over other territories (Porter 1991). With this objective, an attempt has been made to combine the development of endogenous resources with the promotion and incorporation of foreign resources and activities.

Traditionally, however, educational systems have generally been characterized by failing to foster entrepreneurial skill and capacity in such a way that students have not acquired the attitudes that make up an enterprising mentality and therefore, upon completing their studies, set their sitghts on working for large firms or in Public Administration.

1.2 Justification for Investigation

There is consensus in the literarura which affirms that the entrepreneurial initiative, or certain aspects of it, can be taught. In this sense, most of the empirical studies conclude that entrepreneurial initiative can be encouraged through the promotion of entrepreneurial spirit in educational programs. Currently, the challenge is to develop an entrepreneurial culture in a more effective way. (Drucker 1985; Klandt 1993; Vesper and Gartner 1997; Gorman et al. 1997; Varela 1997; Veciana 1998; Reynolds 1999; Charney and Libecap 2000; Kuratko 2004; Petrakis and Bourdelitis 2005).

In this same field of investigation, several researchers focus their analysis on when to impart knowledge concerning entrepreneurial behavior, and the opinions of various researchers do not coincide. Researchers such as (Gasse 1985; Fortin 1992; Fillion 1994, 1995; Abella et al. 2000), consider that it is precisely in secondary education where the entrepreneurial potential of students should be encouraged.

Recent research by Weaver, Dickson and Solomon, collected in the report "The Small Business Economy" (2006), on the impact of the promotion of entrepreneurship in general education suggest three key elements: First, there is consistent evidence pointing to a positive relationship between education and entrepreneurial performance; Second, although the relationship between entrepreneurial education and business creation is somewhat ambiguous, there is evidence to show that when the entrepreneurial culture "by necessity" and the entrepreneurial culture "by chance" are considered separately and the differences between countries are taken into account, the relationship is less ambiguous; Third, that the relationship between Entrepreneurship and entrepreneurial intention is not of a linear nature, in the sense that the highest levels of business intention are related to individuals who have secondary education levels and yet this relationship is not found in the case of individuals with higher education. However, the fundamental conclusion which seems to impregnate most of the research in Entrepreneurship is the positive relationship between Education and business activity. (The Small Business Economy 2006).

1.3   Investigation Questions

Does it make sense to develop entreprising attitudes in university students through Entrepreneurial Education as a preliminary step to the identification and exploitation of business opportunities and the creation of companies?

Is it possible to apply the concept of the entrepreneurial ecosystem to University Entrepreneurial Education and to Entrepreneurial Education in general?

1.4   Expected Contribution

We believe that this research work can be useful for all groups involved in Entrepreneurial Training and Skill Building at the University level and at all educational levels, including Business Creation teachers, as it provides sate of the art knowledge for University Entrepreneurial Education (EEU), from the basics of the issue to the methodological combination best suited to this specific type of training, including a fundamental, operational model.

1.5   Methodology

In order to reach the proposed objectives in this work, different variants of the scientific method are addressed:

- Analytical—a synthetic method for the review of the literature.
- Hermeneutics
- Systemic methodology, in the formulation and management of the theoretical framework and in the proposal of the model that leads to the conclusions.
  - Need to flee from the search for an absolute and perfect system.
  - Prior knowing that the model, in the version in question, will always be incomplete.

2   Basis

2.1   The Question at Hand. *the What*

Entrepreneurial activity has become a popular subject in the agendas of many governments as well as in the field of education. One attempt to channel this interest so that new companies appear has been through educational systems of higher education. This comes as a result of the focus in academic literature on the study of the existing relationship between levels of training and the creation of companies (Álvarez and Urbano 2012). Moreover, more and more frequently, the university has been given a role and an entrepreneurial dimension which go beyond the traditional missions of the institution (Clark 1998), based on the process of the commercialization of the technological resources of a university and the creation of value for society (Bueno and Casani 2007).

The research carried out on the incidence of education in entrepreneurship shows that education provides students with a better knowledge of their possibilities for creating a company and for having greater determination in their intentions (Von Grae-venitz et al. 2010). In the specific case of university students, most of the approaches taken to analyze their entrepreneurial initiative have been developed from a psycholog-ical perspective which conceives intention as a predictor of planned, goal-oriented behavior, especially when it it developed over a long period of time (Ajzen and Fishbein 1980; Azjen 1991, 2002).

For all of the above reasons, it seems that entrepreneurial training in the education of university graduates is a relevant and necessary activity that should be offered to all university students, regardless of their degree or postgraduate study. In addition, there is consensus in the scientific literature that one of the important, if not the most important, niches of new entrepreneurs is the university.

Universities are called to be the great agents of change. One way to meet this chal-lenge is through the promotion of entrepreneurship in students.

Social and cultural environments condition behavior and the decisions made by individuals, and therefore will influence the perceptions of the desirability and capa-bility, as well as the final intention of creating a new company (Bruno and Tyebjee 1982; Kent 1984; Burch 1986; Birch 1987; Dubini 1988), as cited by Díaz et al. (2007).

In order to meet society's demand, an entrepreneurial university must behave according to this new role. Within this context, one of the objectives of an entrepreneurial university is to motivate entrepreneurship and support business development.

Public and Private Promotion Policies and Programs in the EEU: Several initiates have been taken on the part of international, national and regional organizations, both public and private, regarding this subject. As an example, in Europe, the Entrepreneurial Spirit 2020 Plan of the European Commission (2013), has the objective of structuring and putting into action the guidelines on what should be done to promote entrepreneur-ship in the EU, a plan which is contained in the Green Book of Entrepreneurship in Europe (2003).

In the case of Spain, as noted by Batista et al. (2016), *Law 14/2013 in support of entrepreneurs and their internationalization* has placed special emphasis on aligning the education system with the requirements of an entrepreneurial society, which aspires to be more flexible and adaptive, following the guidelines that the European Union has been dictating since the publication of the Delors White Paper (2003), updated and completed in the Entrepreneurial Spirit 2020 Plan (European Commission 2013).

Educational reconversion. Approximately one year ago, Lawrence Summers (2016), former secretary of the American Treasury and former rector of Harvard University, gave a conference in New York on the future of education: "It is probable," he declared before the masters of the city, "that In the next twenty-five years we will see more changes than in the last seventy-five." The world is rapidly changing and with it the expectations that the economy places on the education system. Society is moving towards an extremely flexible and competitive global labor model, whose emphasis will be placed on a series of skills - cognitive and non-cognitive - different from those that were fundamental in the past century. Summers acknowledged that it is more difficult "to reform an academic curriculum than to move a cemetery." However, he ventured to outline a series of measures that should be implemented in the classrooms to move towards the school of the future.

The Charleston College Case, *Charleston Changing the face of education*. There are several national and international movements that defend a change in the educational system, such as the one led by Jimmy Freeman (Berry 2012), at the College of Charleston (South Carolina, USA), which strongly criticizes the current system, which, according to them, prepares students for the past, instead of preparing them for the future. This group defends the idea that a school should give a differentiated treatment to its students, seeking to promote creativity, independence, collaboration, critical spirit and the ability to connect. The concept of customization should be applied to education.

2.2   Understanding Where the World Is Going. *The Why*

Many authors agree that the world economy is moving towards a new model, which could be called the Entrepreneurial Economy, in which small and medium-sizes businesses and the enterprising spirit play a central role (Birch 1979 and 1987). Thus, the universal, transversal and exponential impact of ICT (information and communication technology) on business models, families and public administration, the extension of globalization, new forms of outsourcing and offshoring, the growing presence of a society of knowledge, make the prosperity of countries and regions dependent on the entrepreneurial phenomenon. According to Audretsch and Thurik (2001), and Thurik, Carree, Van Stel, and Audretsch (2008) due to this approach, the current economy up to the end of the twentieth century can be named a directed economy; while the entrepreneurial economy, according to these same authors, is characterized by the following: Flexibility, Novelty, Originality, Creativity and Diversity.

It is evident that the new society which is being formed and the entrepreneurial economy require enterprising spirit, both in its business and social aspects, along with intrapreneurship which relies on the pillars of entrepreneurial training and capability, and must be present throughout the education systems of countries and regions.

2.3   Entrepreneurial Ecosystem the Entrepreneurial University. *"The Where"*

Entrepreneurial ecosystem. The Program of Scientific Research (Lakatos 1976) for the Creation of Companies, following Veciana (1998, 1999) and taking into account the sociocultural or institutional focus block based on the macro level, tells us that:

- Their surroundings determine the ecology of a company (natural ecosystem).
- What environmental factors cause variations in the creation-disappearance rates of companies?

There is a consensus on the need to apply the Entrepreneurial Ecosystem approach or Entrepreneurial Ecosystem Theory when addressing entrepreneurship issues with rigor and long-term vision.

From the paradigmatic model of the Silicon Valley Ecosystem (Munroe and Westwind 2008 and 2009), to the most recent studies on Neuroeconomics (Giordano et al. 2017), including the works of the Triple Helix (Etzkovitz 1983, 2002 and Etzkovitz et al. 1997, 2000a, 200b, 2007 and 2010), and those of Isemberg (2010, 2011), there is a prominent current of research which is headed in this direction, and very much supported by reality (Fig. 1 and 2).

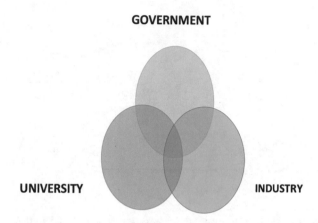

Fig. 1. Model of the triple helix. H. Etzkowitz. Source: Solé Parellada, 2016

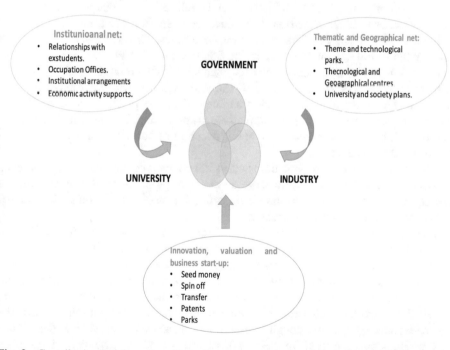

Fig. 2. Contributions of the university to the ecosystem in the triple helix model Source: Solé Parellada, 2016

In Fig. 3 we illustrate how the model of Tappan Munroe based on its study of how the ecosystem of the Silicon Valley is complementary to that of the triple helix and useful to analyze the position of the university in the different elements that make up an innovative ecosystem.

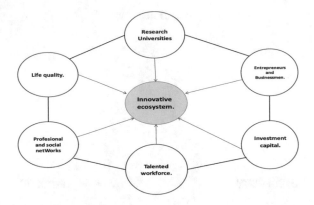

Fig. 3. Elements of the innovative ecosystem Source: Solé Parellada, 2016

The birth of entrepreneurial initiatives can be related to the entrepreneurial ecosystems that occur in different territories because they can offer conditions that allow or inhibit the creation and development of new companies. Therefore, as mentioned in a report of the World Economic Forum on the Education of the Next Generation of Entrepreneurs, it is necessary to know the ecosystems in which entrepreneurship arises, to know what its components are, what their shortcomings are, how they behave and how they interact with each other. This understanding will suggest practices and/or conditions to improve the process, and to plan alternatives to stimulate entrepreneurship and in this way, provide important information to public decision-makers (Wilson et al., 2009), and to private entrepreneurs.

The existence of an entrepreneurial ecosystem which is conducive to the emergence of business initiatives can help to solve a variety of problems which affect a certain territory. In the case of Extremadura, Spain, with 90% rural municipalities, the need for an Ecosystem approach is even greater.

Furthermore, as Solé (2016, pp. 77 and 78) points out, "The university, in addition to its position as agent in the ecosystem, can contribute to improving the other elements to a greater or lesser extent. Universities, as agents of the innovative ecosystem with a research-based and demand-driven model, contribute to the creation of human capital and the attraction of talent. By fostering innovation and the entrepreneurial culture, universities make a significant contribution to the creation and consolidation of the science-technology system through patents, spin-offs and start-ups, such as the science park, which in turn contribute to creating the fabric of productive quality, which puts a powerful global network of science at the service of local networks, and finally, as in the case of MIT, Harvard, Stanford, and Oxford, among others, results in a prestigious global image of the region."

This idea is specified in the driving elements of the model EMFITUR-W2E, which we will see later.

However, it is a mistake to confuse Context, Environment, or External Conditions with Ecosystem, which is something alive, dynamic and with systematic dynamics.

2.4   The Method of Teaching Entrepreneurship. *the How*

According to Varela (2012) what should be provided to the students of EEU is as follows:

- They have an opportunity to fail.
- They allow the mind to measure the consequences.
- They remember that learning, in its structure, should not be equivalent to an appointment at the dentist.
- A change in life's direction is brought about.
- Games and simulations are included.
- It allows for relating concepts to other courses and other learning.
- It goes beyond repeating things to pass an exam.
  - It allows for new forms of learning.
- The teaching work is excellent and dedicated.
- The teacher manages to enthuse them in the subject of the course.

Neck and Green (2011), provide a model which intersects, on the one hand, the figure of the entrepreneur, the Process and Method, and on the other hand, the field, the focus, the level of analysis, the pedagogical elements, language and pedagogical implications.

The Effectual Logic. Sarasvathy (2001, 2008) asks, what is the mental process that entrepreneurs follow when they create new companies? After interviewing 27 successful entrepreneurs, he arrives at the following key elements of the entrepreneur's mind map:

Proven business ideas, creation of a network of trusted stakeholders, decisions, action.

Thus, Sarasvathy (2001, 2008) sees the entrepreneurial process as a set of given means that can be combined in a range of different possible effects (Estrada de la Cruz et al. 2017).

The automobile simile would be the 1st and the 2nd speeds (Sarasvathy et al. 2008), so it only serves the entrepreneur in the start up of his new business or activity. It provides a way to control a future which is inherently unpredictable.

This new approach is disruptive to the traditional way of understanding the entrepreneurial process, which is based on the business plan. As shown in Fig. 4, 3 types of reasoning can be established in companies:

1. Directive Thought: Causal Thought.
2. Strategic Thinking: Creative Causal Thought
3. Entrepreneurial Thought: Effective Thinking.

What is and what is not the Theory of Effectuation: A framework of thought + A set of heuristics, as techniques of inquiry and discovery of opportunities in search of results + Possibilist + How to make the established products and services sellable

What is not the Theory of Effectuation: A system that says what needs to be done + An algorithm + Plan + A way to completely launch a new company.

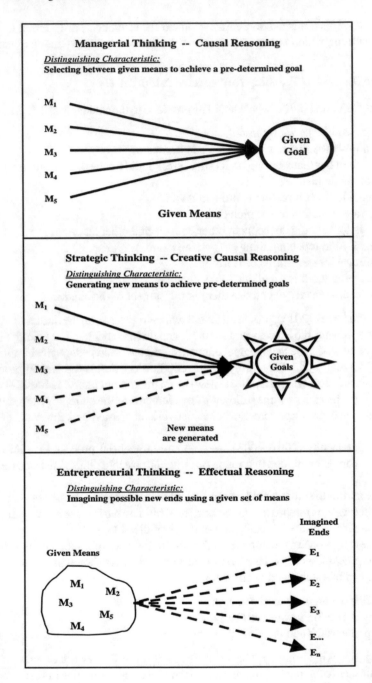

Fig. 4. Different forms of reasoning Source: Sarasvathy (2001)

This theory gives meaning and surpasses the previous ones, and so the working and operational model of EMFITUR M2E is based on this theory.

Teaching models of entrepreneurship. Béchard and Gregoire (2005, 2007), Fayolle and Gailly (2008), cited in Nabi et al. (2015), establish three patterns found in entrepreneurial teaching models:

- Offer Model: behaviorist paradigm (behaviorist). Transmission and reproduction of knowledge, application of procedures.
- Demand Model: subjectivist paradigm. Personalized approach based on exploration, discussion and experimentation.
- Competition model: interactionist paradigm. Students self-organize the available resources to develop skills.

In this subject of the teaching method in entrepreneurial education, one must not fall into the common mistake of confusing managerial training with entrepreneurial training. The first one based on convergent thinking, while the second is based on creative, divergent/convective thinking (Varela 2001, 2010).

One line of work has been developed by Nabi, G.; Liñán, F.; Fayolle, A.; Krueger, N.; Walmsley, A. (2015) and measures the impact of Entrepreneurial Training in Higher Education

2.5   The Entrepreneurial University. *the Who*

The necessary change in the social role of universities is based on the objective importance of the universities as dynamic elements, the importance of their connection with the environment for the university itself and the challenge in assuming a Model of Value Contribution (MAV) and an organization which allows it. (Solé 2016).

The University Entrepreneur Model by Clark (1997, 1998) and Etzkovitz (several).

This American researcher has carried out an in-depth study on the quality of universities in Germany, Great Britain, France, USA and Japan and identifies the factors that lead to excellence in these universities. According to Clark, the five key factors are the following:

1. Solid core management.
2. A plan with a defined objective.
3. A diversified funding base in both public and proven fields.
4. Adequate incentives to stimulate the culture of research and transfer/valorization.
5. An assumed entrepreneurial culture that extends to the entire organization.

The universities that fit the five principles are called "entrepreneurial universities" (Solé 2016). Etzkovitz (1983, 2002, 2003), Etzkowitz, H. and Leydesdorff, L. (1997, 2000); Etzkowitz, H., Webster, A., Gebhardt, C. and Cantisano Terra, B.R. (2000); Etzkowitz H, Zhou C. (2007); Etzkowitz, H, Solé Parellada, F. Pique, J.M. (2007), Etzkowitz, H. and Ranga, M. (2010) and they specify their commitment to the territory. Thus, an entrepreneurial university would be one that meets the following criteria:

1. That it is based on good research with teams that work in a similar way to real companies and have access not only to competitive public investment funds but also to private ones.
2. It produces basic quality research, but with market potential.
3. That it has developed organizational mechanisms that transfer research to the market in terms of protected industrial property (technology transfer and patent offices).
4. That can create companies within the university.
5. That achieves the integration of academic and business elements through new operational units such as the Mixed University - Business Centers in the buildings which are located on the university campus.
6. At this point it is appropriate to point out an aspect of great importance to understanding the evolution of a university towards the model of entrepreneurial university (Clark 1997, 1998) which it is the birth and consolidation of Research Groups (Solé Solé Parellada, F.: The contribution of the university to economic and social development as an object of study in the economic and management sciences. Ed. Royal Academy of Economic and Financial Sciences. Barcelona 2016). These groups are the ones who obtain the funds for research either from competitive public bids or from companies. These organized groups are formed mainly by professors, teachers and entrepreneurial researchers.

3 Micro Entrepreneurial Ecosystem Model EMFITUR, M2E

Conceptual model. This model is part of the bottom up philosophy, and its nature is the ecosystem in its micro version, and it has a catalytic and driving purpose. Its origin dates back to 1992, with the implementation of the first subject of Business Creation that is taught, as an elective, at the University of Extremadura, Spain.

As we pointed out at the beginning of this paper, the final goal of this work is to share our experience, with the idea that it can be useful to different interest groups working in entrepreneurial capacity and training.

All simplification is reductionist. The only pretense that guides us in this part of our work is to provide a teaching model of entrepreneurial initiative which is scientifically based, and inseparably connected to another model which is operational and dynamic, based entirely on research and on the available scientific knowledge. This teaching model has been tried and has evolved over several years owing to our own experience, and with the contribution, although not exhaustive, of the good practices of some universities and companies that stand out in this field.

We will do this by mixing the levels: macro (the context in all its aspects); meso (the educational system, the University, especially public, the research group, the entrepreneurial ecosystem, the entrepreneurial micro ecosystem, the business ecosystem); and micro (the student, the teacher, the classroom, course, degree). All the elements relevant to the entrepreneurial initiative contain these 3 levels, especially in the field of education.

In times where many different attempts are being made to integrate the training and capacity building of the entrepreneurial initiative in the University and in the entire education system in general, it seems appropriate to model, generate micro-ecosystems of entrepreneurship, minitecnopolis, that favor facing this real challenge for the university and educational system in general with guarantees of success because, to a large extent, this approach could well be applied, not only to the university environment, but to the entrepreneurial formation in primary, secondary, college, vocational and private education. We believe this contribution is particularly relevant, since, as Solé (2016) points out, "Some of the inefficiencies observed in the management of the university organizations are derived from the slowness in the perception of the change and the slowness in the application of the remedy according to the logic of the science of management."

This evidence contrasts with what an organization of knowledge is and functions within a society of knowledge (North and Pöschl 2003).

The model proposed below is a well-founded model, and results from several kinds of sources, all of which we understand to be relevant for our intended purpose. The following are cited, not exhaustively, and classified into several groups:

Theories and Basic Models:

Theories:

- Theory of the Ecosystem: Munroe and Westwind (2008, 2009); Isemberg (2010, 2011), Wilson et al. (2009).
- Theory of Neuroeconomics: (Giordano et al. 2017).
- Level of Planned Behavior and Reasoned Action: Krueger (1993), Krueger and Brazael (1994), Krueger and Casrud (1993).
- Theory of Effectuation: Sarasvathy (2001, 2008), Sarasvathy, Dew, Read, & Wiltbank (2008), Sarasvathy and Venkataraman (2011).
- Theory of the Triple helix: Etzkovitz (1983, 2002) and Etzkovitz et al. (1997, 2000a, 2000b, 2007 and 2010).
- Network Theory

Models:

- Entrepreneurial Education Models: Béchard and Gregoire (2005, 2007), Fayolle and Gailly (2008), cited in Nabi, Liñán, Fayolle, Krueger, Walmsley (2015), Neck and Greene (2011).
- Pedagogical Models: Naumes and Naumes (2014); Whitehead (2008).
- University models: Solé (2016), Clark (1998), Eztkovitz (op.cit.).

Collateral theories:

- Theory of Resources and Capabilities: Aaker (1989), Barney, J.B. (1991), Grant, R. G. (1991), Hitt, Hoskisson, Ireland (2001), Penrose (1959), Rummelt (1991) and Wernerfelt (1984).
- Theory of Stakeholders: Friedman and Miles (2006); Harrison and John (1994); Ricardo (1917); Peteraf (1993).
- Theory of Institutional Economics: North (North 1984a, 1984b, 1990, 1991, 1992, 1993, 1994, 2005).
- Theory of the New Institutionalism: Selznick (1948, 1992), Broom and Selznick 1955).

- Theory of the Open System: Scott (1987, 1995, 2014).
- Theory of the management of knowledgeKnowledge Management: Bueno and Salmador (2000); Hernández (2003); Nonaka and Takeuchi (1995); Nonaka, Takeuchi and Umemoto (1996).

Education system. Legal Framework:

- Worldwide Institutions: OECD (2005), UNESCO, WEF (2009).
- European Commission: (1993, 2003, 2006, 2009, 2013).
- Spain: LOCE (2002), LOMCE (2013); Law 14/2013 in support of entrepreneurs;
- Extremadura: Education Law of Extremadura (2011).
- University Regulations: Bologna: ANECA (2009); Alonso, L.E., Fernández Rodríguez, C.J., Nyssen, J. M. (2009); Pulido Trullén, J. I. (2008): VERIFICA.
- Internal regulations of the University of Extremadura (various provisions).

Evidence, Reports, studies:

- GEM Report: http://Gemconsortium.org.; www.emturin2020.gemextremadura.es
- GUESSS Report: http://www.guesssurvey.org. www.emturin2020.gemextremadura.es.

Good practices: Here we have selected universities and experiences that resemble the concepts of Entrepreneurial University and Entrepreneurial Ecosystem, previously studied (Fig. 5).

- Miguel Hernández University, Elche, Spain: Gómez (2013, 2014); http://umh.es/.
- Technological Institute of Monterrey, Mexico: http://www.itesm.mx/
- University of Guadalajara, Mexico: www.udg.mx.
- Universidade da Beira Interior, Portugal: www.ubi.pt
- ICESI-Rodrigo Varela: https://www.icesi.edu.co/.
- University of Mondragón: http://www.mondragon.edu/es.
- Babson College: www.babson.edu/.
- Ecosystem W: http://conectoride.com/.
- Business Institute, Business School: www.ie.edu/business.
- *The Laboratory* of the Faculty of Economics, Business and Tourism ULPGC: https://www.ulpgc.es.

The concept of institutionalization according to Selznick (1948, 1992) (Broom and Selznick 1955: 238) is applied to this conceptual model.

Thus, institutionalization is a neutral idea which can be defined as "the emergence of an orderly, stable, socially integrated model separate from activities: unstable, freely organized, or narrow techniques" (Broom and Selznick 1955). To which we must add the theories of the "open system" (Scott 1987, 1995, 2014).

In other words, what is intended to be designed and implemented is a stable model, open and accepted by interest groups and society, or, in other words, a legitimized and institutionalized model and, of course, useful. It must be borne in mind that according to the Systems Theory, a closed system tends to disappear, while an open system tends to grow.

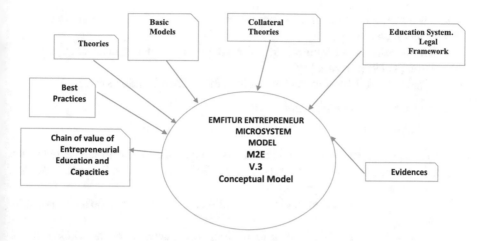

Fig. 5. The emfitur entrepreneur microecosystem model: M2E. conceptual model V.3.0 Source: Prepared by the authors, June 2017

In addition to the above, the conceptual model has been endowed with structure, elements, dynamics, and procedures, and name-entity, following the strategy-structure-operation approach of Chandler (1990).

The Emfitur-M2E model, pursues the <u>following main objective:</u> To get the students to discover their entrepreneurial qualities, consider them valuable, thanks to the simulation they can act in situations of uncertainty like those they would find in real life and standardize behaviors by transforming them into habits, and finally, they learn that entrepreneurial behavior has to be associated with humanistic values (Solé 2016).

In short, help them discover that they are entrepreneurs.

<u>With the following co-objectives:</u>

1. To provide the micro ecosystem with:
    i. Stability over time, continuity.
    ii. Structure and unique elements,
    iii. Dynamics,
    iv. Ente + entity
2. Corporate recognition, with the following dynamics of progression:
    Subject—degree—campus—University, and influence over strategic changes throughout the education system, and at the geographical (regional) level.
3. This implies, above all, overcoming all kinds of barriers.
    Why? Because of everything explained so far in this work, because a change is being brought about in the nature of the observed reality, in the teaching of entrepreneurship at university due to several important reasons:
    - The abundant amount of research, stocks and the flow of knowledge about the initiatives and the entrepreneurial intentions of countries, regions, and university students.
    - The change in the students, who, more than previous generations, are more oriented toward a Milenian profile.

- The existence of movements for disruptive change in the education system.
- General consensus:
- The benefits to be gained and the need to train and stimulate entrepreneurs, through entrepreneurial skills.
- Need to work long-term and throughout the entire education system.

Operational and Dynamic Model V 3.0

Characteristics:

- The Entrepreneurial Ecosystem, in its micro version, open and well-connected to the outside world, business reality, and research. Light ecosystem, well-connected to the existing or emerging ecosystem.
- Not coppied, but adapted to the specific context and local and regional culture, void of "siliconization".
- Fundamental Theory: Theory of Effectuation, which determines the whole process and the micro ecosystem itself.
- Open innovation.
- More closely identified with the models of the Tecnopolis University than with those of the Matricial and Convencional Modern Universityies, which are still dominant in the Spanish university system.
- Management model of diversity and complexity, in line with the descriptions and designs of the Tecnopolis University.
- Intensive use of Communication and Information Technology.
- Intervention of multiple agents.
- The proposed model would be a mixed model of Demand and Competence, according to the classification of Béchard & Gregoire (2005, 2007), Fayolle & Gailly (2008), cited above, since it is behavioral and at the same time interactionist, and applies a combination of methodologies.

The Fundamentals of the EMFITUR-M2E Model, in addition to the Conceptual Model, described above, are comprised of the following three components:

1. Concrete Economic-Social Environment. This is very relevant to the strategy, structure and operation of the Model.

Context:

- Extremadura, in the year 2016, was the Spanish region with the lowest GDP per capita, of only 16,369 euros, 31.71% below the national average in Spain, and 50% less than the region of Madrid (INE 2017).
- Extremadura has an unemployment rate of 28.3% of the active population (4th quarter of 2016), higher than the national rate, with the Autonomous Community having the highest percentage of unemployment in Spain (INE 2017). Unemployment of persons under 25 in that period was 52%.
- Business density does not reach 60 companies per 1,000 inhabitants, compared to 70 at the national level in Spain, with very few medium and large companies.
- Also, as shown in Table 1, it is an eminently rural region:

<u>Transactional University of Extremadura:</u>

- Public University. (www.unex.es)
- There is no model, strategy or ecosystem of Entrepreneurship, neither at a national or regional level, nor at the University of Extremadura.
- Degree: Business Administration.
- Type of subject:
  - elective: 6 credits.
  - Data concerning the number of enrolled (course 2016-2017): 79 with waiting list. Ideal number of students according to the methodological combination to be employed: 20-25.

Being the University where the model under study is generated and applied, the University of Extremadura, Spain, a public, generalist university, and unique in its region, we understand that it entails a territorial commitment which seeks a balance between research and professionalized teaching and that both must be directed toward territorial needs.

At present, the University of Extremadura, like the vast majority of Spanish public universities, can be assimilated to a model of Modern Conventional University, which is far from resembling the Tecnopolis University (Solé 2016) or the Entrepreneurship University model of Clark (1997, 1998) and Etzkovitz (1983, 2002, 2003), but it does have elements on which to base an evolution to the aforementioned models.

2. EMTURIN research group

The corporate framework that supports the genesis and application of the EMTIFUR-M2E.

Model, is the Emturin Research Group (GI Emturin) (http://www.emturin2020.gemextremadura.es/) specialized in Entrepreneurship and with several live and well-connected programs, in addition to research in entrepreneurship, and a vision-mission-operation which is similar to the model of the Entrepreneurial University (Clark 1997, 1998). Synergies

3. Experience in the Business Creation Course (CDE)

The remote origin of the EMFITUR-M2E Model dates back to the year 1992, with the implementation of the first subject of Business Creation, which was taught in the Faculty of Economic and Business Sciences, at the campus of Badajoz, Spain and is currently taught (March 2017) in the Faculty of Business, Finance and Tourism, Degree in Business Administration, as an elective, at the University of Extremadura, Spain. Therefore, the program boasts 25 years of practical experience (Fig 6):

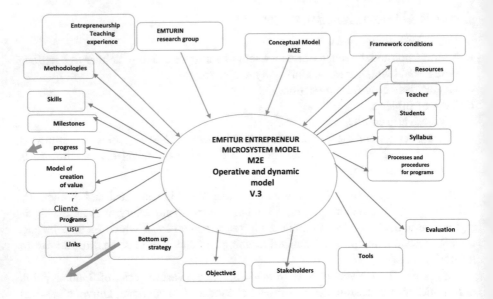

Fig. 6. The emfitur entrepreneur microecosystem model: M2E. operating model V.3 Source: Prepared by the authors, June 2017

EMFITUR-M2E Operational and Dynamic, components of the Model:

1. Model of Value Contribution	10. The Student
2. Objectives	11. Resources
3. Stakeholders	12. Common Errors
4. Instruments	13. Skills-Competencies
5. Teaching Methodologies	14. Milestones, rythm of the course
6. The Teacher	15. Connection-Progress
7. Progress Tests	16. Brakes-Barriers
8. Processes and Procedures by Programs	17. Programs
9. Syllabus	18. Promotive Strategy
	19. Evaluation of the Model

4   Conclusions and Recommendations

A summary of the content of the work carried out, the development of the research questions posed and the specific objectives detailed at the beginning of this article lead us to present the following conclusions. As a synthesis or exponent of the contribution obtained and offered in this paper, we will make 2 blocks of conclusions:

(a) General Conclusions on University Entrepreneurial Education (EEU):

Regarding the research question posed, does it make sense to develop entrepreneurial attitudes in university students, through Entrepreneurial Education, as a previous step to the identification and exploitation of business opportunities and the creation of companies? Through the work we have carried out, we have arrived at the conclusion that the answer is yes, absolutely. Nevertheless, it is necessary to determine *the how* (the method to be followed) and *the when* (at what level of education).

1. There is a wealth of knowledge, of scientific and dynamic basis, generated by the large international scientific community working on entrepeneurialship, in good practice cases, and periodic reports that monitor entrepreneurial activity, as well as programs and public policies, foundations and other entrepreneurship agents that should be used in the present society of knowledge to improve the reality of entrepreneurship at local, regional and national levels, always starting with the education system itself and at all levels.

2. It is increasingly clear that the entrepreneurial initiative, or certain aspects of it, can be taught.

3. A strong reason to address the implementation of entrepreneurial education throughout the education system, and especially in the university system, is the transition that the world economy is experiencing, moving from an economy directed toward an entrepreneurial economy, in an environment of a Society of Knowledge which requires knowledgeable workers.

4. There is a positive relationship between Entrepreneurial Education and Entrepreneurship, although there is a need for more empirical studies that relate both issues.

5. There are indicators of entrepreneurial activity with a scientific basis, such as the GEM project, or the GUESSS project which provide data, information and knowledge about the entrepreneurial phenomenon and its context, in a harmonized, comparative and periodic form, including the education system and, especially, post-secondary education, which can help improve decision-making concerning entrepreneurial education in different countries and regions.

6. The design of the educational spaces can be a decisive factor in the development of the entrepreneurial intention and should be conducive to creating feelings that motivate and emotions that overcome barriers, create positive attitudes, and above all, support the encouragement of positive mental states.

7. The initiatives for promoting entrepreneurialship and its formal training are becoming greater and better all time, both at the public level and in the private sector.

8. In the case of public education, these initiatives come from international, national, regional and local institutions.

9. A common error in practice is to apply convergent thinking in entrepreneurial training rather than creative or divergent thinking, at least in the initial stages.

10. The advisable thing is to use a methodological combination in entrepreneurial training, and teachers with a specialized profile, all within an entrepreneurial ecosystem.

11. The EMFITUR model advises to start with the Skills, with a very practical, yet challenging approach for the student with a systematic plan and procedure which engages the interest of the student and facilitates training and encourages their entrepreneurial skills to stand out.

12. The University should reflect on the strategic of its model of entrepreneurship, or lack of a set strategy, and facilitate training for entrepreneurship in all the courses to be offered. The ideal platform for developing micro-ecosystems of entrepreneurship within universities could be research groups.
13. Research on intentions clearly demonstrates that research groups are the best means of predicting planned behavior.
14. The teachers and trainers in the EEU are key elements and must have entrepreneurial attitudes and aptitudes.
15. The EEU may be the main innovation yet to be incorporated in the University, especially public universities.

On the second research question: Is it possible to apply the concept of an entrepreneurial ecosystem to University Entrepreneurial Education and to Entrepreneurial Education in general? The EMFUTIR-M23 Model is provided as a foundation based on this question.

(b) Conclusions derived from the EMFITUR Entrepreneurial Microecosystem Model, M2E:

This model, which is original and experienced, establishes University Entrepreneurial Education from the approach of the ecosystem, which represents an innovation in this educational field.

1. Since it is necessary to work in the long term and on a systematic basis, the EMFITUR-M2E Model is provided, both in its Conceptual version and in its Operational and Dynamic version 2.0, which is original and adaptable.
2. Double bottom-up dynamics: A model for the contribution of value-objectives-contents-action on the one hand and, on the other hand, Subject/Degree-Faculty-Campus-University and the Educational System in general, although not in sequential order.
3. From M2E, the importance of an Ecosystem approach is derived:
   a. The proposed Model is not possible if you do not have an Ecosystem, if there is no chain equilibrium.
   b. The relationship between education and business is the weak chain.
   c. The relationship between education and public programs and private programs is also necessary in the model and is usually not very strong or stable in reality.
4. An approach that moves from skills to attitudes, from ordinary skills to dynamic or adaptive capacities is necessary.
5. The Model is an organization that learns, and is open to learning.
6. The student must be integrated into the Microcosm of the Entrepreneurial Ecosystem (M2E).
7. Several pitfalls of University Entrepreneurial Education must be avoided:
   a. Working without a long-term model
   b. Siliconization (copying the Silicon Valley model without adapting it to the specific context).
   c. Priority or exclusive use of the Business Plan.
   d. Egosystem Vs Entrepreneurial Ecosystem.

e. It is not just another subject or course material. This must be well understood and work must focus on this so that it is understood and assumed by the "educational system" as a whole.

f. Inadequate teaching staff; demonstrated commitment to Entrepreneurship is required on the part of the teacher.

g. Confusing and erroneously applying entrepreneurial thinking (divergent-convergent thinking, the patchwork quilt) Vs management, managerial thinking (convergent thinking, the puzzle).

h. Failing to consider the context of entrepreneurship.

i. Ignoring research and periodic reports on Entrepreneurship.

j. It would be good to see training for entrepreneurship through the prism of an ecosystem. EMFITUR-M2E, can serve as inspiration and help for this purpose.

8. Solving problems in one corner of the world can be the solution to a global problem.

9. It is a Smart model, experienced, adaptable to other degrees of the University, and to other educational levels, and it is scalable.

References

Aaker, D.: Managing assets and skills: The key to a sustainable competitive advantage. Calif. Manag. Rev. 31(2), 91–106 (1989). Cmr.ucpress.edu

Abellá, S., from Luis, P., Pérez, M.: "Business training and gender". VII Conference on Critical Economics. Albacete, February 2000

Acs, Z.J., Audretsch, D.B.: Innovation in large and small firms: an empirical analysis. Am. Econ. Rev. 78(4), 678–690 (1988)

Ajzen, I.: The theory of planned behavior. Organ. Behav. Hum. Decis. Process. 50(2), 179–211 (1991)

Ajzen, I.: Perceived behavioral control, self-efficacy, locus of control, and the theory of planned behavior. J. Appl. Soc. Psychol. 32(1), 1–20 (2002)

Ajzen, I., Fishbein, M.: Understanding Attitudes and Predicting Social Behavior. Prentice Hall, Englewood Cliffs, NJ (1980)

Alonso, L.E., Fernández Rodríguez, C.J., Nyssen, J.M.: The debate on competencies. Qualitative research on higher education and the labor market in Spain. National Agency for the Evaluation of Quality and Accreditation (2009)

ANECA: Support guide for writing, putting into practice and evaluating learning outcomes. National Agency for the Evaluation of Quality and Accreditation (2009)

Audretsch, D.B., Thurik, A.R.: Capitalism and Democracy in the 21st Century: From the Managed to the Entrepreneurial Economy. Springer, New York, NY (2001)

Barney, J.B.: Firm Resources and continued competitive advantage. J. Manag. 17(1), 99–120 (1991)

Batista, R.M. Fernández-Laviada, A., Medina, M.d P., Esteban, N. N.; Rueda, I.; Sánchez, L.: Education in entrepreneurship. In: Fernández-Laviada, A., Peña, I., Gerrero, M. and González-Pernía, J.L. Global Entrepreneurship Monitor: GEM Spain 2014 Report. Editorial of the University of Cantabria (2015)

Batista-Canino, R.M.: The Process of Business Creation in Cases of Creation of Companies in Spain, pp. 15–18. McGraw Hill, Madrid (2011)

Batista-Canino, Rosa: Report on "Teaching in entrepreneurship in the non-university education system: Recommendations for the Canary Islands." Applicant: General Directorate for Planning, Innovation and Educational Promotion, Ministry of Education, Universities and Sustainability. of the Canary Islands (2014)

Birch, D.G.W.: The Job Generation Process. MIT Press, Cambridge, MA (1979)

Birch, D.G.W.: Job Creation in America. The Free Press, New York (1987)

Birley, S.: New ventures and employment growth. J. Bus. Ventur. 2(2), 155–165 (1987)

Clark, B.: Creating Entrepreneurial Universities in Europe. 19th Forum EAIR. Warwick (1997)

Clark, B.: Creating Entrepreneurial Universities: Organizational Pathway of Transformation (Issues in Higher Education). Pergamon, UK (1998)

Coduras, A., Levie, J., Kelley, D., Saemundsson, R., Schott, T. and GERA: Global entrepreneurship monitor special report: A global perspective on entrepreneurship education and training gem consortium. Babson College, Wellesley, MA (2010)

European Commission: Green Book of the Entrepreneurship in Europe (2003)

European Commission: Growth, competitiveness, employment. The challenges and ways forward into the 21st century. Bulletin of the European Communities 3/93 (1993)

European Commission: Commission staff working document: Towards a European qualifications framework for lifelong learning. Bulletin of the european communities (2005)

European Commission: Recommendations of the european parliament and the council of 18 December 2006 on key competences for lifelong learning (2006)

Official Journal of the European Union 30.12.2006. Available at: http://eurlex.europa.eu/LexUriServ/site/en/oj/2006/l_394/l_39420061230es00100018.pdf

European Commission: Entrepreneurship education at school in Europe. National strategies curricula and learning outcomes (2012). Available in: http://eacea.ec.europa.eu/education/eurydice/documents/thematic_reports/135en.pdf

European Commission: Entrepreneurship 2020 Action Plan. Reigniting the entrepreneurial spirit in Europe. DOE, 09.01.2013 (2013). Available at http://eur-lex.europa.eu/LexUriServ/LexUriServ.do?uri=COM:2012:0795:FIN:EN.Pdf

Dainow, R.: Training and education of entrepreneurs: The current state of the literature. J. Small Bus. Entrep. 3(4), 10–23 (1986)

Díaz Casero, J-C., Hernández, R., Roldán, J.L.: A structural model of the antecedents to entrepreneurial capacity. ISBJ, Int. Small Bus. J. 30(8), 850–872 (2012)

Diaz-Casero, J.C., Ferreira, J.J., Hernández, R., Raposo, M.L.: Influence of institutional environment on entrepreneurial intention: a comparative study of two countries' university students. IEMJ Int. Entrep. Manag. J. 8, 55–74 (2012). https://doi.org/10.1007/s11365-009-0134-3

Diego Rodríguez, I., Vega Serrano, J.A.: Education for entrepreneurship in the Spanish education system. Valnalón. Ed. Ministry of Education, Culture and Sports. Collection Eurydice Spain-Redie (2015)

Do Paço, A., Ferreira, J., Raposo, M., Vaz, J.: How to foster young scientists' entrepreneurial spirit? evidence from scient Project. In the book; XXVI Conference Luso-Espanholas de Gestao Cientificas. Competitividade das Regioes Tranfronteriças. p. 60. Castelo Branco, Portugal (2016)

Drucker, P.: Managing for Results. Harper and Row, New York (1964)

Drucker, P.: Innovation and Entrepreneurship. Harper and Row, New York (1985)

Drucker, P.: Innovation and the Innovative Entrepreneur. Practice and Principles. Edhasa, Barcelona (1986)

Dubini, P.: The Influence of motivators and environment on business start-up: some hints for public policies. J. Bus. Ventur. 4(1), 11–26 (1988)

Dubini, P.: The Influence of motivation and environment on business starts-ups. Some hints for public policies. J. Bus. Ventur. 4(1), 11–26 (1989)

Estrada de la Cruz, M., Gomez Gras, J.M., Verdú Jover, A.J., Alos Simó, l.: The influence of the entrepreneur's social identity on business performance through effective logic. In Rey Juan Carlos (ed.) Strategy, change and business networks. Proceedings 27 congress of Acede. University and Complutense, University of Madrid (2017)

Fayolle, A.: Personal views on the future of entrepreneurship education. ERD, Entrepreneurship & regional development. (2013). http://dx.doi.org/10.1080/08985626.2013.821318

Fayolle, A., Liñán, F.: The future of research on entrepreneurial intentions. J. Bus. Res. 67, 663–666. (2014)

Fillion, J.L.: Entrepreneurship and management: Differing but complementary processes. Eighth Latin American Conference on Entrepreneurial Spirit, Cali, Columbia, 23–25, March 1995

Fillion, L.J.: Vision and relations: keys to success for the entrepreneur. Éditions de l'entrepreneur. Montreal, Qc (1991)

Fillion, L.J.: From Entrepreneurship to Entreprenology. Paper presented at the conference of the United States association of small business and entrepreneurship- USASBE (1997)

Fishbein, M., Ajzen, I.: Belief, Attitude, Intention, and Behavior. An Introduction to Theory and Research. Addison-Wesley, New York (1975)

Fortin, P.: Where we were, where we are: The First Eight CEA Meetings and the Last Four. Cahiers de recherche du département des sciences économiques, UQAM 9215, Université du Québec à Montréal, Département des Sci (1992)

Fostering Entrepreneurship in Europe: Priorities for the Future (European Commission, 1998). Promotion of entrepreneurship and Competitiveness (European Commission, 1998)

Fostering for Entrepreneurship (OECD, 1998)

Freeman, J.: Changing the Face of Education. College of Charleston Magazine. XVI(2), 38–45 (2012). Website: http://www.magazine.cofc.edu

Friedman, A.L., Miles, S.: Stakeholders: Theory and Practice. Ed. Oxford University Press on Demand (2006)

Prince of Girona Foundation: White Paper on Entrepreneurship in Spain. Executive Summary. FPdGI (2011)

Gasse, Y.: A strategy for the promotion and identification of potential entrepreneurs at the secondary school level. Frontiers of Entrepreneurship Research, Babson College (1985)

Genescá, E., Urban, D., Capelleras, J.L., Guallarte, C. and Vergés, J. (coords.): Creation of companies-Entrepreneurship. Tribute to Professor José María Veciana Vergés. Manuals d'Economia. Servei de Publicacions of the Universitat Autònoma de Barcelona (2003)

Giordano, J., Benedikter, R., Flores, N.: Neuroeconomics. An Emerging Field of Theory and Practice-Space (2017). europeanbusinessreview.com

Gnyawali, R.D., Fogel, D.S.: Environments for Entrepreneurship Development: Key dimensions and Research Implications. Entrep. Theory Pract. Summer 18(4), 43–62 (1994)

Gómez, J.M., C. Van-der Hofstadt R.: Competencies and professional skills for university students (2013). books.google.com

Krueger, N., Casrud, A.: Entrepreneurial intentions: applying the theory of planned behavior. Entrep. Reg. Dev. 5, 315–330 (1993)

Liñán, F., Fayolle, A.: A systematic literature review on entrepreneurial intentions: citation, thematic analyses, and research agenda. Springer Science + Business Media, New York (2015)

Hernández Mogollón, R.: Directorate of Knowledge. Theoretical developments and applications. Ediciones La Coria, Trujillo Spain (2003)

Hernández, R., Pérez, P.: An approach to entrepreneurial culture and education in secondary school and in higher-level professional training. IJBE, Int. J. Bus. Envi. 3(1), 120–134 (2010)

Hisrich, R.D., Peters, M.: Entrepreneurship: Starting, developing and managing a new enterprise, pp. 3–23. Boston: Richard D. Irwin, Inc (1989)

Hisrich, R.D.: Entrepreneurship past, present and future. J. Small Bus. Manage. 26(4), 1–4 (1988)

Hornaday, J.A.: Thinking about entrepreneurship: A fuzzy set approach. J. Small Bus. Manage. 30(4), 12–23 (1992)

Iracheta, J.A., Hernández, R., Díaz, J.C., Sánchez, M.C.: The educational space as a factor in the development of the intention to undertake. A theoretical model in the book discovering new horizons in Administration Ed. Esic Editorial. p. 85. Madrid (2013)

Jáuregui, F., Carmona, L., Carrión, E.: 1001 tips to undertake. Ed. Almuzara (2014)

Jáuregui, F., Carmona, L., Carrión, E.: University and employment, instruction manual. Ed. Almuzara (2016)

Johannisson, B.: In search of a methodology for entrepreneurship research. Document presented in RENT VI Workshop. Barcelona (1992)

Juaneda, E., Medrano-Sáez, N., Mosquera, A.: Less + what you want but learn self-learning as a key to student motivation. In the book; XXVI Conference Luso-Espanholas de Gestao Cientificas. Competitividade das Regioes Tranfronteriças. Pág 60. Castelo Branco, Portugal (2016)

Kelley, D., Singer, S. & Herrington, M.: Global Entrepreneurship Monitor 2015/2016 Global Report. Global Entrepreneurship Research Association (GERA). (2016)

Nonaka, I., Takeuchi, H. Umemoto, K.: A theory of organizational knowledge creation. Int. J. Tech. Manage. 11(7–8), 833–845 (1996). inderscienceonline.com

North, D.C.: Structure and Change in Economic History. University Alliance, Madrid (1984)

North, D.C.: Towards a theory of institutional change. Q. Rev. Econ. Bus Perform. through Time 3(4), Winter (1990)

North, D.C.: Towards a theory of institutional change. Q. Rev. Econ. and Bus. Perform. Time 3(4), Winter (1991)

North, D.C.: Institutions and economic theory. Am. Econ. 36(1), Spring (1992)

North, D.C.: "Institutions, institutional change and economic performance". Fund of Economic Culture. Mexico (1993). Original edition: "Institutions, Institutional Change and Economic Performance". Cambridge University Press, Cambridge (UK) (1990)

North, D. C.: Economic performance through time. The American Eco. Rev. 84(3), 359–368 (1994)

North, D.C.: Understanding the Process of Economic Change. Princeton University Press, Princeton, NJ (2005)

North, K., Pöschl, A.: An intelligence test for organizations. In: Ricardo Hernández Mogollón, La Coria, Trujillo (eds.) Knowledge Management: Theoretical Developments and Applications, pp. 183–192 (2003)

Nueno, P.: Undertaking. The Art of Creating Companies and Their Artists. Deusto editions, Bilbao (1994)

Nueno, P.: Undertaking towards 2010. A renewed global perspective of the art of creating companies and their artists. Deusto editions, Bilbao (2005)

OECD: The definition and selection of key competencies. Executive Summary OECD (2005)

Ortega, I.: Obama, Spain and the entrepreneurs. Expansión newspaper of June 27. Madrid (2016)

Ortega, I., Soto, I., Cerdán, C.: Generation z. The last generational leap. Atrevia, Barcelona (2016)

Osborne, D., Gaebler, T.: Reinventing government. Reading, Massachusetts. JSTOR (1993)

Osterwalder, A., Pigneur, Y.: Business Model Generation. Ed. Wiley, USA (2010)

Peng, M., Shekshnia, S.: How entrepreneurs create wealth in transition economies. The Academy of Management Executive 15(1), 95–121 (2001)

Penrose, E.: The Theory of Growth of the Firm. Blackwell, Oxford (1959)

Peteraf, M.A.: The cornerstones of competitive advantage: A resource-based view. Strat. Manag. J. 14(3) (1993)

Petrakis, P.E., Bourletidis, C.A.: Creating a curriculum to teach entrepreneurship in secondary education. IntEnt 2005, Surrey (2005)

Pulido Trullén, J.I.: "Generic skills, what are they?" In generic and transversal competences of university graduates. ICE of the University of Zaragoza (2008)

Raposo, M.L., Ferreira, J.J., Do Paço, Ar., Gouveia, R.: Propensity to firm creation: empirical research using structural equations. IEMJ, Int. Entrep. Manag. J. 4, 485–504 (2008). https://doi.org/10.1007/s11365-008-0089-9

Reynolds, P.D., There is, M., Camp, R.M.: "Global entrepreneurship monitor. 1999 executive report". Babson College, Kauffman center for entrepreneurial leadership, London School Business (Eds.) (1999)

Reynolds, P., There is, M., Camp, R.M.: Global entrepreneurship monitor. 2000 executive report. Babson College, Kauffman center for entrepreneurial leadership, London School Business (Eds.). London (2000)

Reynolds, P., There is, M., Camp, R.M.: Global entrepreneurship monitor. 2001 executive report. Babson College, kauffman center for entrepreneurial leadership, London School Business (Eds.). London (2001)

Reynolds, P., There is, M., Camp, R.M.: global entrepreneurship monitor. 2002 executive report. Babson College, Kauffman Center for Entrepreneurial Leadership, London School Business (Eds.). London (2002)

Reynolds, P., Bosma, N., Autio, E., Hunt, S., De Bono, N., Servais, I., López-García, P., Chin, N.: Global entrepreneurship monitor: data collection design and implementation 1998–2003. Small Bus. Econ. 24(3), 205–231 (2005)

Ricardo, D.: Principles of Political Economy and Taxation. J. Murray, London (1917)

Ries, E.: The lean startup: How today's entrepreneurs use continuous innovation to create radically successful businesses (2011). books.google.com

Rummelt, R.: How much does Industry mater? Strat. Manag. J. 12(3) (1991)

Schumpeter, J.A.: Theorie der Wirtschaftlichen, Entwicklung edn. Verlag Dunker & Humboldt, Munich (1912)

Sarasvathy, S.D.: Cause and affect: Toward a theoretical shift from economic inevitability to entrepreneurial contingency. Acad. Manag. Rev. 26(2), 243–263 (2001)

Sarasvathy, S.D.: Three views of entrepreneurial opportunity. N Dew, SR Velamuri … - Handbook of …, - Springer (2003)

Sarasvathy, S.D.: Effectuation: Elements of Entrepreneurial Expertise. Edward Elgar, Cheltenham (2008)

Sarasvathy, S.D., Dew, N., Read, S., Wiltbank, R.: Designing Organizations that Design Environments: Lessons from Entrepreneurial Expertise. Organization Studies 29(3), 331–350 (2008)

Schumpeter, J.A.: Capitalism, Socialism and Democracy. Ed. Harper & Row 1975, New York (1942)

Senor, D., Singer, S.: Startup Nation: the story of Israel's economic miracle. McClelland & Stewart Ltd, London (2009)

Scott, W.R.: The Adolescence of Institutional Theory. Adm. Sci. Q. 32(4), 493–511 (1987)

Scott, W.Richard: Institutions and Organizations. Sage, Thousand Oaks (1995)

Scott, W. R.: Institutions and Organizations. Ideas, Interests, and Identities Forth. Sage, Thousand Oaks, CA (2014)

Selznick, P.: Foundations of the Theory of Organization. Am. Sociol. Rev. 13(1), 25–35 (1948)

Selznick, P.: Leadership in Administration: A Sociological Interpretation. Quid Pro Books, New Orleans (1957)

Selznick, P.: The Moral Commonwealth. University of California Press, Berkeley (1992)

Sexton, D.L.: Role of entrepreneurship in economic development. In: Hisrich, R.D. (ed.) (nineteen ninety six): Entrepreneurship, Intrapreneurship, and Venture Capital. Lexington Books, Lexington, MA (1986)

Solé Parellada, F.: The contribution of the university to economic and social development as an object of study in the economic and management sciences. Ed. Royal Academy of Economic and Financial Sciences. Barcelona, 2016

Stevenson, H.H., Harmelin, S.: Entrepreneurial management's need for a more chaotic theory. J. Bus. Ventur. 5, 1–14 (1990)

Storey, J.: Impact on the local economy. In: Storey, D.J.: Entrepreneurship and the New Firm, pp. 167–180. Cromm Helm, London (1982)

Storey, J.: Employment. In: Storey, D.J.: Understanding the Small Business Sector, Cap. 6, pp. 160–203. Routledge, London (1994)

Summers, L.H.: What you (really) need to know. The New York Times (2012). stat.wisc.edu

Trulsson, P.: Constraints of Growth-oriented Enterprises in the Southern and Eastern African Region. J. Dev. Entrep. 7(3), 321–339 (2002)

Thurik, A.R., Carree, M.A., Van Stel, A.J., Audretsch, D.B. Does Self-Employment Reduce Unemployment?. Journal of Business Venturing, 23 (6), 673–686. (2008)

Varela, R., Jimenez, J.E.: The effect of entrepreneurship education in the universities of Cali. Frontiers of Entrepreneurship Research (2001). researchgate.net

Varela, R.: Business education based on business skills. In: Varela, R Development, Innovation and Business Culture (2010)

UNESCO and ILO: Career guidance: A resource handbook for low and middle-income countries E Hansen—2006—voced.edu.au. ILO, Geneva, Switzerland (2006). http://www.ilo.org/skills/pubs/WCMS_118211/lang–en/index.htm

Veciana, J.M.: Entrepreneur or Entrepreneur?. Innovating Entrepreneurship Development Newsletter. Icesi University, Cali. Colombia (1997)

Veciana, J.M.: "The Family Business as a program of Scientific Research: Approaches and Current Status". VIIII National Congress of ACEDE. The Gran Canarian palms (1998)

Veciana, J.M.: Creation of companies as a scientific research program. Eur. J. Bus. Manag. Econ. 8(3), 11–36 (1999)

Vesper, K., Gartner, W.: Measuring progress in entrepreneurship education. J. Bus. Ventur. 12(5), 403–421 (1997)

Wernerfelt, B.: A resource-based view of the firm. Strat. Manag. J. 5(2), 171–180 (1984)

Westhead, P.: Exporting and non-exporting small firms in Great Britain. A matched pairs comparison. Int. J. Entrep. Behav. Res. 1(2), 6–30 (1995)

Westhead, P., Wright, M.: Novice, portfolio and serial founders in rural and urban areas. Entrep. Theory Pract. 22(4), 63–100 (1998)

Electronic References

http://www.guesssurvey.org. Retrieved on 25 July 2016

http://Gemconsortium.org. Retrieved on 5 Aug 2016

http://www.fundacionxavierdesalas.com. Retrieved on 5 Aug 2016

http://www.emturin2020.gemextremadura.es. Retrieved on 5 Aug 2016 and 5 Apr 2017

EcosistemaW: http://conectoride.com/. Retrieved on 7 Apr 2017

http://umh.es/.Recuperado. Retrieved on 7 Apr 2017

https://www.icesi.edu.co/. Retrieved on 7 Apr 2017
http://www.mondragon.edu/es. Retrieved on 7 Apr 2017
www.babson.edu/. Retrieved on 7 April 2017
http://www.itesm.mx/ Recovered on 7 April 2017
http://www.udg.mx. Retrieved on 4 Nov 2017
http://www.ie.edu/business. Retrieved on 7 Apr 2017
http://www.dbs.deusto.es Retrieved on 7 Apr 2017
https://www.ulpgc.es Retrieved on 4 Nov 2017
Facebook. The Laboratory feet ulpgc. http://www.ubi.pt. Retrieved on 7 Apr 2017
http://www.ine.es. Retrieved on 30 May 2017
http://www.unex.es. Retrieved on 3 May 2017

Does the Performance of the Company Improve with the Digitalization and the Innovation?

Antonio Fernández-Portillo[✉], Ricardo Hernández-Mogollón,
Mari Cruz Sánchez-Escobedo, and José Luís Coca Pérez

European Academy of Management and Business Economics (AEDEM),
Universidad de Extremadura, Cáceres, Spain
antoniofp@unex.es

Abstract. The use of Information and Communication Technologies (ICT) has allowed for the creation of new business models since the last century. There is a tendency nowadays to digitalize everything that surrounds the company due to the ICT revolution. These actions are generating a great difference in the performance between digitalized and non-digitalized companies, causing great differences between them. Therefore, this paper analyzes how the company's digitalization influences and the innovation its performance, providing a new business performance factor. To do so, the literature on the digital company was reviewed, as well as the business models used by digital companies. In addition, research has allowed us to detect the level of digitalization of the company, its strengths and weaknesses, and even show that as a result of this work, we can conclude that it is necessary to incorporate the digitalization of the company into its performance models, since it represents more than twenty percent of its economic performance.

1 Introduction

In the last decades we have been living the digitalization of everything that surrounds us thus, we are experiencing how the economy and companies are being digitalized. In the case of companies, despite the speed of digitalization, we can still observe the existence of traditional and digital companies, which is why e-business is an alternative to conventional systems (Serarols and Urbano 2007).

At this point, we can consider business digitalization as innovation of the use of e-commerce and e-business. This progress allows the company to be more competitive, provided that technological resources are used adequately (García 2011).

In addition, it is necessary to point out that there are different implementation levels of e-business in European Union countries (Eurostat 2015), which is a problem that generates a limitation when it comes to less digitalized companies competing (García 2011).

Therefore, the interest of this research is justified by the need to know how the digitalization level of companies and innovation affects their performance, focusing on the case of Spanish companies, as this country is in a competitive and globalized economy (Eurostat 2015), located within the European Economic Community.

© Springer Nature Switzerland AG 2019
J. Gil-Lafuente et al. (Eds.): AEDEM 2017, SSDC 180, pp. 276–291, 2019.
https://doi.org/10.1007/978-3-030-00677-8_22

Due to all of the above, the following research question arises: does the digitalization level and innovation of the company influence the company's economic performance? In order to respond to this question, the objective is to verify if the digitalization level of the company and the innovation are determinants factors of its economic performance, for which the present work is structured as follows: first, the theoretical framework of the research; then the field study necessary to try to respond to the objective will be presented, the results will be analyzed and discussed, and finally the conclusions, limitations and future lines of research.

To conclude this section, we must note that this work can be very useful for research, since this may lead to the need to include this items in future business performance models and for entrepreneurs to take it as a key factor of improvement in the competitiveness of their company, being possible to use it as reference in the decision-making of business leaders.

2 Conceptual Framework

The development of ICT has generated a wide range of economic activities in a short period of time. In addition, it is necessary to note that these activities have effects on organizations, causing changes in their structure and functions, improving managerial practices and decisions, productivity and effectiveness, without forgetting that they enable to develop competitive advantages, as well as the simplification of processes and procedures (Ganga and Aguilar 2006; Melville et al. 2004). In addition, ICT in the company helps to generate new business models and markets, such as the exploitation of e-commerce, which enables to diversify traditional channels of access to the market, to provide goods and services to customers (Albornoz et al. 2002). This paradigm makes it possible to use information technology as a means to favor decisions and to serve consumers, thereby gaining significant advantage over competitors.

Continuing with the theoretical review, we will take a step forward that leads us to focus on the concept of digital company and its origin, so we must go back to 1999, when the first authors discussed the digital economy as that economy that makes use of the internet for its performance (Kling and Lamb 1999; Orlikowski 1999; Zimmerman and Koerner 1999). Shortly after that year, we find the first author discussing the digital Company, Slywotzky (1999), understanding it as a company that makes its sales through e-commerce. But we can consider the article by Navas and Breeze (1999) as the most advanced, which deals for the first time with the digitalization of all the administrative and financial systems of the company. On the other hand, the studies performed by Del Aguila et al. (2001), DeLone and McLean (2003), Serarols and Veciana (2003), and Serarols (2003) allow us to focus on the line of research of the digital company, where the determinants for companies in this digital framework are studied. Padilla and Serarols (2006), begin to distinguish characteristics of entrepreneurs and successful companies in digital companies, but focus purely on the digital ones that study the characteristics of the entrepreneur, the environment and the product, leaving aside other also very important factors. On the other hand, Serarols and Urbano (2007) apply part of the model of the study by Padilla and Serarols (2006), but this time making a comparison between digital companies and traditional companies, analyzing the characteristics of the entrepreneur, as well as his professional and family success.

Once the situation of the development of digitalization of the company in Spain has been studied, we will proceed to explain the hypotheses that will be contrasted in this research work, all with the aim of testing, in what way the digitalization of the company influences its economic performance.

Hypothesis 1: The innovation is related directly to the performance of the company.

This hypothesis is based mainly for the works of Despas and Mao (2014), Fernández-Portillo et al. (2015) and on the success model of purely digital companies developed in Serarols´ Doctoral Thesis (2003) and which served as a reference for further research (Padilla and Serarols 2006; Serarols and Urbano 2007).

Following this justification (Padilla and Serarols 2006; Serarols 2003; Serarols and Urbano 2007), we considered it appropriate to also measure the influence of this hypothesis through the "Level of digitalization of the company". This led to the following hypothesis:

Hypothesis 2: The innovation is directly related to the level of digitalization.

As for the justification of the influence of the digitalization of the company on its performance, there are many authors that show the importance of this relationship, but the level of influence is unclear, nor whether all levels of digitalization/application of ICT provide the same performance, so therefore, this study is appropriate (Brynjolfsson and Hitt 2003; Hitt and Brynjolfsson 1996; Powell and Dent-Micallef 1997; Weill and Aral 2006).

Hypothesis 3: The level of digitalization of the company is related directly with the performance of the company.

The hypotheses are shown in the following Figure, together with the theoretical model to be applied in the field work (see Fig. 1).

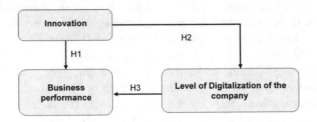

Fig. 1. Research hypotheses. Source: Own elaboration

Once the hypotheses of analysis are presented, we will start the Empirical Framework of this research, which will explain the methodology used in the empirical part and later analyze the research results.

3   Methodology Framework

In this section we will explain the methodology used in this work, so in order to analyze how the different variables influence the performance of the company. Thus, in order to test the hypotheses we have formulated, we used a Multivariate Analysis, all within a Hypothetical-Deductive Methodology.

For this task we used Structural Equations Models (SEM), using the technique of Partial Least Squares (hereinafter PLS) based on the variance. In this case, SmartPLS version 3.2.6 was the tool used.

3.1   Field Study Design

For the study of the field, the sample was considered to be in a homogeneous space, in order to avoid uncontrollable variables (Riquel 2010), therefore commercial companies located in Spain were analyzed. We must mention that the entrepreneurs were surveyed through a digital questionnaire, using mailing and tools provided by Google Drive. In order to obtain the e-mail addresses, it was necessary to use the SABI database, and in this case the mails of the target population of 805,588 commercial companies were obtained, which were reduced to 4041 companies that had their mail available in SABI (Table 1).

Table 1. Population and sample data

Active companies		Sample significance	
Sample	Population (SABI)	Confidence level	Confidence interval
150	805,588	95%	8%

After 4 months of mailing, a sample of 150 surveys was obtained, resulting somewhat small for the objective of the study, but that shed light on the research to be performed.

The questionnaire used was made following the recommendations of Podsakoff et al. (2012), having been developed mainly from previously validated questionnaires such as e-business, e-wach by García (2011) and Digitalization level of the company by Bonnet and Ferraris (2011).

To study the digitalization level of the Spanish company, the study of Bonnet et al. (2015) was followed, which shows that the application of this method should be done by weighting the responses obtained from the set of questions of the questionnaire and according to the results of the study, companies will have either one level of digitalization or another (see Fig. 2). In addition, the variables proposed in this research were measured using a five-point Likert scale, in order to observe and generate ICT adoption profiles in Spanish companies.

To measure the performance of the company, the variables recommended by Serarols and Veciana (2003) were used, such as: EBITDA, financial results and economic and financial ratios, all of them obtained from SABI.

Fig. 2. Level of digitalization of the company according to the score obtained. *Source*: prepared by the authors based on Bonnet et al. (2015).

3.2 Multivariate Analysis

In this section, we will focus on how our study establishes relationships of dependence between variables, measured on metric scales and, as we will see in the next section, there is more than one relationship between the variables that make up the proposed model. Thus, following the studies by Uriel and Aldás (2005) and Jiménez-Naranjo et al. (2016), we understand that the most adequate technique for our research is structural equations, and in addition, structural modeling will be carried out using Partial Least Squares (PLS).

In our case, based on the recommendations of Edwards (2001), the two-step approach will be taken through the latent variable scores (Fig. 3).

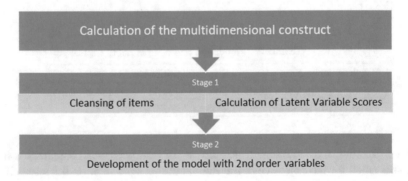

Fig. 3. Two-step approach *Source*: prepared by the authors based on Jiménez-Naranjo et al. (2016), and Roldán and Cepeda (2016).

Once the calculation of the indicators of the 2nd order variable is carried out, the PLS technique is applied, which will lead to the results that we can see in the following section.

4   Results

Following the above methodology, we will proceed to processing the multidimensional variable; then the measuring instrument will be analyzed. Once the validity of the indicators has been determined, we will analyze the proposed structural model where we will test the significance of the hypotheses, in order to contrast them. Finally, we will study the predictive effect of the proposed model, as indicated by the statistical method used.

Then, these first-order factors will perform in the model as the second-order construct they represent (Fig. 4).

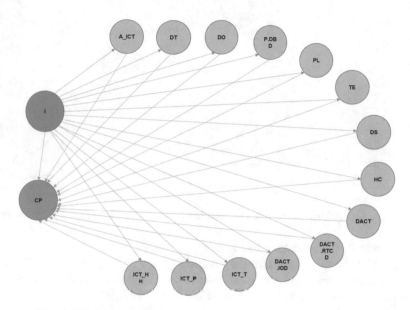

Fig. 4. Model approach in stage 1 *Source*: compiled by the authors.

Once the program to obtain the data has been implemented, and taking into account that the indicators that we will evaluate are reflective, we proceed to evaluate the individual reliability of the indicators of the first order variables. To accept an indicator as a component of a construct, some researchers (Barclay et al. 1995; Chin 1998) suggest that, this indicators with loads between 0.4 and 0.7 can be eliminated from a scale if their elimination leads to an increase in the average variance extracted (AVE), or composite reliability (CR), above the threshold suggested for these parameters (Hair et al. 2011). These weak indicators are sometimes maintained on the basis of their contribution to content validity. The very weak indicators, with values ≤ 0.4, should always be eliminated (Hair et al. 2011), making it necessary to adjust the model and it will be necessary to re-execute the PLS algorithm and obtain new results (Urbach and Ahlemann 2010). To finalize the first stage, the scores of the latent variables are calculated with the model, once the items are cleansed.

In the next stage of PLS analysis, the model is estimated using the latent variable scores obtained by the program for each of the first order components. Next, we will represent the model, including the second-order variable (Fig. 5).

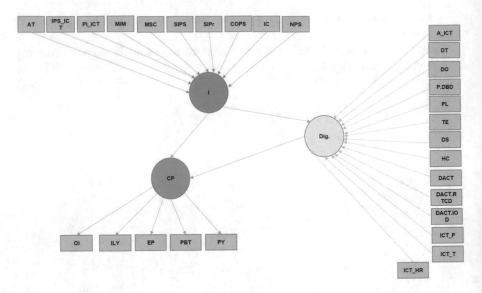

Fig. 5. Latent variables scores with cleansed dimensions *Source*: prepared by the authors.

Evaluation of the measuring instrument

In the first place, we perform the analysis on the validity and reliability of the measuring instruments of the reflective variables. For this we will follow the steps explained in the PLS methodology and which is defined below:

1. Evaluation of the individual reliability of the item.
2. Evaluation of the internal consistency or composite reliability of the construct.
3. Evaluation of convergent validity.
4. Evaluation of discriminant validity.

Next, we will carry out each of the evaluations, showing the required values in a summary table (Table 2).

Table 2. Justification of parametric values

Analysis	Parameter	Values greater than	Justification
Individual reliability	Loads (λ)	0.4	Hair et al. (2014)
Composite reliability	Crombach's Alpha (α)	0.7	Nunnally and Bernstein (1994)
	Composite Reliability (Cr)	0.6	Bagozzi and Yi (1988); Nunnally and Bernstein (1994)
Convergent validity	Average variance extracted (AVE)	0.5	Fornell and Larcker (1981)
Discriminant validity	Compares the AVE with the correlations between constructs	AVE > Correlations	Barclay et al. (1995); Henseler et al. (2009); Hair et al. (2011)

Source: Fernández-Portillo (2016).

Individual reliability is checked by testing the Loads (□) of the indicators with their respective construct. For this purpose, we execute the PLS algorithm and cleanse those items that do not exceed the predetermined values (Hair et al. 2014). This process is iterative. Once the procedure is expressed, the following table shows the load matrix, showing the correlation of each indicator with its construct. This table only shows those indicators that have exceeded the predetermined values of 0.4 (Hair et al. 2014) (Tables 3 and 4).

Table 3. Individual reliability

	Label	Cronbach's Alpha	Composite reliability	Average variance extracted (AVE)
Company Performance	CP	0.770	0.842	0.524

Source: compiled by the authors.

Table 4. Discriminant validity

	Label	Company performance	Innovation	Level of digitalization of the company
Company Performance	CP	0.724	–	–
Innovation	I	0.366	–	–
Level of digitalization of the company	Dig.	0.512	0.675	–

Source: compiled by the authors.

In addition to the above, we carried out the analysis on the validity and reliability of the measurement instruments of the formative variables, which in our model are: "Company performance" and "Level of digitalization of the company"; all following the criteria of MacKenzie et al. (2005).

The first aspect to be analyzed is the existence of multicollinearity of the indicators that make up the formative construct. In order to perform this analysis we will use the VIF (variance inflation factor) indicator, which according to authors like Diamantopoulos and Siguaw (2006), its value should be less than 3.3, and others such as Hair et al. (2011) indicate that values lower than 5 will be valid.

Finally, after carrying out the multicollinearity analysis, we will apply the bootstrapping process, in order to analyze the load-weight relationship and its significance (Hair et al. 2014).

After cleansing the indicators for reflective and formative variables, the model is shown in the following figure, where only those indicators that exceed the values established for initial stages of scale development are shown (Fig. 6).

After the evaluation process of the indicators that estimate the different latent variables of our model, we proceed to evaluate the model as a whole; performing the analysis of the structural model in the next item.

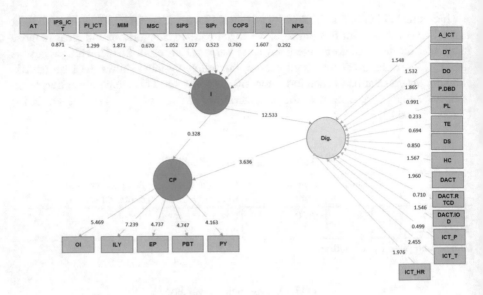

Fig. 6. Final model with cleansed indicators *Source*: compiled by the authors.

Structural Model Analysis

Firstly, since the estimation of the path coefficients is based on OLS regressions, as in a multiple regression, we must avoid the presence of multicollinearity between the antecedent variables of each of the endogenous constructs (Cassel et al. 1999). Following this requirement, according to Hair et al. (2014, p. 170), there will be signs of multicollinearity when FIV > 5 and tolerance levels <0.20 (Table 5).

Table 5. Evaluation of colinearity of constructs

	Label	Company performance	Innovation	Level of digitalization
Company performance	CP	–	–	–
Innovation	I	1.836	–	1.000
Level of digitalization	Dig.	1.836	–	–

Source: compiled by the authors.

Next, by following the manual of Roldán and Cepeda (2016), we will proceed to evaluate the "Coefficient Path" of the hypotheses. They can vary between –1 and 1, so the higher their absolute value, the greater the importance of this hypothesis (Table 6).

Secondly, an analysis of the significance of structural relationships through Boostrapping was carried out using 10,000 subsamples (Hair et al. 2014), so if they are significant there will be empirical support to support the relationships contained in the hypotheses. Next, we must use the critical values of the t-statistic for the one-tailed test, for which we will follow the guideline suggested by Hair et al. (2014). As a result of this process, we will obtain the standard errors, the t statistics and the confidence intervals of the parameters, which will allow us to validate the hypotheses (Roldán and Cepeda 2016) (Table 7).

Table 6. Hypothesis contrast according to "path coefficient"

Hypothesis	Path coefficient	Acceptance of the hypothesis
H1. Innovation - > Company perf.	0.038	YES
H2. Innovation - > Dig. level	0.675	YES
H3. Dig. level - > Company performance	0.486	YES

***p(0.01); **p(0.05); *p(0.1). One-tailed T-Student.
Source: compiled by the authors.

Table 7. Contrast of hypothesis based on its level of significance

Hypothesis	T statistic	Significance
H1. Innovation - > Company perf.	0.328	Non-significant
H2. Innovation - > Dig. level	12.533	***
H3. Dig. level - > Company performance	3.636	***

Source: compiled by the authors.

Henseler et al. (2009) have proposed reporting confidence intervals together with t-values. Confidence intervals have the advantage of being a completely non-parametric approach and are not based on any type of distribution. "If a confidence interval for an estimated path coefficient does not include the value zero, then the hypothesis that it is equal to zero is rejected" (Henseler et al. 2009, p. 306) (Table 8).

Table 8. Hypothesis contrast with confidence levels

Hypothesis	5.0%	95.0%	Acceptance of the hypothesis
H1. Innovation - > Company perf.	0.014	0.373	YES
H2. Innovation - > Dig. level	0.636	0.812	YES
H3. Dig. level - > Company performance	0.289	0.726	YES

Source: compiled by the authors.

The data obtained after contrasting the model indicate the following in relation to the established hypotheses:

H1. It is rejected. It establishes a relationship of the innovation on the performance of the company, which is rejected because it does not achieve the minimum significance. This fact indicates that there is a direct influence, although it is too weak to accept the hypothesis.

H2. It is accepted. This hypothesis indicates the influence of the level of digitalization on the business model, which influences directly; also having a major influence in the proposed model. As for the level of digitalization, the most important indicators are the ICT human resources of the company, closely followed by Technological Experience and Digital Abilities. The result of this relationship indicates that the level of digitalization changes depending on the innovation of the company.

H3. It is accepted. In this case it shows that the level of digitalization of the company influences directly over the performance of the company. When evaluating this hypothesis we must note that it obtains the highest level of significance. This result confirms the theory, where many authors discuss the advantages that ICTs bring to the company, but it is now also empirically corroborated.

Next, the explained variance of dependent latent variables was examined by the constructs that predict them (R2). The required value should not be less than 0.1 (Falk and Miller 1992), although this depends on the study area and context (Sanz et al. 2008). Finally, we performed the analysis of the predictive relevance of the model (Q2) using the "blindfolding" technique. In this case it is necessary to obtain values greater than 0 (Hair et al. 2014). The results are shown in Table 9.

Table 9. Evaluation of the effect of the model

Hypothesis	R2	Q2	Path	Correlation	Explained variance
H1. Innovation -> Company performance			0.038	0.366	2.65%
H3. Level of digitalization of the company -> Company performance			0.486	0.512	24.88%
Performance of the company	0.262	0.086			
H2. Innovation -> Level of digitalization of the company			0.675	0.675	45.56%
Level of digitalization of the company	0.455	0.041			

Source: compiled by the authors.

As for the data obtained, it should be noted that the R2 of the "company's performance" takes a value of 0.262, which is higher than 0.1 (Falk and Miller 1992), and shows a low level $(0.33 < R2 > 0.16)$ (Chin 1998). In addition, we observe that the "level of digitalization of the company" explains 24.88% of the variance in its performance, coinciding with the strong significance shown by hypothesis 3. In addition, R2 of the construct "level of digitalization of the company" is 0.418, obtaining a moderate level $(0.67 < R2 > 0.33)$.

Once we have shown the importance of the value provided by R2, we must emphasize the importance that the predictive capacity of the model acquires through the Q2 value, noting that in relation to this parameter for the company performance construct it has a value (Q2) of 0.086, indicating that the model has predictive relevance (Hair et al. 2014).

In conclusion, we can say that we have developed a model that contains the "level of digitalization of the company" as a factor that influences the "performance of the company".

Based on the model presented with the hypotheses analyzed, the following table shows a summary of the results obtained in relation to the hypotheses for each of the groups of interest that have been analyzed (Table 10).

Table 10. Results of the hypotheses formulated in the model

Hypothesis	Evaluation of the hypothesis	Confidence levels	T statistic	Path coefficient
Hypothesis 1	REJECTED	YES	$p > 0.10$	YES
Hypothesis 2	ACCEPTED	YES	$p < 0.01$	YES
Hypothesis 3	ACCEPTED	YES	$p < 0.01$	YES

Source: compiled by the authors.

By analyzing the results of the contrast of the hypotheses proposed, we can confirm that the proposed model is validated empirically, achieving a contribution to the research under consideration.

5 Discussion, Conclusion and Implications

While conducting this work, we were able to respond to the objectives presented in the introduction, obtaining a series of data that are of interest, which we will conclude with in this section.

Firstly, regarding the review of the state of the art digital company, e-commerce is a fundamental part. In relation to the term e-business, we have proposed the unification of the term and the parts that form it. Finally, it was detected that the concept of digital company arises from the term digital economy.

Secondly and finally, the contrast of the hypotheses shows information that will be decisive in the decision-making of the leaders of companies and public administrations. In this case, there is a slightly positive relationship of the innovation on the performance of the company. This fact shows that, having established this hypothesis with a direct impact, and despite the fact that it is fulfilled, we cannot accept it because it is not significant. For this reason, we cannot say that the innovation influences the performance of the company.

In relation to the influence of the innovation level on the digitalization, in this case we must say that it influences directly, and that it is also the hypothesis with the greatest influence of the proposed model. As for the level of digitalization, the most important indicators are the ICT human resources of the company, closely followed by experience with ICT and ICT skills.

The third and final hypothesis indicates that the level of digitalization of the company directly influences the performance of the company. In addition, this is accepted with a maximum level of significance. This allows us to state that the higher the company's digitalization level, the better the company's performance; thus confirming its improvement in performance.

Finally, we can confirm that the level of digitalization should be part of future performance models of the company, since the structural model has been validated through the analysis, and so has the variance explained by the dependent latent variables, explaining 26.2% ($R2 = 0.262$) of the company's performance, and 45.5% ($R2 = 0.455$) of the digitalization level, results considered to be low ($0.33 > R2 > 0.16$) for the first one and moderate ($0.67 > R2 > 0.33$) for the second one (Chin 1998).

As a final conclusion, we can say that within the factors that influence the performance of the company, the level of digitalization of the company has great importance, since it represents 26.2% of the variance explained in the proposed model.

It is shown that this research topic is current and of great relevance, since companies are being forced to digitalize so as not to lose competitiveness, and for this is necessary to apply the innovation.

5.1 Limitations and Future Lines

We must point out that the main limitation found in this work was collecting the primary data to perform the empirical study, and it may be interesting to increase the sample in future research.

Regarding future lines, it is proposed to include the digitalization level factor in already tested company performance models, and to perform an empirical study to test its behavior together with other business performance factors.

Annex I. Formative Indicators Grouped by Latent Variables, Used in the 2nd Order Model of SmartPLS

Latent variable	Label	Meaning
Innovation (I)	AT	Age of technology used
	IPS_ICT	Innovation of products or services directly related or enabled by ICT
	PI_ICT	Process innovation directly related to or enabled by ICTs
	MIM	Most important market: International, national or regional
	MSC	Market share of the company
	SIPS	Substantial improvement of a product or service in the last year
	SIPr	Substantial improvement of processes in the last year
	COSPS	% of companies offering the same products or services
	IC	% of international customers
	NPS	% of new products or services for customers
Company perfomance (P)	OI	Operating income thousand EURO last year avail SABI
	ILY	Indebtedness (%) % last year avail SABI
	EP	Economic profitability (%) % last year avail SABI
	PBT	Ordinary profit before taxes thousand EURO last year avail SABI
	PY	Profit for the year thousand EURO last year avail SABI

Source: compiled by the authors.

Annex II. Reflective Indicators Grouped by Latent Variables, Used in the 2nd Order Model of SmartPLS

Latent variable	Label	Meaning
Digitalization level of the company (Dig.)	A_ICT	Accessibility of ICT tools and infrastructures
	DACT.RTCD	Access to collaboration data and tools
	DACT	Access to company data and tools
	DACT.IOD	Access to customer data and tools
	HC	High commitment to the digitalization of the company
	PL	Collaborative learning
	P.DBD	Data-based decisions
	TE	Technological Experience
	DS	Digital Skills
	ICT_T	ICT tools available in the company
	DO	Digitalized operations in the company
	DT	Digital Thinking
	ICT_P	Perception of ICT by the employer
	ICT_HR	ICT Human Resources available in the company

Source: compiled by the authors.

References

Albornoz, P., Vergara, E., Failla, F.: Tecnologías de Información en la Pequeña y mediana empresa y el papel del Estado. Tesis, Universidad de Chile. Santiago de Chile, Chile (2002)

Bagozzi, R.P., Yi, Y.: On the evaluation of structural equation models. J. Acad. Mark. Sci. **16**(1), 74–94 (1988)

Barclay, D., Higgins, C., Thompson, R.: The Partial Least Squares (PLS) approach to casual modeling: personal computer adoption and use as an illustration. Technol. Stud. (Special Issue on Research Methodology) **2**(2), 285–309 (1995)

Bonnet, D., Ferraris, P.: Transform to the power of digital: digital transformation as a driver of corporate performance. Digital Transform. Rev. **1**, 14–29 (2011)

Bonnet, D., Deora, A., Buvat, J., Subrahmanyam, K.V.J., Khadikar, A.: Organizing for digital: why digital dexterity matters. MIT Center for Digital Business and Capgemini Consulting (2015)

Brynjolfsson, E., Hitt, L.M.: Computing productivity: firm-level evidence. Rev. Econ. Stat. **85**(4), 793–808 (2003)

Cassel, C., Hackl, P., Westlund, A.H.: Robustness of partial least-scuares method for estimating latent variable quality structures. J. Appl. Stat. **26**(4), 435–446 (1999)

Chin, W.W.: The partial least squares approach to structural equation modeling. In: Marcoulides, G.A. (ed.) Modern Methods for Business Research. Erlbaum, Mahwah (1998)

Del Aguila, A.R., Padilla, A., Serarols, C., Veciana, J.M.: La economía digital y su impacto en la empresa: bases teóricas y situación en España. Boletín Económico de ICE **2705**, 7–24 (2001)

DeLone, W.H., McLean, E.R.: The DeLone and McLean model of information systems success: a ten-year update. Diario De Información De Gestión De Sistemas **19**(4), 9–30 (2003)

Desplas, N., Mao, N.: Análisis paralelo entre e-turismo y e-gobierno: evolución y tendencias. Investigaciones Turísticas **7**, 1–22 (2014)

Diamantopoulos, A., Siguaw, J.A.: Formative versus reflective indicators in organizational measure development: a comparison and empirical illustration. Br. J. Manag. **17**(4), 263–282 (2006)

Edwards, J.R.: Multidimensional constructs in organizational behavior research: an integrative analytical framework. Organ. Res. Methods **4**(2), 144–192 (2001)

Eurostat: (2015). 27 July 2015. Recovered at: http://epp.eurostat.ec.europa.eu/portal/page/portal/eurostat/home/

Falk, R.F., Miller, N.B.: A primer for soft modeling. University of Akron Press, Akron (1992)

Fernández-Portillo, A.: FACTORES DETERMINANTES PARA LA ELABORACIÓN DE UN MODELO DE ÉXITO DE LA EMPRESA EN EL MEDIO DIGITAL. Tesis Doctoral, Universidad de Extremadura. Departamento de Economía Financiera y Contabilidad, Cáceres, España

Fernández-Portillo, A., Sánchez-Escobedo, M.C., Jiménez-Naranjo, H.V., Hernández-Mogollón, R.: La importancia de la innovación en el comercio electrónico. Universia Bus. Rev. **47**(3), 106–125 (2015)

Fornell, C., Larcker, D.F.: Structure equation models: LISREL and PLS applied to customer exist-voice theory. J. Mark. Res. **18**(2), 39–50 (1981)

Ganga, F., Aguilar, M.: Percepción de los proveedores del sistema electrónico "Chilecompra" en la Xª Región-Chile. Enl@ce: Revista Venezolana de Información, Tecnología y Conocimiento. Venezuela **3**(1), 27–48 (2006)

García, M.B.: Factores condicionantes en la adopción del negocio electrónico en la empresa europea. Tesis Doctoral, Universidad Rey Juan Carlos, Facultad de Ciencias Jurídicas y Sociales, Departamento de Economía de la Empresa (Administración, Dirección y Organización), Madrid, España

Hair, J.F., Ringle, C.M., Sarstedt, M.: PLS-SEM: indeed a silver bullet. J. Mark. Theory Pract. **19**(2), 139–152 (2011)

Hair, J.F., Sarstedt, M., Hopkins, L., Kuppelwieser, V.G.: Partial least squares structural equation modeling (PLS-SEM). An emerging tool in business research. Eur. Bus. Rev. **26**(2), 106–121 (2014)

Henseler, J., Ringle, C.M., Sinkovics, R.R.: The use of partial least squares path modeling in international marketing. Adv. Int. Mark. **20**, 277–320 (2009)

Hitt, L.M., Brynjolfsson, E.: Productivity, business profitability, and consumer surplus: three different measures of information technology value. MIS Q. **20**(2), 121–142 (1996)

Jiménez-Naranjo, H.V., Coca-Pérez, J.L., Gutiérrez-Fernández, M., Fernández-Portillo, A.: Determinants of the expenditure done by attendees at a sporting event: the case of World Padel Tour. Eur. J. Manag. Bus. Econ. **25**(3), 133–141 (2016)

Kling, R., Lamb, R.: IT and organizational change in digital economies: a socio-technical approach. Paper presented at the conference understanding the digital economy: data, tools and research, mayo 25 & 26, Department of Commerce, Washington, DC

MacKenzie, S.B., Podsakoff, P.M., Jarvis, C.B.: The problem of measurement model misspecification in behavioral and organizational research and some recommended solutions. J. Appl. Psychol. **90**(4), 710–730 (2005)

Melville, N., Kraemer, K., Gurbaxani, V.: Information technology and organizational performance: an integrative model of IT business value. MIS Q. **28**(2), 283–322 (2004)

Navas, D., Breeze, B.: Scientific-Atlanta integrates its global supply chain and ensures accurate shipments. ID Syst. **19**, 34–39 (1999)

Nunnally, J.C., Bernstein, I.H.: Psychometric Theory, 3rd edn. McGraw-Hill, New York, (USA) (1994)

Orlikowski, W.J.: The truth is not our there: an enacted view of the digital economy. Paper presented at the conference understanding the digital economy: data, tools and research, mayo 25 & 26, Department of Commerce, Washington, DC (1999)

Padilla, A., Serarols, C.: Las características del empresario y el éxito de las empresas puramente digitales. Tribuna de economía. ICE. 833. 155–176 (2006). 7 Aug 2015. Recuperado de. http://www.revistasice.com/CachePDF/ICE_833_155-176__907B253D8894AB5C361D5588E62DCFD6.pdf (2006)

Podsakoff, P.M., MacKenzie, S.B., Podsakoff, N.P.: Sources of method bias in social science research and recommendations on how to control it. Annu. Rev. Psychol. **63**, 539–569 (2012)

Powell, T.C., Dent-Micallef, A.: Information technology as competitive advantage: the role of human, business and technology resources. Strateg. Manag. J. **18**(5), 375–405 (1997)

Riquel, F.J.: Análisis institucional de las prácticas de gestión ambiental de los campos de golf andaluces. Tesis Doctoral, Departamento de Dirección de Empresas y Marketing, Universidad de Huelva, Huelva, España (2010)

Roldán, J.L., Cepeda, G.: Modelos de Ecuaciones Estructurales basados en la Varianza: Partial Least Squares (PLS) para Investigadores en Ciencias Sociales (III Edición). Universidad de Sevilla, España (2016)

Sanz, S., Ruiz, C., Aldás, J.: La influencia de la dependencia del medio en el comercio electrónico B2C. Propuesta para un modelo integrador aplicado a la intención de compra futura en internet. Cuadernos de Economía y Dirección de la empresa **36**, 45–75 (2008)

Serarols, C.: Factores de éxito de las empresas puramente digitales. Tesis Doctoral, Universitat Autònoma de Barcelona, Bellaterra (2003)

Serarols, C., Urbano, D.: El empresario y los factores de éxito. Estudio de casos de empresas tradicionales y digitales. Revista de Contabilidad y Dirección **5**, 139–167 (2007)

Serarols, C., Veciana, J.M.: El empresario digital como determinante del éxito de las empresas puramente digitales: Un estudio empírico. Document de treball núm. 03/4, Departament d'economia de l'empresa, Universitat Autònoma de Barcelona, Bellaterra (2003)

Slywotzky, A.J.: La empresa digital. Gest. **4**(3), 14–15 (1999)

Urbach, N., Ahlemann, F.: Structural equation modeling in information systems research using partial least squares. J. Inf. Technol. Theory Appl. **11**(2), 5–40 (2010)

Uriel, E., Aldás, J.: Análisis multivariante aplicado. Thomson Editores Spain. Paraninfo SA, Madrid, España (2005)

Weill, P., Aral, S.: Generating premium returns on your IT investments. MIT Sloan Manag. Rev. **47**(2), 39–48 (2006)

Zimmerman, H.D., Koerner, V.: New emerging industrial structures in the digital economy-the case of the financial industry. Presented at the 1999 Americas conference on information systems (AMCIS'99), agosto 13–15, Milwaukee (1999)

Strategic Functions Manufacturer-Distributor in Marketing Channels

Bertha Molina-Quintana$^{(\boxtimes)}$ ⓘ, María Berta Quintana-León ⓘ,
and Jaime Apolinar Martínez-Arroyo ⓘ

Universidad Michoacana de San Nicolás de Hidalgo,
UMSNH, Morelia, Michoacán, México
bettymolinaq@gmail.com, mariber0320@gmail.com,
corredor42195@hotmail.com

Abstract. This paper deals with the study of commercial relationships between manufacturers and distributors through a dyadic approach where both parts of the exchange are considered from a perspective based on the components of the relationship, leaving in the background the perspective based on the functions performed the relationship taking as a point of reference the conceptual framework of direct and indirect value creating strategies which has been adapted to the strategic relationships of today due to the commercial needs to strategies that create commercial value and research and development (Walter et al. 2003), which makes it possible to estimate how value can be created for current and potential clients.

1 Introduction

For Webster (1994), the strategic relationships that are maintained as a way to achieve a competitive advantage, are increasingly narrow and lasting. Relationship marketing studies have focused more on the sale of personnel and on distribution channels that have a direct relationship with the end customer (Weitz and Jap 1995), since they pay more attention to customer service.

For any company it is necessary to organize its distribution in order to release the products in the market. The relationships between economic agents are not limited to the exchange of goods and money. But the services, financing, and information, among others, are also part of the flow of the relationship. Then the management of the channel is one of the most decisive strategic aspects for the success and survival of the company. The configuration of the channel and the specific companies with which they carry out the exchanges will determine, then, the profitability of the business, in addition to the characteristics of the product or the environmental factors (changes in consumer preferences, competition, technological evolution). For this the company manages, among others, two critical aspects: the choice of companies with which to carry out the exchanges and the direction of the relationship between the two parties.

J. Gil-Lafuente et al. (Eds.): AEDEM 2017, SSDC 180, pp. 292–305, 2019.
https://doi.org/10.1007/978-3-030-00677-8_23

Some of the background of the manufacturer-distributor relationships dates back to 1900 how it is broken down below:

Sheth and Gardner (1982) and Sheth et al. (1988) proposed a classification of marketing thinking approaches in which the importance given to the consumer by each of them is appreciated (Table 1):

Table 1. Clasification of marketing thinking approaches

Dimensions	Not interactive	Interactive
Economic	- Product school - School of functions - Geographical school	- Institutional school - Functionalist school - Management school
Not economic	- School of consumer behavior - School activist - Macro-marketing school	- School of organizational dynamics - School of systems - School of social exchange

The institutionalist approach of the economic and interactive quadrant of the school of marketing thinking that began to develop in 1900 (Sheth and Gardner 1982; Sheth et al. 1988) centers the interest in commercial organizations, from the producer to the wholesaler and from this to the retailer. And they consider that the discipline of marketing would benefit enormously if the attention in the organizations that distribute the goods from the producer to the consumer is centered in the distribution channel.

The School of Organizational Dynamics dates from 1957, has its direct antecedent in the Institutional School, since both try to explain the functioning of the distribution channels; however, they differ in that the Institutional School relies on economic principles while the School of Organizational Dynamics focuses its attention on consumer welfare through the analysis of the objectives and needs of the different agents participating in the channel.

The following table presents some of the most frequent problems that arise in buyer-seller relationships in marketing channels (Tables 2 and 3):

Table 2. Conflict sources between manufacturers-distributors

Sources of Conflict between manufacturers-distributors	
1.	Lack of professionalism of the distributor
2.	Poor knowledge of the products by the distributor
3.	The distributor does not care about growing and/or looking for new product applications
4.	The distributor sees himself more as an independent company oriented towards the end customers than as a part of the company's distribution system
5.	Lack of involvement with the company's products
6.	Distributors are reluctant to provide information
7.	Your company wants distributors not to sell competing products while they want a broad assortment

(continued)

Table 2. (*continued*)

Sources of Conflict between manufacturers-distributors	
8.	Incorporation of distributors and duplication of areas
9.	Do not worry about innovating and applying new techniques of doing business
10.	Responsibility for large customers (especially when they have been developed by the intermediary). Competition between sellers and distributors
11.	Resistance on the part of the distributors to support the levels of inventory that the company considers adequate
12.	Continuous complaints to the distributor regarding the conditions of the offer
13.	Slowness of the distributors when it comes to reporting cancellations and/or changes in orders
14.	Succession problems in the intermediary business
15.	Problems derived from establishing who is responsible for the returns
16.	Complaints by the distributors about the slowness with which the manufacturer informs them about changes in the issues of interest

Source: Sanzo et al. (2003)

Table 3. Sources of conflicts with government in commercial relations

Sources of conflicts before government in commercial relations	
1.	The problem of the safeguard, derived from the concern of the company for the possible opportunistic behavior of the agent with which it interacts, against the specific investment made in assets for the ultimate realization of the exchange
2.	The problem of adaptation, which arises as a result of the difficulties that the company must face in order to modify its contractual agreements in the face of changes that may occur in the environment (uncertainty relative to the environment), derived from its limited capacity
3.	The problem associated with the evaluation of performance, caused by the limitation of rationality that characterizes the company and that consequently makes it difficult to assess the degree of compliance or satisfaction achieved with the exchange by the other party (uncertainty regarding the behavior)

Source: Williamson (1985) and Rindfleisch and Heide (1997).

2   Conceptual Framework

Understanding the creation of value for the client from the perspective of relationship marketing can be identified as one of the most recent developments in the investigation of the concept of value (Eggert et al. 2006). Although the number of both theoretical and empirical works in this area is limited (Payne and Holt 2001), it is now possible to recognize two important currents of thought (Lindgreen and Wynstra 2005): one that directs its efforts to examine the balance or trade off between the multiple components of benefit and sacrifice that make up a relationship and another that, even recognizing this balance, has more impact on the tasks or functions that relationships can perform for the parties involved. Next, each of them is commented:

First Stream: Focused on the Balance of the Components of the Relationship

The offer of a product of a company can be considered the means through which current value is offered (Zeithaml 1988). A classic way of understanding this offer is taking into account the different levels of the product: generic, expected, increased and potential (Levitt 1980). In this line, the competition increasingly focuses on the increased product, on what exceeds the customer's expectations. However, in a relational context, such conceptualization of the product may be restrictive since no attention is paid to the value of a relationship and how it affects the client's perception of value (Eggert and Ulaga 2002).

In this way, when reference is made to the value received by the client in an episode of a relationship, in addition to contemplating the essential product and complementary services or goods, the effects deriving from the fact of maintaining a relationship of exchange must be incorporated (Lindgreen and Wynstra 2005). Therefore, at the theoretical level, in the delimitation of the concept of value, the balance between the benefits and the perceived sacrifices should not be reduced, solely and exclusively, to a specific episode of buying and selling in a given relationship, but should be extended. To the set of successive transactions over time that allow you to incorporate the benefits and sacrifices of a relationship (Ravald and Gronros 1996).

In the case of consumers of services, we can highlight Gwinner et al. (1998), who maintain that in order to positively assess a relationship with a long-term orientation, at least three types of benefits must be present: of security, social and special treatment. On the other hand, in the sector of information and communication technologies (ICT) and in the financial sector, the work of Lapierre (2000) represents a turning point in the purely conceptual literature that prevailed up to that date, being the first of a series of recently published empirical studies where not only the term value for the client is measured, but also a construct that distinguishes between domain and scope is specified.

A categorization closer in time is that of Ulaga and Eggert (2005) who, in their purpose of limiting and measuring the value concept of a relationship in manufacturing sectors, identify five dimensions of benefit (product, service, know-how, market time and social) and two sacrifice (price and process costs). To summarize it is necessary to emphasize that although some authors may agree on certain components of the value concept of a relationship, the characteristic note at this time is the diversity of proposals, without any of them prevailing over the others.

Second Stream: Focused on the Functions That the Relationship Performs

There is a line of research focused on the analysis of the creation of value in a relationship that is recognized as a functionalist perspective (Lindgreen and Wynstra 2005). This is because it is based on the premise that buyer-seller relationships can perform a series of tasks or functions that create value for the actors involved (Anderson et al. 1994). Specifically, these allow the parties to connect activities, unite their resources and develop affective bonds between people, thus allowing the accumulation of knowledge, the creation of new resources and the development of new activities (Harris and Wheeler 2005).

The functions of relationships can be classified primarily into direct functions or commercial strategies and indirect functions or research and development functions (Walter et al. 2003):

(a) The direct, primary or commercial strategies represent the positive and negative effects of the interaction in a main dyadic relation on the companies involved. It is these functions, it is intended in relation to resources, activities and actors, seeking efficiency through the interconnection of activities, creativity based on the heterogeneity of resources, and mutuality based on the self-interest of the actors.

(b) Indirect, secondary or research and development functions, also known as network functions, include the positive and negative indirect effects of a relationship, when they are directly or indirectly connected to other relationships. This type of functions includes the chains of activities that involve several organizations, the portfolio of resources controlled by more than one company, and the shared perceptions of the network by more than two companies.

Among the most significant and recent contributions within this perspective of analysis we can highlight the Walter et al. (2001) who understand the value creation functions for the client as the activities developed and the resources used by the supplier company, that is, what the provider contributes to both the main dyadic relationship and the network of relationships connected to it.

Development of the Analysis Perspective

In this research, the functionalist perspective has been chosen to study the concept of value for the client. This is due to the novelty of this approach, together with the lack of a widely accepted concept on the creation of value in a relationship, which makes it necessary and relevant to apply proposals of these characteristics in areas of study such as the current one. Next, each of the value creation functions that the supplier develops for the client is described:

1. Commercial Strategies Functions
   (a) Benefit function. The fact that the supplier in its relationship with a given customer is able to offer products or services at the best prices or, at least, at competitive prices, without the client having to renounce minimum standards of technical quality, can be a determinant or significant factor of the client's profitability (Walter et al. 2001). This function reflects the importance that the supplier has in generating economic benefits for the client. Precisely, it helps the client, through the cash generated by the purchase of products where it obtains a high margin of profit, to stimulate relational exchanges with different suppliers that can perform complementary functions in their business strategy (Kalwani and Narayandas 1995).
   (b) Volume Function. The supplier develops this creative function of value for the client when it satisfies the high volume of demand experienced by its counterpart (Walter et al. 2001). This circumstance implies that, on occasion, customers need to maintain stable relationships with their suppliers from a strategic point of view over the criterion associated with obtaining high profits in successive transactions, which in certain circumstances may be in the background. For a client, the supplier developing the volume function means having the certainty that the supplier will supply him, in time and form, with an important amount of product when necessary (Turnbull 1982).

(c) Safeguard function. The supplier performs this function in a relational exchange from the moment in which it guarantees the client a certain level of supply, even though from an economic point of view it is not as favorable, as a reduction in the dependency it has on other suppliers (Hakansson 1982). The selling party acts therefore as an emergency supplier for the business partner. The argument is that, as a result of market uncertainties, it is necessary for clients to maintain relationships with suppliers other than the usual ones that, although they may not have a profit function and/or volume function, acts as an insurance in the face of possible crises or difficulties that may arise in other exchange relationships (Walter et al. 2003).

(d) Trade Marketing function. This means for the manufacturer to see the distributor as a customer, rather than as a distribution channel, so that Trade Marketing would be oriented to satisfy the consumer through the integration of the marketing activities of the manufacturer with those of the distributor, thinking together in the needs of market development. The three most important variables to consider in the trade marketing are: The management by categories, assortment and promotion, however since we talk about the assortment and management by categories in the previous functions that is why for the trade marketing function In this study we will talk about price and promotion.

2. Research and Development Functions

(a) Innovation function. Customers tend to establish cooperative relationships in the development of new products and processes with suppliers that are at the forefront of technology and where product experience is high (Walter et al. 2001). From this point of view, considering the existence of a network of interdependent relationships, both product and process innovation have a prominent role, since it is assumed that both types of innovation can be the result of the interaction between two or more actors (Parkinson 1985; Hakansson 1987), which may mean that, in certain circumstances, the client sacrifices the economic benefits for the long-term usefulness of a cooperation of these characteristics. That is, some companies may choose to work in relative isolation, while others may be inclined to develop part of these innovations with other actors (Walter et al. 2003).

(b) Market function. Clients can use prestige or good credit in the daily work of some of their suppliers and use it, as references, as an important support to access markets and establish new business links (Walter et al. 2001). In this sense, the market function can be found when working with suppliers that use strict criteria and conditions in the selection of their business partners (Corstjens and Merrihune 2003). For this reason, interacting with this type of suppliers can have a reference effect in other exchange relationships, although these suppliers are not the most important in the market in economic terms (for example, in sales volume).

(c) Exploratory function. It can be argued that a large part of the success of a company can originate, to a large extent, in obtaining significant information from other actors that are located outside it (Walter et al. 2001). Thus, a possible criterion to be used in the selection of new suppliers may be their ability to provide relevant commercial information to the client. These providers can become useful channels of business information relevant to the customer. These providers can become useful information channels, given their extensive knowledge about the sector itself on other markets with which they are related (Walter et al. 2003). As pointed out, it can be pointed out that commercial information of utility obtained through suppliers include, among others, details about the offers of competing companies and issues related to market trends.

(d) Exploratory function. It is important to obtain relevant commercial information from other actors, about market trends and possible rival companies and third organizations to support their growth and analysis of their strengths, weaknesses, advantages and disadvantages of the way in which the company with its commercial relationships unwrap.

(e) Technology function. Nowadays, the use of technology facilitates business and commercial relations, which is why it is important to make use of ICT, since with the implementation of computer systems monitoring, review, analysis, streamlines processes and improves relationships.

Below is the diagram of variables described in this part for a clear perception of the study (Diagram 1).

3   Methodology

The present study is basically the structure for the application of said research, which is in its first planning phase; will be made to manufacturing companies located in Michoacán, Mexico of medium and large size, belonging to more than one sector of economic activity, registered in the Industrial Association of the State of Michoacán. The measurement instrument will be through Likert surveys of 5 points where 1 means totally in disagreement and 5 totally agree, directed to the General Manager, Manager or Commercial Director.

Because there are four items per latent variable to be able to achieve the necessary degrees of freedom with which to calculate the adjustment indices of the measurement model, it is possible to evaluate each of the measurement scales in isolation using a confirmatory factor analysis (Tables 4, 5 and 6).

Universe
Table 7 shows the conceptual model proposed intercepted with the measurement scale, where the latent factors are measured from the perspective of one of the members of the main dyadic relation, that is, the manufacturer, which values, by virtue of the exhaustive knowledge that it presupposes from this relationship, how it develops a series of direct and indirect functions that provide value to its main distributor.

The scale proposed by Walter et al. (2001) as a reference point. These functions were adapted in our study to obtain the perception of the manufacturer about how it contributes to the creation of value for its most important distributor by performing a series of tasks, both direct and indirect (Table 8).

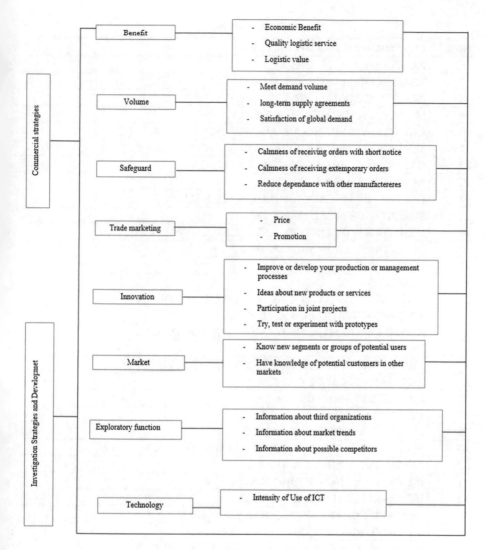

Diagram 1. Variable diagram. Source: Own Elaboration based on Walter et al. (2001).

Table 4. Manufacturing companies of Michoacan

Num.	Manufacturing companies in Michoacan
1	AAK MÉXICO, S.A. DE C.V.
2	CHECOLINES, S.A. DE C.V.
3	CHOCOLATERA MOCTEZUMA, S.A. DE C.V.
4	CONGELADORA MORELIA, S.A. DE C.V.
5	GEUSA DE OCCIDENTE, S.A. DE C.V.

(*continued*)

Table 4. (*continued*)

Num.	Manufacturing companies in Michoacan
6	PATYLETA, S.A. DE C.V.
7	TEAM FOODS MÉXICO, S.A. DE C.V.
8	ACEITES ESPECIALES TH, S.A. DE C.V.
9	MEXICAN AVOCADOS, S.A. DE C.V.
10	INDUSTRIALIZADORA DE AGUACATE HASS, S.A. DE C.V.
11	SERVICIOS REFRESQUEROS DEL GOLFO Y BAJÍO, S. DE R.L. DE C.V.
12	CAMPO ALEGRE ALIMENTOS, S.A. DE C.V.
13	ALKEMIN, S. DE R.L. DE C.V.
14	DERIVADOS MACROQUÍMICOS, S.A. DE C.V.
15	EUROTON DE MÉXICO, S.A. DE C.V.
16	INDUSTRIAL DE PINTURAS VOLTON, S.A. DE C.V.
17	INDUSTRIAL LA FAMA, S.A. DE C.V.
18	PINOSA, S. DE R.L. DE C.V.
19	QUIMIC, S.A. DE C.V.
20	RESINAS NATURALES Y ARTIFICIALES, S.A. DE C.V.
21	RESINAS SINTÉTICAS, S.A. DE C.V.
22	ACEROS Y TREFILADOS DE MORELIA, S.A. DE C.V.
23	DEACERO, S.A. DE C.V.
24	ARCELORMITTAL LÁZARO CÁRDENAS, S.A. DE C.V.
25	ARTIFIBRAS, S.A. DE C.V.
26	INDUSTRIAS MARVES, S.A. DE C.V.
27	JANESVILLE DE MÉXICO, S.A. DE C.V.
28	SEELE INDUSTRIAL, S.A. DE C.V.
29	EQUIPO INDUSTRIAL Y SERVICIOS, S.A. DE C.V.
30	ASTRO EMPAQUES, S.A. DE C.V.
31	GRUPO PAPELERO SCRIBE, S.A. DE C.V.
32	INDUSTRIAS JAFHER, S.A. DE C.V.
33	KIMBERLY CLARK DE MÉXICO, S.A.B. DE C.V.
34	LA UNIVERSAL IMPRESORA, S.A. DE C.V.
35	MADERAS Y SINTÉTICOS DE MÉXICO, S.A. DE C.V.
36	AGROVIM, S.A. DE C.V.
37	ALSTOM MEXICANA, S.A. DE C.V.
38	ANDRITZ HYDRO, S.A. DE C.V.
39	FUNDIDORA MORELIA, S.A. DE C.V.
40	PROYECTOS INDUSTRIALES Y MANTENIMIENTO DE MORELIA, S.A. DE C.V.
41	INDUSTRIAL CORDELERA MEXICANA, S.A. DE C.V.
42	INDUSTRIAS POLIPLÁSTICOS DE MICHOACÁN, S.A. DE C.V.
43	PROMOTORA ZACAPU, S.A. DE C.V.
44	TREOFAN MÉXICO, S.A. DE C.V.

(*continued*)

Table 4. (*continued*)

Num.	Manufacturing companies in Michoacan
45	AMBIENTAL MICHOACANA
46	CENTRO DE ESTUDIOS EN MEDIO AMBIENTE, S.C.
47	ECOLOGÍA 2000, S.A. DE C.V.
48	GAS NATURAL URUAPAN, S.A. DE C.V.
49	GLOBAL SOLUTIONS IN AUTOMATION, S.A. DE C.V.
50	GRUPO ERÉNDIRA DE PROYECTOS INDUSTRIALES, S.A. DE C.V.
51	KANSAS CITY SOUTHERN MÉXICO, S.A. DE C.V.
52	LOGÍSTICA ARQUITECTÓNICA, S.A. DE C.V.
53	GRUPO INVISA
54	SANIRENT MÉXICO, S.A. DE C.V.
55	INTEGRACIÓN DE NORMAS CON APORTACIÓN DE VALOR

Source: Asociación de Industriales del Estado de Michoacán, A.C.

Table 5. Number of manufacturing companies of Michoacan located by sector of economic activity

Manufacturing companies in Michoacan

Sector	Num. companies
Food	12
Chemistry	9
Steel	3
Automotive	3
Ceramic	1
Electric	1
Wood-Paper	7
Metal mechanic	4
Plastic	4
Textiles	0
Services	11

Source: Own Elaboration.

Table 6. Research technical sheet

Target population	Medium-sized companies and large manufacturers belonging to the sectors of food (12), chemistry (9), steel (3), automotive (3), ceramics (1), electrical (1), wood-paper (7), metalworking (4), plastic (4), services (11).
Sample unit	General Director, Manager, or Commercial Director
Scope	State owned

(*continued*)

Table 6. (*continued*)

Target population	Medium-sized companies and large manufacturers belonging to the sectors of food (12), chemistry (9), steel (3), automotive (3), ceramics (1), electrical (1), wood-paper (7), metalworking (4), plastic (4), services (11).
Information Collection Method	Structured surveys like Likert scale.
Population census	55
Sample size	49
Margin of error	5%
Level of trust	95%; z = 1.96; p = q = 0.5
Sample Procedure	Simple, random in excel

Source: Own Elaboration

Table 7. Operationalization of variables

Variable	Dimension	Indicator	Question
Commercial strategies	Benefit	− Economic benefits − Quality of logistic service − Logistic value	1–2 3–6 7–8
	Volume	− Satisfy demand volumes − Long-term supply agreements − Satisfaction of global demand	9 10 11
	Safeguard	− Peace of mind to receive orders at short notice − Tranquility of receiving extemporaneous orders − Reduce dependency with other manufacturers	12 13 14
	Trade Marketing	− Price − Promotion	15 16
Investigation and development	Innovation	− Improve or develop your production or management processes − Ideas about new products or services − Participation in joint projects − Test, try out, or experiment with prototypes	17 18 19 20
	Market	− Know new segments or groups of potential users − Have knowledge of potential customers in other markets	21 22
	Exploratory function	− Information on third organizations − Information on market trends − Information about possible competitors	23 24 25
	Technology	− Intensity of Use of TIC's (Technology of information and communication)	26–28

Source: Own Elaboration based on Walter et al. (2001).

Table 8. Function measurement scale

	Our most important distributor has achieved …
1	A high profit margin per product as a consequence of the nature of our offer
2	High economic benefits
3	Quality of service because the knowledge and experience of our staff are adequate
4	Speed when placing orders as the ordering process is effective and easy to use
5	Attention to the service, should any problem arise, is it selected in a satisfactory manner
6	Are you happy with the level of logistics service we offer
7	A Competitive Advantage because logistics adds value to the relationship with this distributor
8	Increase orders when the level of logistics service offered exceeds your expectations
9	Can you meet high demand volumes because we want and can meet the supply of such orders
10	Important long-term supply agreements with our company
11	The satisfaction of the global demand that you experience of the category/s of product/s that you buy from us
12	The peace of mind of knowing that you can receive orders made at short notice
13	The peace of mind that you can receive orders of an extraordinary nature
14	Reduce dependence with other manufacturers
15	That you improve the price that is in the market
16	Offers and promotions from us
17	Any ideas about new products or services that you can market to develop your business
18	Suggestions about how to improve or develop your production or management processes
19	Participate in joint projects that have allowed you to increase your technical knowledge about the benefits and applications of our offer
20	Try, test or experiment with prototypes that allow you to anticipate your marketing strategy
21	Know the existence of new segments or groups of potential users that had not previously been considered
22	Have knowledge of potential customers in other markets
23	Information on third organizations (for example: research centers, business associations
24	Information on market trends
25	Information about possible competitors
26	Information technologies for transport optimization
27	Information technologies for tracking/traceability of products (GPS)
28	TIC's for the management and optimization of inventories/warehouses

Source: Own Elaboration based on Walter et al. (2001).

4   Conclusions

This research aims to achieve an understanding about the consequences of creating value in the business markets. Starting from the premise that the buyer-seller or manufacturer-distributor relations fulfill a series of tasks or functions, both direct and indirect, or commercial strategies or research and development strategies, which generate value for the business partners involved. The novelty of this study, together with the lack of a definition of the concept of value, makes it pertinent to apply such proposals, since it should be noted that the creation of value in an instrument to promote the commercial relationship that the manufacturer maintains with the distributor.

However, it would be interesting once this research has been carried out since it is the theoretical and planning part, for which it is necessary to carry out the field work, to carry out the counterpart of the dyadic relation; the distributor, since for this work the manufacturer's side is addressed.

References

Anderson, J.C., Hakansson, H., Johanson, J.: Dyadic business relationships within a business network context. J. Mark. **58**(4), 1–15 (1994)

Association of industrialists of the state of Michoacan www.aiemac.org.mx

Corstjens, M., Merrihue, J.: Optimal marketing. Har. Bus. Rev. **81**(10), 114–121 (2003)

Eggert, A., Ulaga, W.: Customer perceived value: A substitute for satisfaction in business markets. J. Bus. Ind. Mark. **17**(2/3), 107–118 (2002)

Eggert, A., Ulaga, W., Schultz, F.: Value creation in the relationship life cycle: a quasiLongitudinal analysis. Ind. Mark. Manag. **35**(1), 20–7 (2006)

Gwinner, K.P., Gremler, D.D., Bitner, M.J.: Relational benefits in services industries: the customer's perspective. J. Acad. Mark. Sci. **26**(2), 101–114 (1998)

Hakansson, H. (ed.): International Marketing and Purchasing of Industrial Goods. An Interaction Approach. Wiley, Chichester (1982)

Hakansson, H. (ed.): Industrial Technological Development. A Network Approach. Croom Helm, London (1987)

Harris, S., Wheeler, C.: Entrepreneurs' relationships for internationalization: functions, origins and strategies. Inter. Bus. Rev. **14**, 187–207 (2005)

Kalwani, M.U., Narayandas, N.: Long-term manufacturer-supplier relationships: do they pay off for supplier firms? J. Mark. **59**, 1–16 (1995)

Lapierre, J.: Customer-perceived value in industrial contexts. J. Bus. Ind. Mark. **15** (2/3), 122–145 (2000)

Lindgreen, A., Wynstra, F.: Value in business markets: what do we know? Where are we going? Ind. Mark. Manag. **34**, 732–748 (2005)

Levitt, T.: Marketing success through differentiation—of anything. Harv. Bus. Rev. **58**, 83–91 (1980)

Parkinson, S.T.: Factors influencing buyer-seller relationships in the market for high-technology products. J. Bus. Res. **13**, 49–60 (1985)

Payne, S.H.: Diagnosing customer value: integrating the value process and relationship marketing. Br. J. Manag. **12**(2), 159–182 (2001)

Ravald, A., Gronroos, C.: The value concept and relationship marketing. Eur. J. Mark. **30**(2), 19–30 (1996)

Rindfleisch, A., Heide, J.B.: Transaction cost analysis: past, present, and future applications. J. Mark. **61**, 54 (1997)

Sanzo M.J., V.R. (s.f.): Fuentes de Conflicto en los Canales de Distribución Industriales: Análisis de los Factores Condicionantes desde el punto de vista de los Distribuidores Industriales del Sector Químico. *D-S 10*, 107–127

Sanzo, M.J., Santos, M.L., Vázquez, R., Álvarez, L.I.: The effect of market orientation on buyer-seller relationship satisfaction. Ind. Mark. Manag. **32**(4), 327–345 (2003)

Sheth, J.N., Gardner, D.M.: History of marketing thought: an update. In: Bush, R., Hunt, D.S. (eds.) Marketing Theory: Philosophy of Science Perspectives, pp. 52–58. A.M.A. (1982)

Sheth, J.N., Gardner, D.M., Garret, D.E.: Marketing Theory: Evolution and Evaluation, pp. 19–22. Wiley, London (1988)

Turnbull, P.W.: Britmet. A marketing case study of a large producer of special steel products. In: Hakansson, H. (ed.) International Marketing and Purchasing of Industrial Goods. An Interaction Approach, pp. 88–101. Wiley, New York (1982)

Ulaga, W., Eggert, A.: Relationship value in business markets: The construct and its dimensions. J. Business-to-Business Mark. **12**(1), 73–99 (2005)

Walter, A., Ritter, T., Gemunden, H.G.: Value creation in buyer seller relationships. Ind. Mark. Manage. **30**(4), 365–377 (2001)

Walter, A., Muller, T.A., Helfert, G., Ritter, T.: Functions of industrial supplier relationships and their impact on relationship quality. Ind. Mark. Manage. **32**(2), 159–169 (2003)

Webster Jr., F.E.: Market-Driven Management: Using the New Marketing Concept to Create a Customer-Oriented Company. John Wiley & Sons, Inc., New York (1994)

Weitz, B.A., Jap, S.D.: Relationship marketing and distribution channels. J. Acad. Mark. Sci. **23**(4), 305–320 (1995)

Williamson, O.E.: Employee ownership and internal governance: A perspective. J. Econ. Behav. Organ. Elsevier **6**(3), 243–245 (1985)

Zeithaml, V.A.: Consumer perceptions of price, quality, and value: a means-end model and synthesis of evidence. J. Mark. 2–22 (1988)

The Italian Approach to Industry 4.0: Policy Approach and Managerial Implications in a SMEs Environment

Maurizio Fiasché[1(✉)] and Francesco Timpano[2]

[1] Department of Electronics, Information and Bio-engineering,
Politecnico di Milano, Milan, Italy
maurizio.fiasche@ieee.org
[2] Department of Economic and Social Sciences,
Università Cattolica del Sacro Cuore di Milano, Milan, Italy
francesco.timpano@unicatt.it

Abstract. The Industry 4.0 Italian plan is the new attempt of the National Government in Italy to adopt a comprehensive industrial policy aimed to adapt the industrial sector to the new trajectories of manufacturing. A variety of policy plans and interventions have been recently developed in order to sustain these processes everywhere in Europe. The 9 enabling technologies identified in the so called 'Calenda plan' (from the Italian ministry of economic development name, coordinated the plan writing) are not just a list of industrial technologies available, but are a chance for (re)create a new business model in the factory 4.0, but more and more in the 'social environment' 4.0 where the whole value chain in a business company need to be re-thought. An analysis of the potential impact of these policy frameworks has still to be implemented. This paper aims to disentangle the relationship between policies and the Italian manufacturing starting from an overview of the industrial policies in Italy and raising the main problems and contradictions potentially stemming out from the Italian plan.

1 Introduction

Industry 4.0 is a comprehensive process of manufacturing transformation. A variety of different definitions have been provided until now by scholars, practitioners, policy makers. A relatively encompassing definition has been provided by the European Parliament "*Industry 4.0 describes the organisation of production processes based on technology and devices autonomously communicating with each other along the value chain in virtual computer models. Industry 4.0 involves a series of disruptive innovations in production and leaps in industrial processes resulting in significantly higher productivity. Challenging preconditions for successful implementation of Industry 4.0 have to be met as regards standards, work processes and organisation, availability of products, new business models, security and IP protection, availability of workers, research, training and professional development and the legal framework*".

© Springer Nature Switzerland AG 2019
J. Gil-Lafuente et al. (Eds.): AEDEM 2017, SSDC 180, pp. 306–315, 2019.
https://doi.org/10.1007/978-3-030-00677-8_24

A large debate has been produced in the last years, most of it is a clear consequence of the increasing interest of policy makers on the "fourth industrial revolution"[1]. It is an interdisciplinary debate that has promoted an interesting confrontation among engineering, production and business both from academia and business. In Hermann et al. (2016) an analysis of literature identifies six design principles for the implementation of Industry 4.0: interoperability, virtualization, decentralization, real-time capability, service orientation, and modularity. Until now, it is quite limited the literature analysing the (forecast or real) effects of the policymaking on the evolution of the Industry 4.0 in the countries where a policy approach has been recently adopted.

A variety of policy interventions have been implemented starting, as an example, from EU Communication (EC) (2012) and from the Communication "For a European Industrial Renaissance" (2014). National and regional plans are flourishing in most of the European and non-European countries. An ex-ante evaluation of the potential changes produced by the introduction of the policies can be done by comparing the guidelines of the policy with the specific features of the industrial sector in Italy (and in Europe).

The fourth Industrial revolution is strictly connected to the evolution of the manufacturing sector and is a process that is developing with the evolution of the need in the instrumental goods sector along with a massive evolution of the digital technologies applied to the production systems. The 3D printing allows one to efficient rapid prototyping and also will allow to customize production. The high volume of data coming from the enforcement of the relationships between physical and digital systems will enable firms to implement new service by exploiting data management. How the 3D technologies and the policies that are implemented in Italy to push their adoption will impact on a system characterized by positive features, linked for example to the wide experience in machinery sector and mechatronics sector of Italian SMEs, and negative features, linked to the limited investment in new knowledge and the adoption of innovative technologies more through investment in machinery than through comprehensive investment on the business models of the firms. Is the policy well designed? What are the potential issues for the future?

In the Sect. 2 an overview of the industrial policy in Europe and in Italy is provided with a short description of the Calenda plan. In the Sect. 3 a preliminary analysis of the potential impact of the plan on the Italian firms, both in the process of adoption and in the process of evolution will be suggested for a debate that will certainly be enriched by monitoring the whole process[2] in Sect. 4 a description how technological and market trends can impact on production processes and vice versa, in Sect. 5 a brief conclusion and next investigation steps are inferred.

---

[1] Klaus Schwab (2016) at World Economic Forum, Davos.

[2] A significant contribution will be certainly provided by the observatories on Industry 4.0 that have been created in Europe. In Italy see http://www.osservatori.net.

2   The Policy Approach to Industry 4.0 in Europe and in Italy

The industrial policy in Italy has been recently characterized by robust changes after years of difficulties and poor implementation of EU pushed initiatives.

Most of industrial policies in Italy have been based on the support to "national champions" like the mechanical sector or the automotive sector and most of the policies have been aimed to sustain the adoption of new machines and new system of productions by sustaining internal demand. More recently, the crisis of internal demand for capital goods has oriented Italian firms producing machineries to serve the global market and most of the policies have been diverted to sustain the innovation process and hopefully the competitiveness of the firms.

As Viesti (2013) pointed out, the new industrial policies are actually oriented towards the support of entrepreneurship, the support of innovation through public demand, the renewal of regulation to support and the integration with urban and regional policies.

More recently a renewed interest towards manufacturing has characterized the EU discussion around industrial policy and the Industry 4.0 paradigm has been introduced by several EU countries in order to push a new industrial revolution based on a strict interoperability between capital goods, information and services through a massive use of a number of enabling technologies.

The "manufacturing renaissance" has been strongly pushed by EU initiatives that emphasized the potential role of new business models coming from the manufacturing sector. For example, the EU platform "Manufuture"[3] has played a significant role after the 2004. The basis of that platform is *"the mission is to propose, develop and implement a strategy based on Research and Innovation, capable of speeding up the rate of industrial transformation to high-added-value products, processes and services, securing high-skills employment and winning a major share of world Manufacturing output in the future knowledge-driven economy."* More recently in Sautter (2016), the ManuFuture Vision 2020 generation and implementation process has been defined by "setting the frame and depicting the ambition as well as the challenge for the upcoming vision building process with the time horizon of 2030 and beyond. This upcoming scenario and vision building process with the aim to create a new vision 2030 for a smart, clean and human-centered EU industry should be adaptive to the various stakeholders' needs and objectives like the ManuFuture 2020 process in the past."

Additional actions based on public-private partnerships have been developed like the important European Factory of the Future Research Association (EFFRA)[4] that started in 2009 where "the overall aim of the partnership is to enable a more sustainable and a more competitive European industry at the centre of Europe's economy – generating growth and securing jobs. The partnership will achieve this by supporting European manufacturing enterprises in strengthening their technological base." More recently new platforms have been activated like Vanguard[5] on advanced manufacturing

---

[3] http://www.manufuture.org.

[4] http://www.effra.eu.

[5] http://www.s3vanguardinitiative.eu.

capabilities in some EU regions and a KIC Manufacturing has about to be launched by the European Institute of Innovation and Technology[6].

All this can be considered on one side as a way to sustain the transformation of the manufacturing sector in the third Industrial revolution and to push the digitalisation of the factory of the future.

Moreover after the last economic crisis, the reaction to the quick and strong reduction of the aggregate demand has been largely based on a demand pulled approach with policy interventions largely based on increasing public demand. In some countries, like US, this has also been directed towards the green sector trying to give active support to the sustainability issues. The sovereign debt crisis has reduced the appeal of demand pull policies and it has re-activated supply side policies, mostly oriented towards tools aimed to sustain research and innovation, where selection of objectives and relatively small scale interventions are used also to enforce the competitiveness of the firms and of the industrial sector.

The relationship between local and national government in the field of industrial policies has been another relevant topic of the last years. R&D and innovation policies have increased their importance at regional level also in the context of the Smart Specialisation Strategies promoted to plan the EU Structural Funds 2014/2020. An increased role for national based policy has been justified by the need of optimising the policies and enlarging the scale of intervention.

Industria 4.0: The Italian Plan

In Italy there have recently been at least three important plans: Industria 2015 (in 2006), the competition Technological Cluster promoted by the Ministry of Research in 2012 and more recently the Plan for Industry 4.0 promoted by Minister Carlo Calenda in 2016. Industria 2015 was considered an ambitious and modern industrial plan but the performance of the management has been poor with a very limited impact at macro level. The Technological Cluster competition promoted in 2012 by the Government Monti was much more ambitious and strictly oriented towards enabling technologies in the specialisation areas of the country. A poor management of the measures have been replied also in this case but there still room for recovery since the project started recently.

Industria 4.0 is a comprehensive plan with a more strategic appeal that is activating public and private actions at different levels. On the side of the financial support, it has implemented a benefit from the over-amortization of 250%, set up under the 2017 Budget Law. This is a maximum increase that allows to increase the deductible cost of all the instrumental goods acquired to transform the company into a technological and digital key 4.0. This is actually investment in intelligent, interconnected machines and includes functional assets for the technological and/or digital transformation of businesses according to the "Industry 4.0" model and intelligent devices, hardware and software instrumentation and components for integration, sensing and/or interconnecting and automatic process control used also in modernization or revamping of existing production systems. This is a disruptive action in a country like Italy, where

---

[6] https://eit.europa.eu/sites/default/files/webinar_presentation_eit_call_manufacturing_20170907.pdf.

the fleet of machineries in manufacturing sector has an average age of 13 years as reported in a research by UCIMU association (2014).

But Industry 4.0 is not only over-amortisation but it is a comprehensive action aimed to different goals mostly linked to new business paradigm like mass customization, servitization, circular economy and networking. These new paradigms are not new for the Italian industry. Most of the Italian SMEs are characterized by an amazing ability in customization, small scale production and more recently by firms supplying services along with products like machines. The processes robotizing are also familiar for example in the machine tool sectors, where the performance of Italian firms has always been very relevant, as well as visual intelligent technologies like Augmented Reality (AR) or Virtual Reality (VR) business sectors, where the Italian system is quite strong. Nevertheless, as observed in Assolombarda (2016), a widespread adoption of the approach is still difficult for the large number of SMEs that do not see an immediate positive feedback from a change of paradigm that is necessary in order to implement the Industry 4.0 approach. In Industry 4.0 the comprehensive investment in the organization of the firm, in her business model, in human resources and in the adoption of new technologies oriented to provide a product-service along a new supply chain, more characterized by efficiency in terms of reduction of waste, information and use of the potential coming from data, circularity in the use of production factors, etc. Industry 4.0 should not be simply the investment in new machinery physically connected to intelligent systems but it should be a change in the entire value chain paradigm, involving also skills training system.

Nevertheless, Italy may play a role to raise advantage from the adoption of the approach, exactly because the industrial structure of the country and the strength of the instrumental goods sector, already trained to largely use digital technologies.

The Calenda plan can be summarized in a comprehensive model of governance that should link national ministries, universities, research centers, Cassa Depositi e Prestiti, firms and trade unions. The main guidelines are oriented towards two key areas (innovative investment and competences) on one side and support areas (enabling infrastructure and public policies). The public financial effort of the plan is estimated in 13 billions of Euro to activate additional 24 billions of Euro of private investment and the key instruments will be iper-amortisation and over-amortisation along with a tax rebate linked to the private R&D expenditure and some additional tools in measures to reduce credit crunch for firms buying new machines (Nuova Sabatini and Fondo di garanzia), venture capital and start-up support, support to patenting[7].

The lack of a precise support to investment in research and in people, along with an uncertainty on the financial resources available and a complete misunderstanding on the role of local governments (actually absent) are some weaknesses of the first Plan.

The plan has been relaunched for the 2018 (now it is called Impresa 4.0) and it now contains new measures sustaining human capital investment, starting from the Competence centers and a tax rebate for new employees involved in the I4.0 investment activity.

---

[7] For a comprehensive descripition see also http://www.sviluppoeconomico.gov.it/index.php/it/industria40.

Some contradictory features of the plan should not reduce the importance of the effort implemented by the Italian government. There are at least three challenges that emerge from the preliminary evidence on the impact of the plan on Italian industrial system:

1. *How to fill the competence gap.* It is certainly emerging that a gap in terms of competences will emerge mainly in the field of the development of the big data potential stemming from the plan. This is also strictly linked with the ability in improving the servitization process;
2. *How to activate the Italian industrial districts and clusters.* Despite the crisis that has reduced the Italian productive basis, Italian districts and clusters are still one the main source of innovation for the Italian industry. The Industry 4.0 Plan should activate the Italian districts (see Mosconi 2016) towards a renewed collaboration between large, medium and small (micro) enterprises;
3. *How to sustain productivity and employment.* The fourth industrial revolution should improve productivity and may reduce employment. The uncertainty about the final outcome of Industry 4.0 will be one of the fields of analysis for the future, nevertheless the operative challenge is in understanding how productivity will actually be increased by the new technologies applied to traditional and new productions and how the employment structure of the country will be changed in this new context.

3   Managerial Implications of the Italian Policies Plans

The context where the Industria 4.0 plan was born is inside a so complex situation.

Demographical sciences provide us provisional studied where in the next 30 years the EU will be destined to see an ever-growing population aging with a stationary education level, while the African continent will be characterized by an enhancement of young people (a trend continued from the past) but with always more high level education, as reported in Samir et al. (2010). This represent for sure a problem to manage for Italy, especially for its front-end position in the Mediterranean Sea but this is also an opportunity, in fact, it means also new available markets (e.g. to be consider the birth of a new dynamic middle class and consider its need).

It's the same for the enhancement of the GDP of China, India and other countries reported in the *United States Energy Information Administration* (2011). *Annual Energy Outlook 2011*, a new middle class (measured also with some parameter like the enhancement of new cars owners per year) developed in these emergent economies means new market, new opportunities and new challenges for business companies. In this representation, a projection of the GDP by Region, 1990–2030 (PPP) Outlook see the overtaking of China on USA and EU zone, starting from 2020 and India, Africa, and Central and South America overtake countries like Middle East, Brazil, Japan etc.

Other aspects like the rise in commodities price in the last 25 years, the provision of an ever-increasing scarcity of water in the next 30 years, is the reason for new and renewed Government green policies in the next years aiming at the reduction of emissions and energy efficiency, also in manufacturing sector weighing more than 30%

over the total of these phenomena, as reported *IEA, Worldwide trends in Energy Use and Efficiency, Energy Indicators, 2008* (2008). This means new policies, business processes and not just technologies for achieving these so important aims.

This represents the context, full of challenges inside the Industry 4.0 paradigm is born following the trend of the smart manufacturing action in EU and in the rest of advances industrial countries worldwide.

4 Technological and Markets Evolution vs Evolution of Business Strategy and Production Processes: Which Should Come First?

In Table 1 is shown as technological trends changed with a certain dynamism in the last three years. In the meanwhile, Markets trends as the Mass Customization, and the continuous search by companies to reduce the Time to Market, are phenomena with a strong interconnection with continuous new customer needs and desiderata, creating new production paradigms and requesting technological innovation for achieving these desiderata, all these implications need of new business strategies (Fig. 1).

Table 1. The first 10 technological Trends in 2014–2016

No	2014	2015	2016
1	Mobile device Diversity (BYOD)	Computing Everywhere	The Device Mesh
2	Mobile Applications	The Internet of Things	Ambient User Experience
3	Internet of Everything	3D Printing	3D-printing Materials
4	Hybrid Cloud	Advanced, Pervasive and Invisible Analytics	Information of Everything
5	Cloud/Client Architecture	Context-Rich Systems	Advanced Machine Learning
6	Personal Cloud	Smart Machines	Autonomous Agents & Things
7	Software Defined Anything (SDx)	Cloud/Client Computing	Adaptive Security Architecture
8	Web-Scale IT (Amazon, Google, Facebook)	Software-Defined Applications and Infrastructure	Advanced Customer Architecture
9	Smart Machines	Web-Scale IT	Mesh App and Service Architecture
10	3-D printing	Risk-Based Security/Self-Protection	Internet of Things

Source: Gartner Group 2015

At least for a decade now, product margins should not be the primary yardstick for launching a product anymore, as the profitability of services towards the customer is an important asset to take into account. One such example is buying a train, which

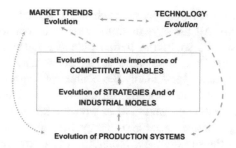

Fig. 1. Cause-effects pseudo blocks-scheme describing the connection loop among trends

represents merely 10% of all costs associated along its life cycle phase, thus leaving the other 90% to costs such as locomotive services, train operations etc. as reported in Wise and Baumgartner (1999). Another point of view could be simply that a product is completely commoditized and services have to be added on top so that the manufacturer regains its competitive edge; for instance the manufacturing enterprise of dynamite ICI-Nobel, had to start selling the services of blasting rocks using its product, because dynamite became completely commoditized in the 1990s, as reported in Thether and Bascavusoglu-Moreau (2012). This successful and pervasive trend of service bundling with products is known as Product-Service (P-S) strategy.

Providing services downstream need to provide a lot of different competences and resources, because the logic and underlying ideas are so different from the one required for a physical product. As it is impossible for manufacturing enterprises to own all of them and as the move downstream the value chain requires close collaboration with other enterprises, new form of collaboration arose, like the Manufacturing Service Ecosystem (MSE). It enables a high number of companies to jointly design, build and provision P-S onto the market in a distributed, dynamic and often non-hierarchical way. Yet before composing and provisioning them to the market, functional managers in manufacturing enterprises have to define a P-S strategy from the operations management perspective, which represents the nexus of the offering's success. Namely, if the strategy is ineffective, there is a strong possibility it will be completely rejected by the market. Hence, designing a P-S strategy can be a complex issue, as it is a process of multiple strata, requiring smart and experienced managers in the fields of product and service positioning. But this could be not enough! As shown in Opresnik et al. (2017), data is the key factor for understanding processes and taking right decision using Decision Support System (DSS) built with smart analytics systems. Machine Learning and Artificial Intelligence are becoming the key for approaching these problems.

The P-S strategy is revolutionizing the product value chain concept, adding a service chain during the Middle of Life (MOL) and the End of Life (EOL) of a product. In a Factory 4.0 data collected during all lifecycle of a product via smart sensors with a Product Lifecycle Management (PLM) system are useful for elaborating strategies useful for improving performances during each lifecycle phase, information about product dismission and recycle, as well as during the usage and maintenance and the service provided, are useful during the Beginning of life (BOL) in the design phase for reducing the time to market or for developing a smart maintenance system for the

MOL. In conclusion, all lifecycle phases exchange information collected and elaborated in a PLM system. All these 'evolutions' require innovation, but also the innovations should be 'smart', so in the Italian plan the Open Innovation scheme is expected also with funding. Innovation is expected also with the data driven horizontal/vertical integration, and finally the context of Industry 4.0 plan need to be inserted in a larger vision, smart manufacturing and smart supply chain have to be considered inside the context of smart cities, with the integration of production sites inside new cities, e.g. sharing place and energy for an optimization of spaces and resources for building the sustainability thought not just how environmental, but in a larger acceptation, how 'social sustainability' implementing another address of EU Factory of the Future community, the human-centric-manufacturing concept, like also reported in Fiasché et al. (2016).

First get in the right direction, seizing the opportunity, this is the challenge for the Italian SME, for being competitive in the worldwide 'markets 4.0'.

5 Brief Conclusions

Risks, challenges and opportunities for the Italian entrepreneurship have been approached in this preliminary analysis of the Industry 4.0 paradigm, of its declination in the Italian use case and of how the nine technological directions presented in the plan are the platform on which develop innovation in all the value chain of a business company. The macro-economic framework has been analysed and some business model has been read inside a context 4.0. Strength and weak points have been introduced in this analysis, and some theoretical aspect (no technological aspects) of how a factory can implement the 4.0 paradigm has been explored. In the next future a larger analysis need to be investigated especially about the Italian manufacturing eco-system.

References

Assolombarda: Industria 4.0, Position Paper n.2/2016 Area Industria e Innovazione e Centro Studi (2016)
EC Communication: For a European Industrial Renaissance, COM/2014/014 final (2014)
EC Communication: A Stronger European Industry for Growth and Economic Recovery Industrial Policy Communication Update, No. 582 final (2012)
European Parliament: Directorate General for Internal Policies: Industry 4.0, Policy department a: economic and scientific policy, February 2016
Fiasché, M., Pinzone, M., Fantini, P., Alexandru, A., Taisch, M.: Human-Centric Factories 4.0: a mathematical model for job allocation. In: 2nd IEEE International Forum on Research and Technologies for Society and Industry Leveraging a Better Tomorrow, RTSI 2016, Bologna, September 2016, pp. 7–9. IEEE (2016). Article number 7740613
Hermann, M., Pentek, T., Otto, B.: Design Principles for Industrie 4.0 Scenarios: A Literature Review. Technische Universität Dortmund Fakultät Maschinenbau Audi Stiftungslehrstuhl Supply Net Order Management. Working paper, n. 1 (2016)
IEA: Worldwide trends in Energy Use and Efficiency, Energy Indicators (2008)

Mosconi, F.: I distretti industriali alla prova della nuova sfida tecnologica: un'introduzione. *L'industria* / n.s., a. XXXVII, n. 3 (2016)

Opresnik, D., Fiasché, M., Taisch, M., Hirsch, M.: An evolving fuzzy inference system for extraction of rule set for planning a product–service strategy. Inf. Technol. Manag. **18**(2), 131–147 (2017)

Samir, K.C. et al.: Projection of population by level of educational attainment, age, and sex for 120 countries for 2005–2050. Demogr. Res. **22**, 383–472 (2010)

Sautter, B.: Futuring European industry: assessing the ManuFuture road towards EU re-industrialization. Eur. J. Futures Res. **4**, 25 (2016)

Tether, B., Bascavusoglu-Moreau, E.: Servitization: the extent and motivations for service provision amongst (UK) manufacturers. In: Proceedings of DRUID, pp. 19–21 (2012)

UCIMU: Il Parco Macchine Utensili E Sistemi Di Produzione Dell'industria Italiana. Centro Studi & Cultura Di Impresa Di UCIMU-SISTEMI PER PRODURRE, V Edizione, update 31.12.2014 (2014)

United States Energy Information Administration: Annual Energy Outlook 2011 (2011)

Viesti, G.: The rediscovery of industrial policy: the return to growth. Rev. Italian Cond. **1**, 79–104 (2013)

Wise, R., Baumgartner, P.: Go downstream. Harvard Bus. Rev. **77**(5), 133–141 (1999)

Review of the Literature About the Incidence of Port Dynamics in the Local Economy

María Berta Quintana-León[✉] [iD], Bertha Molina-Quintana[✉] [iD],
and Marco Alberto Valenzo-Jiménez[✉] [iD]

Universidad Michoacana de San Nicolás de Hidalgo,
UMSNH, Morelia, Michoacán, México
mariber0320@gmail.com, bettymolinaq@gmail.com,
marcovalenzo@hotmail.com

Abstract. In a context of economic integration and globalization with profound changes in the models of industrial organization and increasing competition in international markets, maritime ports are configured as a key piece of competitiveness. The present work analyzes theoretical aspects of the relation of the dynamics of ports with the economic growth of its surrounding region. It is important to characterize these theoretical approaches, because this question is not only important for the intellectual interest but also for the policy objective.

1 Introduction

The regions of ports seem to have always been an advantage compared to the regions that are not in the sea or in the rivers (Vleugels 1969). Since then, the rise of globalization has revealed the fallacy of such deterministic arguments defining ports as natural farming areas. Neoclassical theories at the poles of growth and industrial placement fall short when explaining the regional decrease in the benefits derived from maritime ports, in particular, when observing the scarcity of local impacts of containerization (Vallega 1996). Although ports can be seen as elements of urban structuring within their surrounding region (Wakeman 1996), their economic links with the peripheral regional economy seem to diminish (Boyer and Vigarié 1982; Grobar 2008).

The most striking problems of the economic geography "new economic geography" or (NEG) theories that follow the approach presented in Krugman's book in 1991 and, in particular, his article in the Journal of Political Economy are the concentration of economic activity in cities. Generally, in many countries, their dominant cities have developed mostly in ports (including sea, river, and lake ports). Many geographers said that the port represents the most convenient location for export and import. But, many economists argued that port cities seem to remain an unresolved question. The investigation of this question is important not only for the intellectual interest but also for the policy objective (Fujita and Mori 1996).

According to Fujita (Fujita and Mori 1996) the formation of cities is equally likely to begin in the interior and location of the port. In this context, port cities have the additional advantages of access to transport, are more likely to grow predominantly than non-port cities, and eventually some non-port cities can be absorbed by port cities.

© Springer Nature Switzerland AG 2019
J. Gil-Lafuente et al. (Eds.): AEDEM 2017, SSDC 180, pp. 316–326, 2019.
https://doi.org/10.1007/978-3-030-00677-8_25

This phenomenon can explain the relationship between the location of ports and cities and open our perspective on port cities.

The definition of the port industry has varied between studies, although it has generally incorporated the activities necessary to move ships, cargo and passengers through the port. The type of region used to estimate the flow of effects has also varied, with the region ranging from a town or city to a state.

Measurements of impacts reported in the studies have included production, added value, household income, employment, and payments to governments. The impact has been identified by the function of the port, type of cargo/basic products, port area and industry sector (on the effects of flow only). Several review articles have highlighted the general limitations of port impact studies and the need to interpret the results adequately.

2 Conceptual Framework

The port cities have a cause of advantages for the strategic geography for the trade that eventually generates its growth. Rodrigue (1999) described the typology of port cities to explain the size of port cities determined by their port traffic (see Fig. 1). In the figure you can determine the type of port cities. It also shows the relationship between the port traffic that we assume as the economic activity in port that can be used to determine the type of port cities. Using Rodrigue's typology, we can assume that port traffic is a factor that explains the development of cities.

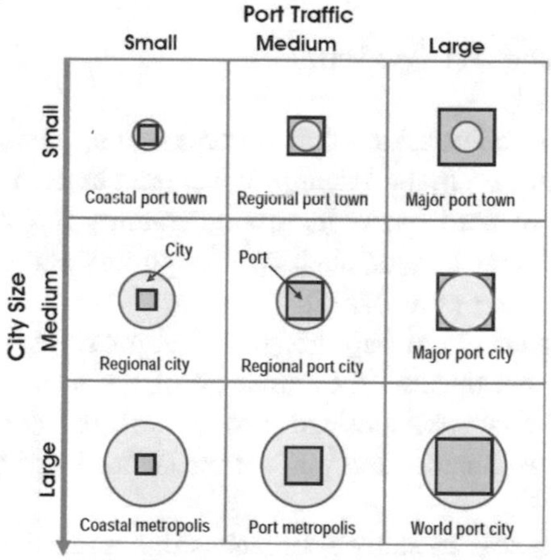

Fig. 1. Typology of the type of port cities of Jean-Paul Rodrigue. Source: Rodrigue 1999.

The ports have been associated with the public interest for the concept that they should be accessible to all potential users. Historically, countries such as France have maintained that ports must remain in the public sector, while other countries such as England have allowed private port companies to establish ports.

Many port cities in general can be described as a center of transportation and commerce. The following statement can be visualized using Rodrigue's model (Rodrigue 1999) that describes the function of the port (Fig. 2). Two types of port are appreciated. The first is the main port. The second is the regional port. The function of the main port is as a center that serves the regional port and another activity that has connectivity with economic activity. The Port is an infrastructure that creates service to its users. The import or export activity needed the service of a port. And because of this the city with port generates a lot of services in its economy.

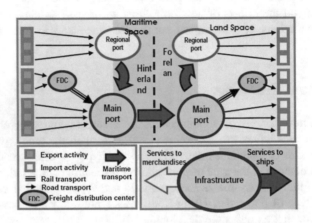

Fig. 2. Role of the Port of Jean-Paul Rodrigue. Source: Rodrigue 1999.

3   Port Regions as Port Systems

The port region can also be considered as a port system, or a system of two or more ports (and terminals), located in the vicinity of a certain area. A three-decade review of the analysis of the port system shows increasing stability or even a decrease in traffic concentration due to several factors, such as transporters and congestion strategies in large cargo centers (Ducruet et al. 2009).

The enjoyment of ports is not only geographical proximity, but also the functional interdependence of the sea through the exchange of services and land.

Local and regional characteristics are also crucial, but are often ignored by port specialists, who tend to consider the port as an isolated entity connected by cross-border networks.

Reference is made below to study approaches that relate economic growth to ports and their surrounding regions.

4 Ports as Facilitators of Economic Growth

The literature on ports and regional development can be classified into two categories. The optimist sees the port as the engine of local and regional economic growth, while for the pessimistic approach, ports simply respond to demand through the physical transfer of goods flows. This echoes the intense debate about whether infrastructures foster or follow development (Rietveld 1989).

The optimistic approach defines ports as the poles of growth for the development of economies of scale for production and trade, therefore, that offer comparative advantages to the regions and cities where they are located (Fujita et al. 1999; Clark et al. 2004). This general statement based on location theories implies that the efficiency of the ports generates more economic benefits, since it will allow more cargo processing, while inefficient ports can put a country or a region further away from the sources of cheaper inputs for markets or good production (Haddad et al. 2006).

The first empirical studies have shown the importance of multiplier effects at the local and regional level in developing countries (Omiunu 1989) and developed countries (Witherick 1981).

The pessimistic approach calls into question the local and regional benefits of port investments (Goss 1990), and the structuring of the effects of transport infrastructure (Offner 1993). This is particularly true in the case of ports located outside the main economic regions (Stern and Hayuth 1984; Fujita and Mori 1996). The improvement of the internal connectivity and the efficiency management can accentuate the tunnel effect that is defined by the diminution of the local benefits and a greater yield of volumes destined to distant areas.

Several scholars have observed the negative effects of traffic growth at the local level, such as congestion and lack of attractiveness (Mc Calla 1999; Rodrigue 2003; Rozenblat et al. 2004; Grobar 2008). A moderate approach proposed by Vallega (1983) interprets the development of ports and regional development as two distinct processes and with episodes of indirect interactions. This approach has been much more complemented by the works of Langen (2003) on ports as groups of economic activities. Far from establishing a direct line between port activity and industrial development, the concept of a port cluster depends on institutional arrangements and on the presence of leading companies in particular, economic activities.

5 The Measurement of Port-Region Interdependence

Several studies have addressed benefit measurements related to the local and regional port, using a wide variety of methodologies, but at the same time the regional planner is interested in the benefits that a port brings to a city or region, the difficulty lies in the quantification of benefits (Bird 1971). Studies of port impacts have flourished since the 1950s in the United States and elsewhere (Hall 2004), with many case studies on measuring the multiplier effects of port activities in surrounding areas (Taylor 1974; Witherick 1981; Omiunu 1989). The literature on regional development has largely focused on results (Porter 2003) but with few works related to transport infrastructures or port activities (Rietveld 1989).

Ports are not isolated entities of virtual connection to value chains. They are part of a regional economy, and the evolution of the regional economy affects the performance of the port. In particular, industrial specialization constitutes a weakness for the performance of ports in the era of globalization.

6 The Model of Fujita and Mori

The neoclassical model of FUJITA and MORI (1996) suggests that the port related to urban growth is more likely to occur in remotely located regions where specialized industries develop local resources in comparative advantages.

The geographical distance of the essential economic regions plays an important role. Maritime ports located in regions beyond the center enjoy an equivalent portfolio of merchandise as a remedy for distance and internal transport costs. This confirms the model in which the remoteness of the main advantageous economic regions if converted to the interior offers sufficient comparative advantages. Some ports in the capitals of the countries are more diversified, but do not generate large volumes of urban traffic due to limitations for the expansion of the local port.

As we see the importance of distance in terms of the port and the main city, there is a hierarchical relationship within the system, as well as the comparative advantages.

As reflected in this model, the impact depends on the organization, the hierarchical union between the port city and the main city.

7 The Economic Situation of the Areas Surrounding the Main Ports

Historically, ports have been considered as engines of economic development of the cities and regions where they are located.

Companies that wish to export or import merchandise by sea have found it advantageous to locate themselves near a port to minimize land transport costs. Traditional methods for moving and handling the load were labor-intensive, creating important effects on direct local employment. Therefore, according to Campbell, ports have traditionally been the centers of economic and cultural activities in cities.

However, with the latest advances in transport technology, the role of ports in local economic development has changed. Containerization has made the merchandise movement process much more capital intensive, thus reducing the benefits of local employment to have a port. The relatively low cost of overland transport has reduced the advantage of locating export companies near a port. Export companies are now more likely to be located in areas where land is relatively cheap and where there is good access to transportation services, which allows them to send their exports to the district port. Container imports increase their flow to distribution centers located inland instead of coastal locations.

While the benefits of proximity to the port have diminished over time, costs have increased. The most important ports now process millions of containers per year. These containers must be loaded onto trucks and transported by rail from the port area or their destinations. As a consequence, traffic congestion and pollution from port activities is becoming a growing concern in areas adjacent to ports.

The literature on ports districts has grappled with the rapidly changing reality of relations between ports and their cities. Ducruet and Jeong (2005) suggest that, although there is no consensus in the literature on the precise definition of a port city, on a local scale, it is the urban zone and the mixture of jurisdiction and functions, the transition zone (Hayuth 1982; Hoyle 1989), on a broader scale, is the node of the system as a whole, including several cities and ports within a regional area (serial port, country, continent), in the assumption of land-sea connection.

As Hesse (2006) points out, the globalization of manufacturing has been an important factor in the expansion of port activities, as worldwide interwoven production systems have led to an increasing need for the trade of raw and semi-finished materials, as well as finished products. However, increases in trade volumes in ports have not led to an increase in employment in water transport, as the increase in productivity in transport water has offset the impact of increased trade volumes. In addition, Hesse considers a growing trend of logistics and distribution activities to find further away from ports, which reduces a second area of economic benefit to port cities.

One of the first studies to document the decline in local economic benefits from port cities was Campbell (1993). In a case study of the Port of Oakland, Campbell documented the dispersion of functions related to the port and industries that depend on the port throughout the Bay Area. He considers that, while the Port of Oakland acquired control over cargo handling over the San Francisco port during the 1970s and 1980s, maritime employment services remained concentrated in San Francisco. In addition, he considers that the ports that depend on industries in the bay area, which he defines based on the percentage of his business that is based on maritime trade, is not geographically concentrated in the counties that contain the ports. Therefore, he concludes, the general result of containerization, therefore, seems to be the change of the benefits of a port from the local to the regional and national scales.

In a more recent article, Poister and Helling (2000) present evidence on the weakening of economic links between ports and their cities and the decline of direct employment in water transport.

Defining the geographic concept of "district port", we chose to define the concept as the zone located within the radius of 7.5 miles of the port. This choice of geographical boundaries reflects a desire to collect not only an economic activity that occurs directly in the port, but also in the areas immediately adjacent to the port.

In most cases, the per capita income is lower in the district port than it is in the metropolitan area.

In contrast to the per capita income indicator, which indicates that some district ports compare favorably with their surrounding metropolitan areas, we find a much more negative picture when comparing unemployment and poverty rates of the district port with the statistics of the metropolitan area.

Unemployment rates are higher in the port districts than in the metropolitan areas that surround it. When we compare the percentage of families that are below the poverty level, we find that, without exception, poverty rates, in many cases, are significantly higher, higher than the poverty rates in the metropolitan area.

In almost all cases, the prevalence of low-income families is greater in the district port than in the metropolitan area, if the low-income threshold has been set for families earning less than $ 10,000 or less than $ 25,000. (Grobar 2008).

In a review of the literature on the subject of environmental inequality and environmental justice, Brulle and Pellow (2006) find extensive literature documenting the existence of environmental inequality in the United States. This literature considers that people of low socioeconomic status and minorities are disproportionately affected by the environmental risks in their communities and the adverse health effects derived from this exposure. It is found that the population of the port districts are poorer than the population in general and, in many cases, they have a greater proportion of minorities than in the metropolitan areas that surround it. This would imply that the burden of environmental hazards created by the largest US container ports, is disproportionately in charge of low income and minority populations.

Despite the high unemployment rate in the port districts, as measured by household statistics, the data reveal that job creation is relatively abundant in the port district. Of course, residents of port districts can often find employment outside the narrow geography of their place of residence, so this statistic may underestimate the employment opportunities available to residents of the port district, especially at large metropolitan areas.

Therefore, port operations can serve to cause surrounding areas to become increasingly undesirable places for housing. As a result, real estate prices may fall or increase more slowly than the values in the metropolitan area. Therefore the poorest residents can move to the area, attracted by relatively low prices or rents. Ironically, then, the same ports that serve as "economic engines" of the region and the nation can be the cause of the economic decline and deterioration of the immediate areas that surround them.

The high rate of unemployment among residents of the port district is also consistent with the literature on the spatial distribution of the disadvantaged situation in metropolitan areas. This literature began with the work of Kain (1968), but was later expanded and refined by a large number of researchers (including Houston 2005; Raphael 1998; Stoll et al. 2000; Thomas 1998).

The indicators of disadvantage, such as high unemployment and poverty rates, are not evenly distributed in metropolitan areas, but are spatially concentrated in large cities. They also have a high minority percentage of residents, and thus adapt to the description of the port districts described above. One possible explanation for the high unemployment rates observed in these areas is that there is a "territorial imbalance" between job seekers and jobs, which are often found in suburban areas. This explanation has come to be known as the "spatial mismatch hypothesis".

The city center tends to have a lower proportion of low-skilled jobs as a proportion of total jobs compared to the suburbs. If this is the case in our port district, this explains why it observes a high rate of unemployment and poverty rates, even in an area in which the total number of jobs is abundant. Unemployment noted a spatial function of the lack of correspondence between low-skilled workers and low-skilled jobs.

Studies of the economic impact of maritime ports have been increasingly important, since they measure their direct and indirect impact on the patterns of employment, income and tax revenues in the regional economy. Measuring that impact of maritime ports on the local economy becomes even more crucial from the point of view of state and local governments, as it can serve as an important educational tool for the community in understanding the structure of a port, as well as its immediate economic effects.

Several port impact studies have been conducted that attempt to measure the impact of ports in a local economy in terms of employment, sales, income, and taxes (Pearson (1964); Waters (1977); Chang (1978); Hoffman (1980); Davis (1983); Yochum and Agarwal (1988); and Brass and Groseclose Colbert (1989); Warf and Cox (1989); DeSalvo (1994); Gripaios and Gripaios (1995); Verbeke and Debisschop (1996); Castro (1997)). However, there is no standard methodology that accurately measures the economic impact of a seaport. The first studies differ among themselves in their methodology and definition of the economic effects of a port.

Several review articles have discussed the limitations of port impact studies and the proper interpretation of results. They also include proposals to improve the method used in these studies.

Waters proposes the use of cost-benefit analysis to determine the direct effects, with the input-output models that are used to estimate the flow of the effects.

Chang (1978) pointed out that Waters had rightly pointed out many limitations that involve the use of port impact studies.

He proposed a model that linked the expansion of capacity with the existing profitability, the utilization capacity and expected growth of the demand for port services.

Davis (1983) noted that port impact studies had three main weaknesses. First, there was no commonly accepted definition of the port industry. Second, existing studies use at least four alternative methods to estimate the flow of effects, the basis of economic analysis, analysis of income and expenditure, input-output analysis, and the application of a multiplier from a previous study. Third, the studies had some deficiencies that should be used for the evaluation of the economic effects of changes in the volume of port services.

8 Conclusions

The activity of ports, conceived as merchandise distribution platforms, is considered a multi-product process. This is because the heterogeneity of the merchandise served and the services provided in them mean that even a relatively small facility has many features in common with the large industrial companies of multiple production. However, port activity is surrounded by a series of special circumstances that make it different from any other of a strictly industrial or commercial nature since, among other aspects, society tends to consider ports as facilities at the service of the common good.

Despite this, authors of the prestige of Goss or Grossdidier de Matons recognize that, despite the economic impact that this activity may have on its geographical environment, the main objective of its managers must be to try to maximize the performance of the docks, which entails maximizing the traffic served in them and, therefore, the demand for port services performed in them.

Manufacturing labor and export and import activity in these ports may be explaining and influencing the economic growth of these port cities. Partially, the manufacturing industries are not significant, only because the industrial agglomeration is optimal. The functions of these cities are not particular in the manufacturing activity but provide services and commercial activity, so they could develop the industrial performance of the interior.

Cities with higher transportation costs would also be less likely to attract investment in commercial activities, because cities are temporary suppliers of economic activity. Cities in particular have two important roles in their environment. An urban center like the cities serves as the center of its rural environment and as a mediator of the interaction of the outside world. This statement may consider a city to have a strong influence, strength or power in its rural environment.

References

Boyer, J.C., Vigarié, A.: Les Ports et l'Organisation Urbaine et Régionale. Bulletin de l'Association des Géographes Français **487**, 159–182 (1982)

Bird, J.: Seaports and Seaport Terminals. Hutchinson, London (1971)

Brulle, R., Pellow, D.: Environmental justice: human health and environmental inequalities. Annu. Rev. Public Health **27**, 103–124 (2006)

Campbell, S.: Increasing trade, declining port districts: port containerization and the regional diffusion of economic benefits. In: Noponen, H., Graham, J., Markusen, A.R. (eds.) Trading, Industries, Trading Region: International Trade, American Industry, and Regional Economic Development, pp. 212–255. Guilford Press, New York (1993)

Castro, J.: Economic impact analysis of santander port of its hinterland. Int. J. Transp. Econ. **XXIV**(2), 259–277 (1997)

Chang, S.: In defense of port impact studies. Transp. J. **17**, 79–85 (1978)

Clark, X., Dollar, D., Micco, A.: Port Efficiency, Maritime Transport Costs and Bilateral Trade. National Bureau of Economic Research Working Paper 10353 (2004)

Davis, H.C.: Regional port impact studies: a critique and suggested methodology. Transp. J. **23**, 61–71 (1983)

DeSalvo, J.: Measuring the Direct Impacts of a Port. Transp. J. Summer, 33–42 (1994)

Ducruet, C., Jeong, O.: European Port-city Interface and its Asian Application, Korea Research Institute for Human Settlements, Research Report 17 (2005)

Ducruet, C., et al.: Going west? spatial polarization of the north korean port system. J. Transp. Geogr. (2009). (forthcoming)

Fujita, M., Mori, T.: The role of ports in the making of major cities: self-agglomeration and hub-effect. J. Dev. Econ. **49**(1), 93–120 (1996)

Fujita, M. et al., (eds.): The Spatial Economy: Cities, Regions and International Trade. MIT Press, Cambridge & London (1999)

Goss, R.: The economic functions of seaports. Marit. Policy Manag. **17**(3), 207–219 (1990)

Gripaios, P., Gripaios, R.: The impact of a port on its local economy: the case of Plymouth. Marit. Policy Manag. **22**(1), 13–23 (1995)

Grobar, L.M.: The economic status of areas surrounding major U.S. container ports: evidence and policy issues. Growth Change **39**(3), 497–516 (2008)

Groseclose, B., Brass, J.S., Colbert, E.L.: South Carolina State Ports Authority: Economic Impact Study (Charleston, S.C., South Carolina State Ports Authority) (1989)

Haddad, E.A. et al.: 'Port Efficiency and Regional Development' (2006). Unpublished paper, Niteroi Hille, J. et al.: Associação Nacional dos Centros de Pós-Graduação em Economia (1975)

Hall, P.V.: We d have to sink the skips: impact studies and the 2002 West Coast Port Lockout. Econ. Dev. Q. **18**, 354–367 (2004)

Hayuth, Y.: The port-urban interface: an area in transition. Area **14**(3), 219–224 (1982)

Helling, A., Poister, T.: U.S. maritime ports: trends, policy implications and research needs. Econ. Dev. Q. **14**, 300–317 (2000)

Hesse, M.: Global cahain, local pain: regional implications of global distribution networks in the German Nort range. Growth Change **37**, 570–596 (2006)

Hoffman, R.F.: Economic Impact Survey: The Effects of Waterborne Commerce on the Community: A Four-County Standard Metropolitan Statistical Area Survey. Port of Milwaukee, Board of Harbor Commissioners, Milwaukee (1980)

Houston, D.: Employability, skills mismatch and spatial mismatch in metropolitan labour markets. Urban Stud. **42**(2), 221–243 (2005)

Hoyle, B.S.: The port-city interface: trends, problems and examples. Geoforum **20**, 429–435. Journal of Commerce Online, (2007) New York-New Jersey, Rotterdam in environmental agreement, 4 September 1989

Kain, J.: Housing segregation, Negro unemployment and metropolitan segregation. Quart. J. Econ. **82**(2), 175–197 (1968)

de Langen, P.W.: The Performance of Seaport Clusters. Erasmus Research Institute of Management, Rotterdam (2003)

Mc Calla, R.: Global change, local pain: intermodal seaport terminals and their service areas. J. Transp. Geogr. **7**(4), 247–54, et al. (2004), Dealing with globalisation at the regional and local levels· the case of contemporary containerization. The Canadian Geographer **48**(4), 473–87 (1999)

Offner, J.M.: 'Les Effets Structurants du Transport: Mythe Politique. Mystification Scientifique', L'Espace Géographique **3**, 233–242 (1993)

Omiunu, F.G.I.: The port factor in the growth and decline of Warri and Sapele townships in the western delta region of Nigeria. Appl. Geogr. **9**, 57–69 (1989)

Pearson, R.L.: Measuring the impact of the waterborne commerce of the ports of Virginia on employment, wages and other key indices of the Virginia economy, 1953–1962 (University of Virginia, Bureau of Population and Economic Research) (1964)

Porter, M.E.: The economic performance of regions. Reg. Stud. **37**(6/7), 549–578 (2003)

Rodrigue, J.P.: The Port Authority of New York and New Jersey: Global Changes, Regional Gains and Local Challenges in Port Development. Les Cahiers Scientifiques du Transport **44**, 55–75 (2003)

Raphael, S.: The spatial mismatch hypothesis and black youth joblessness: evidence from the San Francisco Bay area. J. Urban Econ. **43**(1), 79–111 (1998)

Rietveld, P.: Infrastructure and regional development. Ann. Reg. Sci. **23**(4), 255–274 (1989)

Rodrigue, J.P.: Transport Geography. Department of Economics & Geography Hofstra University, Hempstead, New York (1999)

Rozenblat, C.: Les Villes Portuaires en Europe, Analyse Comparative. CNRS, Montpellier (2004)

Stern, H.: Developmental effect of geopolitically located ports. In: Hoyle, B.S., Hilling, D. (eds.) Seapaort Systems and Spatial Change, pp. 239–249. Willey, Chichester (1984)

Stoll, M., Holzer, H., Ihlanfeldt, K.: Within cities and suburbs: racial residential concentration and the spatial distribution of employment opportunities across sub-metropolitan areas. J. Policy Anal. Manag. **19**(2), 207–231 (2000)

Taylor, M.J.: The impact of New Zealand s secondary ports in their associated urban communities. N. Z. Geogr. **30**, 35–53 (1974)

Thomas, J.M.: Ethnic variation in commuting propensity and unemployment spells: some U.K. evidence. J. Urban Econ. **43**(3), 385–400 (1998)

Vallega, A.: Fonctions portuaires et polarisations littorales dans la nouvelle régionalisation de la Méditerranée, quelques réflexions. Villes et ports, développement portuaire, croissance spatiale des villes, environnement littoral, pp. 355–367 (1979)

Vallega, A.: Nodalité et Centralité Face à la Multimodalité: Eléments pour un Relais entre Théorie Régionale et Théorie des Transports. In: Calogero Muscara and Corrado Poli (eds.) Transport Geography Facing Geography. International Geographical Union, Rome (1983)

Vallega, A.: Cityports, Coastal Zones and Sustainable Development. In: Hoyle, B.S. (ed.) Cityports, Coastal Zones and Regional Change. Wiley, Chichester (1996)

Verbeke, A., Debisschop, K.: A note on the use of port economic impact studies for the evaluation of large scale port projects. Int. J. Transp. Econ. **XXIII**(3), 247–266 (1996)

Vleugels, R.L.M.: The Economic Impact of Ports on the Regions They Serve and the Role of Industrial Development. Paper presented at the 6th Biennial Conference of the International Association of Ports and Harbors, Melbourne, Australia, March (1969)

Warf, B., Cox, J.: The changing economic impact of the port of New York. Marit. Policy Manag. **16**(1), 3–11 (1989)

Waters, R.C.: Port economic impact studies: practice and assessment. Transp. J. **16**, 14–18 (1977)

Wakeman, R.: What is a sustainable port? The relationship between ports and their regions. J. Urban Technol. **3**(2), 65–79 (1996)

Witherick, M.E.: Port developments, port-city linkages and prospects for maritime industry: a case study of Southampton. In: Hoyle, Brian S., Pinder, David A. (eds.) Cityport Industrialization and Regional Development. Pergamon Press, Oxford (1981)

Yochum, G.R., Agarwal, V.B.: Static and changing port economic impacts. Marit. Policy Manag. **15**(2), 157–171 (1988)

Theoretical Aspects of Creating Customer Value

José Alfredo Delgado-Guzmán[1], Bertha Molina-Quintana[2]([⊠]) [iD],
and María Berta Quintana-León[2] [iD]

[1] Universidad Nacional Autónoma de México, UNAM,
Ciudad de México, CDMX, Mexico
adelgadoguzman@hotmial.com
[2] Universidad Michoacana de San Nicolás de Hidalgo, UMSNH,
Morelia, Michoacán, México
bettymolinaq@gmail.com, mariber0320@gmail.com

Abstract. The concept of value has changed from being seen as simply a balance between benefits and sacrifices towards a process of interaction in which consumers and companies collaborate to create a joint value. The role of marketing is to help the company create value to its customers and to be superior to the competition. It is important to understand the concept of value, value for the client and creation of value for the client in order to analyze the situation of the Company and act strategically in order to be competitive.

1 Introduction

The American Marketing Association has five different definitions of marketing between 1935 and 2007. The definition of 1935 was; Marketing is the performance of business activities that direct the flow of goods and services from producers to consumers (American Marketing Association, (AMA) 2007), through the 4Ps (product, price, place and promotion). In 1985 the definition was; Marketing is the process of planning and executing the conception, pricing, promotion and distribution of ideas, goods and services to create exchanges that meet individual and organizational objectives (AMA 2007). Today it is believed that marketing is fundamentally about value creation. Therefore, the most recent definition of the American Marketing Association states that:

> Marketing is the activity, set of institutions and processes to create, communicate, deliver and exchange offers that have value for customers, partners and society in general. This was the definition in 2007 (AMA 2007).

An alternative definition, proposed by Grönroos (2006), and rooted in marketing services and in the literature of industrial marketing is:

> Marketing is a focus on the client that pervades the processes and functions of organization and is oriented to make promises through value proposals, which allows the fulfillment of the individual expectations created by such promises; Meeting these expectations supports the customer's value-generating processes as well as the creation of value in the company.

© Springer Nature Switzerland AG 2019
J. Gil-Lafuente et al. (Eds.): AEDEM 2017, SSDC 180, pp. 327–339, 2019.
https://doi.org/10.1007/978-3-030-00677-8_26

While there are important conceptual differences between the two definitions, what can be seen is that the creation of value is found in both.

This is in line with other developments in marketing, stating that the role of marketing is to help the company create value to its customers, to be superior to the competition (e.g. Tzokas and Saren 1999; Lindgreen and Wynstra 2005).

The great importance that the construct value has been acquiring, has caused that in the year of 1997 Slater began to outline what he called "theory of the firm based on value for the client". It is important to mention that as a result of the development it has had so far, it cannot be said that it is a theory but rather an approach derived from the perspective of resources and capacities (RBV), with the distinctive feature of mentioning in a more express way the concept of value for the customer.

For its construction it argues that this approach must respond to three basic questions: why companies exist and their main purpose, why are there differences in the scale, scope and types of activities between companies, and why are there differences in their performance. In relation to the first question, the companies have as a fundamental mission or reason the client's satisfaction. This satisfaction is achieved by the delivery of a higher value, which will later be translated into a superior result for the company. In such a way that the companies will exist to provide a product or service, being neither effective nor efficient for the buyers to try to satisfy all their necessities by themselves.

With regard to the differences in scale, scope and types of activities between companies, they will be determined by the value strategy for the customer, which includes the following elements: the establishment of appropriate market objectives, the selection of segments of specific market, creating a value proposition that establishes a position of competitive advantage, and the development of the skills needed to understand the customer's needs and deliver the promised value.

2 Conceptual Framework

It is a fact that companies' competitiveness depends on their ability to create sustainable competitive advantages over time. Currently, the environment they face has undergone major changes, mainly characterized by an increase in the competition and higher levels of market demand. That is why companies, besides competing, must seek to satisfy the needs of the client. Said satisfaction is achieved when the company is able to offer the consumer a value equal to or greater than what it expects (López and Vásquez 2002).

Companies now pay more attention to their markets, since customers will be the true judges of value creation. In this sense, one of the factors that must be analyzed in depth is the perception that customers have about the value of the specific offer of a company, which should be perceived as such.

Value creation presents three key characteristics; it must be strategic, because the core of the company's global strategy must be to create and offer the client what they value; it must be systematic, because, generally, it will imply an organizational and behavioral change, and finally, it must be continuous, given the changing nature of the clients and other elements of the environment (Band 1994).

Any strategy aimed at the market must concern itself with two primary objectives: to provide superior value to the client through the distinctive competences of the organization, create an economic value for the owners of the company, bearing in mind that this objective will be the consequence of the first from them. The creation and delivery of a superior value for the client, especially for those customers of greater value to the company, will increase the value for the organization (Slywotzky 1997).

Despite its importance, there is no consensus on how to define the concept of value and consequently how to measure it. Table 1 shows some definitions that reveal how the notion of value has changed in recent years. The concept of value has changed from being seen as simply a balance between benefits and sacrifices towards an interaction process in which consumers and companies collaborate to jointly créate value.

Table 1. Definitions of value by different authors

Definition of value	Author and year
Is the benefit that a customer obtains from a product or service, less the cost of obtaining it	Carr (1992)
Supposes the estimation by the consumer of the capacity of the products to satisfy their needs	Kottler et al. (1995)
Is a function of what a customer gets, the solution provided by an offering and the sacrifice of the customer to get this solution	Gronroos (1997)
Are the advantages that are perceived in exchange for the loads that are supported	Berry and Yadav (1997)
Is the worth in monetary terms of the technical, economic, service and social benefits that the customer of a company receives in exchange for the price it pays for a market offer	Anderson and Narus (1998)
Is the perceived value of a product or service compared to what has been paid and the opportunity costs incurred	Goodwin and Ball (1999)
Will increasingly have to be created jointly with consumers. Interaction of the consumer-company is the locus of creation of value	Prahalad and Ramaswamy (2004)
Is the perceived exchange between the multiple benefits and sacrifices generated through the relationship with the client by those who make the key decisions in the provider organizations	Walter et al. (2001)
Is a cognitive response that is composed of several basic dimensions, where the benefits received and the sacrifices borne are processed together	Martin (2001)
Can be seen as the sum of the communication processes and the results that the consumer integrates with their brand relationship. Through this process, the role of the consumer is established in the generation of value through their interactive participation	Lindberg-Repo and Gronroos (2004)

(*continued*)

Table 1. (*continued*)

Definition of value	Author and year
Is the client's judgment about the service received, where all the benefits and sacrifices perceived therein are processed simultaneously in the client's mind, which leads to an overall evaluation of the service provider. This judgment affects the response and current and future behavior of the client with respect to the service provider	Martin et al. (2004)
Is not delivered by a company to the clients but is created in the client's processes as support to those processes and through the co-creation of the interactions with the clients	Gronroos (2006)

Source: Own elaboration.

Lindgreen and Wynstra (2005) identify two trends in value research in industrial marketing, the value of goods and services and the value of the buyer-seller relationship. The first research position contemplates multiple components, where perceived benefits are balanced (for example, a combination of physical attributes, service attributes and technical support) against perceived sacrifices (e.g. price). Value is a function of individual perceptions since it is perceived subjectively by customers, and it is relative to the competition as customers to compare different offers and the expected benefits of them (Ulaga and Chacour 2001).

Value as a derivative of the buyer-seller relationship has been the focus in industrial marketing and the relationship of marketing literature. This view on value recognizes that the relationship can have an important influence on how value is perceived (Ravald and Gronroos 1996).

A somewhat different view of the value relationship has developed recently, as marketers recognize the growing importance of customers in the process of value creation. While the focus is still on the value relationship, it is not the value of the relationship per se, but the value created through the relationship. Experts in management and marketing have begun to underline that value is co-created with customers in an interactive process (Ramírez 1999; Prahalad and Ramaswamy 2004), in which companies can offer value propositions (Normann and Ramírez 1993), but ultimately it is the client in the process of experiencing the value offers who determines the value (Ponsonby and Boyle 2004; Gronroos 2006).

The recent resurgence of the concept of value in use, however, broadens this point of view by placing an emphasis on interaction with customers in the process of creating value as a central marketing activity (Ballantyne and Varey 2006).

Value for the Client is considered an essential prerequisite for the company's long-term survival and success (Porter 1996; Woodruff 1997; Payne and Holt 2001; Huber et al. 2001). Understanding the value and how customers judge a service or product is crucial to achieving a competitive advantage. Scientists and practitioners have recognized the power of the concept of Value for the Client in the identification of value for clients and the management of customer behavior (Johnson et al. 2006; Kothari and Lackner 2006; Setijono and Dahlgaard 2007).

A multitude of Customer Value approaches have emerged, and ambiguous empirical results have been presented. So far there is very little consensus in the literature regarding the notation and conception in this field of research. Even the term Value for the Client is used and evaluated in very different ways in the marketing literature (Woodruff 1997). For now there is no consistent definition of "Value for the Client". So in Table 2, some definitions of Value for the Client are presented.

Table 2. Definitions of the concept of customer value by different authors

Author and year	Definition of value for the client
Gale (1994)	Is perceived quality adjusted by the relative price of the product of the market. It is the opinion of its customers of its products (or services) compared to that of its competitors
Holbrook (1994)	Is a realistic preference (comparative, personal and situational) characterizing an experience of the subject of the consumer to interact with some object, that is, any good, service, person, place, thing, event or idea
Woodruff and Gardial (1996)	Is the clients' perception of what they want to happen in a specific use situation, with the help of the offer of a product or services, in order to achieve a desired goal or purpose
Woodruff (1997)	Is a preference and evaluation perceived by the customer of those attributes of the product, of the result and of the consequences derived from the use that facilitates reaching the objectives and purposes of the client when he uses them
Woodruff (1997)	Is a preference perceived by the client and evaluation of the attributes of the product, attribute performance, and the consequences derived from the use that facilitate (or block) the achievement of goals and objectives of the client in situations in use
Goodstein and Butz (1998)	Are the perceptions that said client has about the good that his needs have been met

Source: Own elaboration.

In recent years, the development of relationship marketing has led to a relational approach to value creation having a greater importance than the transactional approach (Ulaga and Eggert 2005). That is, if you have to determine the value that the exchange partner receives through the offer of a product, questions of a relational nature are taken into account or not taken into account (Gronross 2004).

Next, Table 3 shows the different ways in which the concept of value for the client has been studied. From the review of the literature, seven perspectives of analysis on the creation of value for the client are identified and examined.

In the same way as presented in Table 3, the different perspectives of analysis with which the concept of value for the client has been evaluated, now in Table 4, the models of several authors can be observed in different approaches related to the concept of value for the client, from the manufacturer's focus, the distributor's focus, both, etc.

As indicated by Mocciaro and Battista (2005) the creation of value supposes the development of a set of capacities related to the creation and renewal of the roads for

Table 3. Perspectives of analysis on the concept of customer value

Perspective analysis	Concept approach of value for the client	Representative authors
Perspective based on a quantitative approach		
Perspective based on a monetary quantification	The concept of value refers to the relative valuation in monetary units of the set of net benefits (social, service, technical and economic) that a part of the exchange receives, in exchange for the price it pays for the supply of a product	Anderson, Jain and Chintagunta (1993); Anderson (1995); Anderson and Narus (1998); Anderson, Thomson and Wynstra (2000)
Perspective based on a value hierarchy	The concept of value is the perception that customers have regarding preferences and evaluations about the attributes of a product, the behavior of the product in relation to these attributes and the consequences that derive from the use of the product, which facilitate (or block) the achievement of the objectives and purposes of the clients in the situations of use	Woodruff (1997)
Perspective based in marketing de relationships		
Perspective based on joint effort	The creation of value is the process by which the competitive abilities of the parties are improved by being in the relationship. In this way, it is considered that to add more value to the offer of a product must resort to a joint effort that is mutually beneficial, created by the synergistic combination of the forces of the parties	Wilson (1995); Kothandaraman and Wilson (2001)
Dynamic perspective	The basic premise of this approach is that the meaning of value for the client changes as a relationship of exchange develops and evolves. That is, the prolonged use of a product may result in the valuation made on certain aspects of the same does not coincide with that made in previous moments	Parasuraman (1997)

(*continued*)

Table 3. (*continued*)

Perspective analysis	Concept approach of value for the client	Representative authors
Perspective based on the components of the business relationship	The concept of value seeks to collect the balance or trade-off between the multiple benefits and sacrifices that are associated with being part of a relationship of exchange. In other words, value judgments are not evaluated or made on specific or specific product offers, but the relationship as a whole is taken into account	Wilson and Jantrania (1994); Ravald and Gronroos (1996); Grönroos (1997); Gwinner, Gremier and Bitner (1998); Lapierre (2000); Ulaga and Eggert (2005)
Perspective based on the functions performed by the business relationship	The creation of value is manifested through the functions or tasks that a relationship can perform for the actors involved. These functions can be: (a) direct or primary, when they collect both positive and negative effects derived from the interaction in the main dyadic relation; and/or (b) indirect or secondary, if they reflect the effects caused by the existence of connections between more than one relationship, that is, positive or negative indirect effects of a relationship, being connected to other relationships	Anderson, Hakansson and Johanson (1994); Anderson (1995); Walter et al. (2001); Walter and Ritter (2003); Walter et al. (2003); Ryssell, Ritter and Gemunden (2004)
Other perspectives	First case: The value for the client is created when the consumer benefits associated with a product or service exceed the life cycle costs of the offer for the client	Slater and Narver (2000)
	Second case: A conceptual model of value for the client is proposed, where five different types of value are collected: promised value, desired value, designed value, expected value and received value	Van der Haar et al. (2001)

Source: López et al. (2010).

Table 4. Models of literature studied about value

Study	Authors	Focus
Customer value	Cantone model 2004	Approach based on theory
	Menon, Homburg and Beutin 2005 model	Quantitative study, questionnaires to 981 managers
	Flint, Woodruff and Gardial 1997–2002 model	Approach based on Theory
Perceived value For the client	Anderson Jain C. and Chintagunta 1993 model	Approach based on Theory
	Lapierre 2000 model	Quantitative study, questionnaires to 209 managers
	Ulaga and Chacour 2001 model	Qualitative study, 36 respondents in 12 companies
Value in relation to the customer's perspective	Wilson and Jantrania 1994 model	Approach based on Theory
	Ravald Grönroos 1996 model	Approach based on Theory
	Grönroos 1997 model	Approach based on Theory
	Ford and Mc Dowell 1999 model	Approach based on Theory
	Ulaga and Eggert 2001 model	Qualitative approach, interviews with 15 managers
	Ulaga 2003 model	Approach based on Theory
	Gadde and Snehota 2000 model	Approach based on Theory
	Ulaga and Eggert 2005 model	Quantitative study, questionnaires to 221 managers
	Ulaga Eggert and Schultz 2005 model	Quantitative study, questionnaires to 420 managers
	Ulaga and Eggert 2006 model	Quantitative study, 112 questionnaires for the exploration sample and 288 of the validation sample
Value in relation to the perspective of suppliers	Walter, Ritter and Gemünden 2001 model	Quantitative study, 30 interviews and 247 questionnaires
	Baxter and Matear 2004 model	Quantitative study, questionnaires to 318 managers
Value in relation to emphasis on network dimensions	Gummensson 2004 model	Approach based on Theory
	Möller and Törrönen's 2003 model	Approach based on Theory

(*continued*)

Table 4. (*continued*)

Study	Authors	Focus
Value in relation to considerations in both; clients and suppliers	Hogan 2001 model	Individual relationship evaluation
	Biggemann and Buttle 2005 model	Approach based on Theory
	Mandjak and Simon 2007 model	Qualitative interviews and care groups
	Bouzine, Chameeva, Durrieu and Mandjak 2001 model	Approach based on Theory
Value In Relation To The Combined Perspective	No contributions	No contributions

Source: Corsaro (2008).

the competitive advantage. Therefore, they will involve new combinations of resources to develop new skills and new knowledge about the current and future needs of clients (Mocciaro and Battista 2005; Zander and Zander 2005).

Under the Perspective of Resources and Capabilities (RBV), the idea of creating value for the client appears in a large part of the literature, although it is true that little attention has been paid to the question of how resources generate value for the client (Srivastava et al., Zander and Zander 2005), also called the demand perspective (Adner and Zemsky 2006). This has meant little attention to the role and importance of customers in the company's strategy and growth, underestimating many of the possible ways in which companies can generate income and ensure long-term growth (Zander and Zander 2005).

From a business point of view, the creation of value refers to the role played by companies with respect to the economic system in which they operate, and involves the ability to perceive and implement new combinations of resources to develop new combinations of resources to develop new competencies and new knowledge that are capable of increasing the efficiency achieved in the use of the resources present in the economic system (Mocciaro and Battista 2005). Such efficiency can be achieved through two paths; have resources for other uses, so that opportunity costs are reduced, or that products or services with higher levels of performance and new combinations of resources are produced that lead to the production of goods and services that satisfy desires and needs that it was not possible to satisfy or that they had not been previously expressed.

Therefore, it will focus on the appropriation of market rents generated by the possession of certain resources or differential capacities (Mocciaro and Battista 2005). There will be no incentives to create value for the client (Mizik and Jacobson 2003). Among these factors, these authors point out the following: the effects of reputation and brand, exchange costs, advertising, etc.

A definition of the creation of value for the client based on the above definitions would be: set of capabilities that companies have to develop new combinations of resources that result in greater efficiency or greater benefit for the client, through of the products and services that they develop.

3 Conclusions

In the first place, the importance of the role played by the creation of superior value for the client in order to achieve a sustainable competitive advantage has been highlighted in this work. Thus, the competitiveness of companies will depend, among other factors, on the degree to which they are able to satisfy the needs of their customers better than the competition.

The question of how resources generate value for the client has been a forgotten element, despite its importance for business strategy. In relation to this, the need arises to integrate some of the approaches made under the scope of Marketing to the Perspective of Resources and Capacities (RBV) to try to make up for this deficiency, trying to identify which resources are capable of creating a superior value for customers and how these resources should be transformed into new value combinations for customers.

The satisfaction of the client's needs, sometimes even above their expectations, is not an easy task, since it can vary not only between clients, but also between sectors, companies, etc., and change over time. This led us to carry out a preliminary analysis of the value for the client, studying in the first place, its nature. In this way we were able to verify its high subjectivity, its dependence on the context and the situation, the influence of the client's own factors, its dynamic character, etc. Likewise, its components have been analyzed, basically grouped in the benefits perceived by the client and the sacrifices it supports, as well as its main results or consequences, among which loyalty and satisfaction stand out.

There are many more studies that deal with value for the client, than studies that take into account the point of view of the provider towards which, among other things, the analyzes are more fragmentary and are often carried out in a logical economic-financial. There is greater attention to the dimensions of benefits than to those of sacrifices.

A tendency to focus on the value generated within the dyad can be observed, leaving aside the other relationships in the network that can have an impact on it.

Above all, it is possible to observe the scarcity of models included in the analysis of suppliers and customers and the total lack of studies comparing the two perspectives.

It is not even clear whether a reflexive approach (the construct makes its variables) or a formative approach (dimensions make the construct) better, in order to study value.

Finally, we can notice that in marketing there is no well-established theory on the subject of "value" that is capable of integrating construction in the broader landscape of relational marketing studies.

Thus we can see that, contrary to the many models of study of the value of the relationship of suppliers and customers, there is no model that can be considered predominant over others, a perspective that is more credible than the others, in simpler words a theoretical reference.

References

Adner, R., Zemsky, P.: A Demand-based perspective on sustainable competitive advantage. Stategic Manag. J. **27**, 215–239 (2006)

Anderson, J.C., Narus, J.A.: Business marketing: Understand what customers value. Harvard Bus. Rev. **76**(6), 53–65 (1998)

Anderson, J.C., Gerbing, D.W.: Structural equation modelling in practice: a review and recommended two step approach. Psychol. Bull. **103**(3), 411–423 (1988)

Anderson, J.C., Jain, D.C., Chintagunta, P.K.: Customer value assessment in business markets: a state-of-practice study. J. Bus.-to-Bus. Mark. **1**(1), 3–29 (1993)

Anderson, J.C., Hakansson, H., Johanson, J.: Dyadic business relationships within a business network context. J. Mark. **58**(4), 1–15 (1994)

Anderson, J.C., Narus, J.A.: Capturing the value of supplementary services. Harvard Bus. Rev. **73**(1), 75–83 (1995)

Anderson, J.: Relationships in business markets: exchange episodes, value creation and their empirical assessmen t. J. Acad Mark. Sci. **23**(4), 346–350 (1995)

Anderson, E.W.: An economic approach to understanding how customer satisfaction affects buyer perceptions of value. In: Stewart, D.W., Vilcassim, N. (eds.) Proceedings of the AMA Winter Conference: Marketing Theory and Applications. vol. 6, pp. 102–106 (1995)

Anderson, J.C., Narus, J.A.: Business marketing: understand what customer value. Harvard Bus. Rev. **76**(6), 53–65 (1998)

Anderson, J.C., Thomson, J.B.L., Wynstra, F.: Combining price and value to make purchase decisions in business markets. Int. J. Res. Mark. **17**(4), 307 329 (2000)

Arias, L.: Metodología de la Investigación. Trillas, México (2007)

AMA.: Asociación Americana de Marketing en www.ama.org

American Marketing Association. J. Mark. (2007)

Ballantyne, D., Varey, R.J.: Creating value-in-use through marketing interaction: the exchange logic of relating, communicating and knowing. Mark. Theory **6**(3), 335–348 (2006)

Biggemann, S., Buttle, F.: Conceptualising business-to-business relationship value. In: Wynstra, W. (ed.) Proceedings of the 21st Annual IMP Conference: Dealing with Dualities Erasmus University: Industrial Marketing and Purchasing (IMP) Group (2005)

Carr, L.P., Ittner, C.D.: Measuring the cost of ownership. J. Cost Manage. 42–51 (1992, Fall)

Flint, D.J., Woodruff, R.B., Gardial, S.F.: Customer value change in industrial marketing relationships, A call for new strategies and research. Ind. Mark. Manage.**26**(2), 163–175 (1997)

Ford, D., McDowell, R.: Managing business relationships by analysing the effects and value of different actions. In: Ind. Mark. Manage. **28**(3), 429–442 (1999)

Corsaro, D.: Relationship Value in Business Markets: Strategic, Relational and Technological. In: 8th Global Conference on Business & Economics. Universita Cattolica del Sacro Cuore di Milano, Florence (2008)

Gale, B.T.: Managing Customer Value: Creating Quality and Service That Customers Can See. The Free Press, New York (1994)

Goodwin, R., Ball, B.: Closing the Loop on Lotalty, Primavera, 25–34 (1999)

Grönroos, C.: Value-driven relational marketing: from products to resources and competencies. J. Mark. Manag. **13**(5), 407–419 (1997)

Grönroos, C.: On defining marketing: finding a new roadmap for marketing. Mark. Theory **6**(4), 395–418 (2006)

Grönroos, C.: The relationship marketing process: communication, interaction, dialogue, value. J. Bus. Ind. Mark. **19**(2), 99–113 (2004). https://doi.org/10.1108/08858620410523981

Gronroos, C.: From marketing mix to relationship marketing: towards a paradigm shift in marketing. Asia-Aust. Mark. J. **2**(1), 9–29 (1994)

Holbrook, M.B.: The nature of customer value. In: Rust, R.T., Oliver, R. L. (eds.) Service Quality: New Directions in Theory and Practice, pp. 21–71. Sage Publication, Thousand Oaks (1994)

Huber, F., Herrmann, A., Morgan, R.E.: Gaining competitive advantage through customer value oriented management. J. Consum. Mark. **18**(1), 41–53 (2001)

Johnson, M.D., Herrmann A., Huber F.: The Evolution of Loyalty Intentions. Am. Mark. Assoc. **70**, 122–132 (2006)

Kothari, A., Lackner J.: A value based approach to management. J. Bus. Ind. Mark. **21**(4), 243–249 (2006)

Homburg, C., Garbe, B.: Towards an improved understanding of industrial services: quality dimensions and their impact on buyer-seller relationships. J. Bus.-To Bus. Mark. **6**(2), 39–71 (1999)

Kothandaraman, P., Wilson, D.T.: Implementing relationship strategy. Ind. Mark. Manag. **29**(4), 339–349 (2000)

Porter, M.E.: What is strategy? Harvard Bus. Rev. **44**, 61–78 (1996)

Parasuraman, A.: Reflections on gaining competitive advantage through customer value. J. Acad. Mark. Sci. **25**(2), 154–161 (1997)

Martin, D.: El Valor Percibido como Determinante de la Fidelidad del Cliente. Universidad de Sevilla, Tesis Doctoral (2001)

Martin, D., Barroso, C., Martin, E.: El Valor Percibido de un Servicio. Rev. Esp. Investig. Mark. ESIC **8**(1), 47–73 (2004)

Lindgreen, A., Wynstra, F.: Value in business markets: What do we know? Where are we going? Ind. Mark. Manage. **34**, 732–748 (2005)

Mocciaro, A., Battista, G.: The development of the resource-based firm between value appropriation and value creation. Adv. Strat. Manag. **22**, 153–188 (2005)

Mizik, N., Jacobson, R.: Trading off between value creation and value appropriation: the financial implications of shifts in strategic emphasis creation. J. Mark. **67**, 63–76 (2003)

Normann, R., Ramírez, R.: From value chain to value constellation: Designing interactive strategy. Harvard Bus. Rev. **71**(4), 65– 77 (1993)

Payne, A., Holt, S.: Diagnosing customer value: integrating the value process and relationship marketing. Br. J. Manag. **12**(2), 159–182 (2001)

Prahalad, C.K., Ramaswamy, V.: Co-creation experiences: the next practice in valuecreation. J. Interact. Mark. **18**(3), 5–14 (2004)

Ponsonby, S., Boyle, E.: The 'value of marketing' and the marketing of value in contemporary times—A literature review and research agenda. J. Mark. Manage. **20**(3-4), 343–361 (2004)

Ramírez, R.: Value co-production: Intellectual origins and implications for practice and research. Strateg. Manag. J. **20**, 49–65 (1999)

Ravald, A., Gronroos, C.: The value concept and relationship marketing. Eur. J. Mark. **30**(2), 19–30 (1996)

Ritter, T., Wilkinson, I.F., Johnston, W.J.: Managing in complex networks. Ind. Mark. Manag. **33**(3), 175–183 (2004)

Rojas, R.: El Proceso de la Investigación Científica. Trillas, México (2007)

Setijono, D., Dahlgard, J.J.: Customer Value as a key performance indicator (KPI) and a key improvement indicator (KII). Measur. Bus. Excell. **11**(2), 44–61 (2007)

Slater, S.F.: Developing a customer value-based theory of the firm. J. Acad. Mark. Sci. **25**(2), 162–167 (1997)

Slater, S.F., Narver, J.C.: Market orientation, customer value, and superior performance. Bus. Horiz. **37**(2), 22–28 (1994)

Tuominen, M.: Channel collaboration and firm value proposition. Int. J. Retail. Distrib. Manag. **32**(4), 178–189 (2004)

Tzokas, N., Saren, M.: Value transformation in relationship marketing. Australas. Mark. J. **7**(1), 52–62 (1999)

Ulaga, W.: Capturing value creation in business relationships: a customer perspective. Ind. Mark. Manag. **32**(8), 677–693 (2003)

Ulaga, W., Chacour, S.: Measuring customer perceived value in business markets. Ind. Mark. Manag. **30**(6), 525–540 (2001)

Ulaga, W., Eggert, A.: Relationship value in business markets: the construct and its dimensions. J. Bus.-To-Bus. Mark. **12**(1), 73–99 (2005)

Ulaga, W., Eggert, A.: Value based differentiation in business relationships: gaining and sustaining key supplier status. J. Mark. **70**(1), 119–136 (2006a)

Ulaga, W., Eggert, A.: Relationship value and relationship quality. Eur. J. Mark. **40**(3/4), 311–327 (2006b)

van der Haar, J.W., Kemp, R.G.M., Omta, O.: Creating value that cannot be copied. Ind. Mark. Manag. **30**(8), 627–636 (2001)

Walter, A., Muller, T.A., Helfert, G., Ritter, T.: Functions of industrial supplier relationships and their impact on relationship quality. Ind. Mark. Manag. **32**(2), 159–169 (2003)

Walter, A., Ritter, T., Gemqnden, H.G.: Value creation in buyer–seller relationships. Ind. Mark. Manage. **30**(4), 365–377 (2001)

Wilson, D.: An integrated model of buyer-seller relationships. J. Acad. Mark. Sci. **23**(4), 55–66 (1995)

Wilson, D.T., Jantrania, S.: Understanding the value of a relationship. Asia-Aust. Mark. J. **2**(1), 55–66 (1994)

Woodruff, R.: Customer value: the next source for competitive advantage. J. Acad. Mark. Sci. **25**(2), 139–153 (1997)

Woodruff, R.B., Gardial, S.: Know your Customers–New Approaches to Understanding Customer Value and Satisfaction. Blackwell, Oxford (1996)

Zander, I., Zander, U.: Knowledge and the speed of the transfer and imitation of organizational capabilities: an empirical test. Organ. Sci. **6**(1), 76–92 (2005)

Consequences of Poor Accounting Practices

Eladio Pascual-Pedreño[1], Laura Pascual-Nebreda[2]([⊠]),
and Juan Gabriel Martínez-Navalón[2]

[1] University of Extremadura, Madrid, Spain
[2] University Rey Juan Carlos, Madrid, Spain
Laura.pascual@urjc.es

Abstract. Every entrepreneur or professional is obliged to carry out the accounting and that such conduct is carried out in a correct manner. This paper refers, firstly, to such compliance, requirements to be complied with and other accounting obligations such as drawing up the annual accounts, auditing them, publishing them, legalizing them or retaining them. Secondly, the main consequences of improper handling or lack of accounting, which may be of a criminal nature (existence of crimes against the Public Treasury) or of an administrative nature (no crime, but Or the place to apply the method of indirect estimation of tax bases.

1 Introduction

The businessperson is a natural or legal person under private law who acts on its own behalf or on behalf of others and carries out a commercial, industrial or service activity (Broseta and Martínez 2013). From this moment, he or she is subject to an exclusive statute in order to regulate all obligations of the businesspersons, for example, the one set by Article 25 of Code of Commerce on the accounts of businesses: "All businesses must keep orderly accounts, in keeping with the activity of their business activities that allows chronical monitoring of all their operations, as well as periodic preparation of balance sheets and inventories. Notwithstanding the terms set forth in the laws or special provisions, and inventories and annual accounts book and another daybook must necessarily be kept" (Royal Decree, dated August 22nd, 1985).

The Article 29.1 of the Code of Commerce refers to the correct keeping of the accounts as follows: "All the accounting books and documents must be kept, whatever the procedure used, with clarity, by order of dates, without blank spaces, interpolations, crossing out or erasures. As soon as errors or omissions suffered in the accounting annotations are noticed, they shall be noted. No abbreviations or symbols whose meaning is not clear according to the Law, the Regulations or generally applicable business practice may be used".

However, we must consider the requirements of the tax legislation about full compliance with accounting standards are confined to businesspersons whose fiscal regulations demand accurate accounting practices. This concerns, for example, natural persons under the direct estimate tax regime who carry out commercial activities (art.104 of the Personal Income Tax Law; art.68.2 of the PIT Regulations); or legal corporate tax payers (art.120 of the Law 27/2014 of the Corporate Tax Act). In the

© Springer Nature Switzerland AG 2019
J. Gil-Lafuente et al. (Eds.): AEDEM 2017, SSDC 180, pp. 340–348, 2019.
https://doi.org/10.1007/978-3-030-00677-8_27

remaining cases, accounting obligations require to keep the following books: sales and income records, purchases and expense records and investment assets record (Pascual 2016).

Article 10.3 of the Law of Corporate Income Tax states that "the tax base is equal to the accounting profit, once this has been corrected according to law" (Pascual 2009). Likewise, this was also established in the Royal Legislative Decree 4/2004 of March 5, whereby the amended Text of the Corporation Tax Act was approved (BOE dated March 11, 2004) (Báez 2005).

In case of a natural person, all data reflected in the accounting records act as a basis for determining the amount to be paid to the public purse in accordance with tax compliance. In addition to the required corrections, the law also states in Article 120: "Corporation tax payers must keep their accounts in compliance with the Code of Commerce or with the regulations governing them".

Those entrepreneurs under tax regulations and required to keep their accounts, must necessarily keep the daybook and the inventories and annual accounts book. As stated in article 25 of the Civil Code (Royal Decree dated July 24th, 1889), every businessperson is necessarily required to keep the inventories and annual book and a daybook. The daybook is "an accounting document that shall record the daily account of all the operations related to the activity of the business (Article 28 of the Code of Commerce); including either ordinary or extraordinary activities (Alonso 2009). In addition to the above, as stated in the article 34 of the Code of Commerce, every entrepreneur must draw up the annual account of his business at year-end. According to the General Accounting Plan, these documents form a single unit and should be prepared in compliance with accounting regulations (Code of Commerce, the Companies Act and the General Accounting Plan) (Cosín 2009).

Once all accounting and registration requirements are described, we would like to dwell on the possible consequences of carrying poor or lack of accounting practices.

2   Theoretical Framework

Keeping the accounts should not be carried out arbitrarily but entrepreneurs must follow all legal and technical, regulations on accounting and commercial matters to harmonize the accounting information. In the words of Alonso Espinosa (2009), accounting is "a continuous and systematic process of recording, documentation, management, planning, processing, sorting and keeping all business activities that are likely to have a direct, indirect, current, future or past impact on the composition of the business' assets, on its valuation or on its economic outcomes".

This is associated with the newly coined term "accounting manipulation" (Elvira and Amat 2008), which means "range of techniques selected by the businessperson in order to obtain desired profits taking advantage of the flexible Generally Accepted Accounting Principles" (Apéllandiz and Labrador 1995).

As stated in Article 29.1 of the Code of Commerce "all the accounting books and documents must be kept, whatever the procedure used, with clarity, by order of dates, without blank spaces, interpolations, crossing out or erasures. As soon as errors or omissions suffered in the accounting annotations are noticed, they shall be noted.

No abbreviations or symbols whose meaning is not clear according to the Law, the Regulations or generally applicable business practice may be used".

This responsibility for keeping accounting records of the business is not individual, however it can either fall on the entrepreneur (in case of a natural person) or on the shareholders or directors (in case of a legal person), or it may be even delegated to other person linked professionally. It is important to highlight the lack of provisions with punitive nature in commercial law about keeping the accounts. The Article 279 of the Companies Act stipulates the obligations of the company's directors to file annual accounts at the Business Registry within a month since the Board has approved it. However as an exception, the Articles 282 and 283 of the Companies Act describes what are the consequences if annual accounts have not been filed at the Business Registry.

Regarding both penal and tax legislations, the Article 180.1 of the General Tax Act states the non-concurrence of tax penalties, thus when an issue is not considered subject to penalty against Public Treasury by the tax administration it should follow the proceedings by the Prosecution Service. In case of condemnatory sentence, judicial authorities will prevent the administrative penalty.

As noted by Pérez (2016), breaches of the accounting regulations that according to Article 200 of the General Tax Act are subject to penalty will not be a breach as long as this has been taken into account as graduation criteria of the sanction applicable to the infringement of the article 191 of the Law. In case the offence is not appreciated by the competent Court of Law, the last paragraph of the article 180.1 establishes it is possible for the Tax Administration to act "according to proven facts". It is clear that "there can be non-concurrence between administrative and penal infringements" (Mestre and Cervantes 2005).

If we focus on the main consequences of not keeping the accounts or the lack of accounting practices, we should consider them as criminal (crimes against Public Treasury), administrative or tax offences.

1. Criminal offence: crimes against Public Treasury
2. Administrative offence: it refers to tax breaches.
3. Tax offence: possibility of applying the indirect estimation method of tax bases.

3   Results

3.1   Criminal Consequences: Crimes Against Public Purse

There are several types of criminal penalties concerning accounting. In addition to the known as "accounting crime", there are others actions that can lead to crime such as forgery of commercial documents, corporate crimes or even fraud. Therefore, carrying out improper accounting practices can lead to accounting crime as long as besides breaching the accounting obligations, it results in tax fraud. To sum up, accounting crime implies "there is a tax regulation that forces the breached accounting obligation" (Omeñaca 2013).

The accounting crime is the most severe consequence when breaching accounting obligations. Since criminal and tax law regulates it, the commission of the offence is only possible if it is bound by tax legislation as stated in Article 310 of the Criminal Code. The commission of the offence shall not exist unless keeping accounting books and records is required by taxation lax. The Organic Act 5/2010 dated 22nd June on Criminal Code, amended penalties applicable to all offences against Public Treasury and Social Security, except for accounting crime.

The following amendments regarding offences against Public Treasury and Social Security shall apply:

1ª. The article 310 was introduced in order to consider the liability of legal persons with regard to an offence committed

2ª. The maximum penalty rose to 4 or 5 years for offences defined in Title XIV (Crimes against Public Treasury and Social Security) and in accordance with Article 131, the statutory period of limitation of five years shall remain unaffected. However, the increasing sanctions will not be applicable for accounting crimes that are still punishable by jail from 5 to 7 months.

Regarding what the current Criminal Code states about accounting crime, we must differentiate between individuals (Article 310 of the Criminal Code) and corporations (Article 310 of the Criminal Code):

(a) Accounting Crime Committed by Individuals

The Article 310 of the Criminal Code establishes that:

"Whoever is obliged by law to keep corporate accounting, books or tax records shall be punished with a sentence of imprisonment from five to seven months when:

(a) He absolutely fails to fulfill that obligation under the direct assessment of the tax bases regime.

(b) He keeps different accounts that, related to the same activity and business year, conceal or simulate the true situation of the business.

(c) He has not recorded businesses, acts, operations or economic transactions in general, in the obligatory books, or has recorded them with figures different to the true ones.

(d) He has recorded fictitious accounting entries in the obligatory books.

The consideration as a criminal offence of the cases of fact referred to in Sections (c) and (d) above, shall require the tax returns to have been omitted, or for those submitted to provide a record of the false accounting and that the amount, by more or less, of the charges or payments omitted or forged exceeds, without arithmetic compensation between them, two hundred and forty thousand euros for each business year.

As seen above, this regards a special offence since individuals who are enforced by Tax Law can only commit it.

(b) Accounting Crime Committed by Legal Persons

The article 310 of the Criminal Code establishes the appropriate sanctions in case legal persons commit the offence. This article, amended by the Organic Act 5/2010 dated 22nd June on the Criminal Code, introduces a new regulation by which corporations can also be judged for crimes, since only individuals were subject to penalty so far.

Even so, corporations have limited criminal liability and only Law can declare the offences such as fraud, money laundering and breach of confidence.

The punishment of imprisonment shall be imposed on natural persons according to article 310 bis, but in case the offender is a legal person he will be punished with a fine. Criminal liability of legal and natural persons is not alternative but cumulative being legal persons criminally liable of the offence committed as well as natural persons, legal representatives and *de facto* or *de jure* directors of a company who took advantage of it (Article 31 of the Criminal Code).

As to the subjective element, this is a criminal offence in all its forms. According to the new Criminal Code, it is stated that there is no perpetration of the offence prescribed. Furthermore, there is no subjective element on the unjust that, in all cases, the perpetrator acts with the intention of committing tax fraud. Nevertheless, in case of keeping different accounts it is often required to hide information or show a false image of the real situation of the business.

Corporate Offences: Falsification of Annual Accounts and Other Documents

Considering how important the duties of the company director are, it is necessary to regulate its responsibility in case they breach their obligations. Titles XIII of the Criminal Code on corporate offences states in its first article 290 that "The *de facto* or *de jure* directors of a company incorporated or under formation, who falsify the annual accounts or other documents that should record the legal or financial status of the company, in such a way to cause financial damage thereto or to any of its shareholders or partners, or to a third party, shall be punished with a sentence of imprisonment from one to three years and a fine from six to twelve months. If financial damage were actually caused, the penalties shall be imposed in the upper half.

As stated by Omeñeca García (2015), it is important to differentiate the article 290 from the 310 and 310 bis mentioned before. The difference lies in the person who may commit the deed, since the article 290 declares the administrator guilty of falsification of the annual accounts while the articles 310 and 310 bis make reference to "legal or natural person obliged by law to keep corporate accounting, books or tax records".

Offence of Swindling

The Title XIII of the Criminal Code describes the criminal offences against property and against social-economic order. In this framework, in particular in chapter on fraud, the offence of fraud is explained. The Article 248.1 of the Criminal Code states "Those who use sufficient deceit, for profit, to cause lead another into error so as to have him carry out a deed of disposal in his own detriment or that of another, commit swindling".

To be considered a swindling offence, "the existence of a causal relationship between the victim and the swindler that confirms the detriment is caused by misleading behaviors should be noted" (Nuñez 2010) and also that this behavior is motivated by financial gain. This behavior arises when the person acts in order to obtain a patrimonial advantage. Therefore, the attitude is what makes an act punishable and it might occur before the fraud.

The study of the accounting crime should be supplemented with the next chapter referring to the framework of the procedure of penalties of administrative nature. The type of accounting crime and the facts that represent the administrative infringement can be the same.

3.2   Administrative Consequences: Existence of Tax Infringement

Regarding the procedure of administrative sanctions, the breach of the accounting obligations and with the registry can lead not to transfer an specific amount concerning the tax under consideration, even though it is also possible it does not cause an economic damage against Public Treasury. It is important to determine if there is an infringing behavior before defining the act committed.

The Article 183 of the General Tax Law states that tax infringements are negligent acts or omissions registered as infringements in this or other laws, which can be defined as the principle of criminal-law liability. In the words of Mestre and Cervantes (2005), "the Law must incriminate all negligent practices which assumes no criminal responsibility could be laid down which is prohibited by Constitution". It is relevant to consider the Article 179 of the General Tax Law that establishes persons who commit a tax infringement are subject to punishment as long as they result to be liable of the act committed. "There are exceptions for not being punished with sanctions such as incapacity, force majeure, collective decision, necessary diligence (reasonable interpretation of the regulation established in consultations of the Tax Administration) and errors of software tools of the Tax Administration" (Banacloche 2000).

The regulations on the procedure of tax sanctions can be found in Title IV of the General Tax Law in articles 207 and 212. The exercise of the sanction by the Administration is organized independently and separated from the tax liability and independent of both material aspects of the classification of infringements and sanctions and procedural aspects. Regarding the material scope of application of the General Tax Law, it is essential to highlight that besides dedicating an entire chapter to regulations of infringements and tax sanctions, it expressly rules out two categories of infringements and sanctions of its scope of application: sanctions in terms of contraband (regulated by the Organic Law 12/1995 and by the Royal Decree 1649/1998) and in terms of cadastral records (regulated by Royal Legislative Decree 1/2004) (Aníbarro and Sesma 2005).

In this paper we will focus on the most usual assumption: how breach of accounting activities lead to financial damage for the Public Treasury. In this case, this breach cannot be sanctioned as an independent sanction, but it will be evaluated according to graduation criteria of the breach of not paying or improperly obtaining tax refunds. According to the Articles 183 and 184 of the GTL, infringements are categorized in minor, serious or severe offences depending on the degree of culpability of the alleged infringer. An infringement will be serious in case of false or default invoices or there are substantial accounting anomalies that could not exceed a proportion. If fraudulent means are used a severe infringement is committed. In the absence of both situations or in case the Law evaluates it that way; the infringement will be defined as a minor fault

The decisive circumstance in order to evaluate the infringement regards the utilization of fraudulent means. This will define the sanction as a very severe sanction and, therefore, at least a 100% penalty will be imposed. The following instruments are considered fraudulent: anomalies in the account books and tax records, false invoices with more than a 10% of the value of the sanction and the utilization of related persons or corporations.

The following acts are considered substantial anomalies in accounting books or records established by tax regulations (Article 184.3ª) General Tax Law and 4.2 of the General Tax Penalty Regime (from now on RGRST):

1. Total breach of the obligation of carrying accounting books.
2. Regarding the same business activity, different accounting practices are carried out and do not show the true image of the situation of the business.
3. Breach of the required obligation if its impact is greater than 50% through the following circumstances:

   – False accounting entries, records or amounts.
   – Omission of realized transactions.
   – Incorrect booking of accounts so its tax treatment could change.

According to Mestre and Cervantes (2005) "from the sanction regime point of view, as long as the breach is strictly on accounting regulations, the improper booking of accounts will not take place.

Possible Incidence of Accounting Crimes
The type of accounting offence considered in the Article 310 of the Criminal Code and the facts that represent the administrative infringement of substantial anomalies in accounting books and tax records can be the same. This raises the problem of the consequences coming from the complaint of the accounting crime on administrative procedures of inspection and the possible disciplinary regime. This has been solved by the Article 33.1 of the RGRST that stipulates that in such cases "appropriate acts of assessment can be enforced with no possibility of starting or continuing the procedure for the application of sanctions based on identical facts and on the same grounds and with no chance of being considered in order to evaluate the infringement".

4 Discussions and Implications

The article 53.1c of the Genera Tax Law establishes "the substantial breach of keeping the accounts or records" is one of the circumstances that can define the application of direct estimation. This circumstance cannot totally be identified by substantial anomalies of accounting from Article 184.3 of the General Tax Law even though both qualifications can be usually the same. However, there may be instances where they are not the same. For instance, there can be anomalies in accounting without applying indirect estimation, because the data have been collected during tax inspections to determine the tax base with the direct estimation method (invoice delivery, information required to clients and providers, analysis of bank accounts or any other method that allows to establish the tax base through the direct estimation method). Thus, indirect estimation can be applied without committing infringement for breach accounting obligations if the books have been eliminated or lost due to force majeure.

The article 51 of the General Tax Law, indicates that when applying the direct estimation method, the tax administration will use the documents filed, the data recorded in the books and verified records and further documents and data related to tax obligations. In the direct estimate regime and since the assessment should be based on real facts, there should be sufficient evidence. It is obvious that as long as means of evidence are presented, an estimate must not take precedence. The accounts and other elements will represent the means of evidence.

The article 53 of the General Tax Act establishes that in order to estimate the tax base, the indirect estimation method will be applied when the tax authority cannot count on the required data and documents due to the following reasons:

(a) Tax declarations are not submitted or incomplete.
(b) Lack of cooperation, hindering or refusing the tax inspections.
(c) Substantial breach of the accounting obligations.
(d) The books and other supporting documents have been eliminated or lost due to force majeure.

One of the most important notes of the current regime is its subsidiarity regarding the direct estimation method, therefore its application may be limited in cases direct estimation method is not applicable. The Tax Administration shall prove that the taxpayer has breached his obligations and the cause and effect relationship between the infringement and the impossibility of applying the ordinary estimation regime. In addition to this, if the tax administration could determine the tax base using any other method (existing records of the system, invoices, information provided by clients or providers...) it will be impossible to make use of the indirect estimation method.

Concerning the burden of proof, if the taxable person wants to dispute the estimation he will have to prove the estimation is not correct, either because the used means are not suitable to be maintained or because they are not specific enough or because of any other appropriate reason. In this case, in addition to the evidence that can be presented to undermine the inspector's proposal, also means that differ from the ones applied by the Tax Administration such as signs, modules, criminal records etc. can be presented.

References

Alonso Espinosa, F.J.: El deber de documentación de la empresa y de llevanza de contabilidad tras la Ley 16/2007, de 4 de Julio. (G. W. La Ley, Ed.) Revista de Derecho del Mercado de Valores (RMV) (5), 2, pág. 26 y pág. 42 (2009)

Aníbarro, S., Sesma, B.: Infracciones y Sanciones Tributarias, Lex Nova, 1ª Edición, Valladolid, pág. 23 (2005)

Apéllandiz, P., Labrador, M.: El impacto de la regulación contable en la manipulación del beneficio. Estudio empírico de los efectos del Plan General de Contabilidad de 1990. Revista Española de Financiación y Contabilidad 24(82), pág. 13–40 (1995)

Alonso Espinosa, F.J.: "El deber de documentación de la empresa y de llevanza de contabilidad tras la Ley 16/2007, de 4 de Julio". In: G. W. La Ley, (Ed.) Revista de Derecho del Mercado de Valores (RMV) (5), (2009)

Banacloche, C.: El procedimiento sancionador tributario, Universidad Rey Juan Carlos, Madrid, pág. 7 (2000)

Báez Moreno, A.: "Capítulo III: Las consecuencias del sistema de relaciones entre la base imponible del IS y el resultado contable en el procedimiento de gestión tributaria: la potestad de determinación del resultado contable (Art. 143 TRLIS). Normas contables e impuesto sobre sociedades. (Ed.) Aranzadi, 2005.

Broseta, M., Martínez, F.: Capítulo 3: El empresario mercantil y su estatuto jurídico (2013)

Cosín, R.: La nueva contabilidad en el Impuesto sobre Sociedades, Ed. CISS, Madrid, pág. 32 (2009)

Elvira, O., Amat, O.: La manipulación contable: tipología y técnicas. Revista Partida Doble, núm. 203, pág. 48 y pág. 59 (2008)

Mestre García, E., Cervantes Sánchez-Rodrigo, C.: Infracciones y sanciones. Ed. CISS, Madrid (2005)

Omeñaca García, J.: Estudios contables y fiscales, curso de contabilidad (sitio web), 2013. Recuperado el 30 de Abril de (2015) (Disponible en www.omenaca.com)

Pascual, E.: Practicum contable, Ed. Thomson Reuters, Pamplona, pág. 69 (2016)

Pascual, E.: Guía práctica contable y fiscal para juristas. Ed. Lex nova, pág. 299 y pág. 620 (2009)

Pérez, F.: Derecho financiero y tributario parte general, Thomson Reuters, 26ª edición, Navarra, pág. 431 (2016)

Nuñez Castaño, E.: Lección XXIII. Delitos patrimoniales de enriquecimiento mediante defraudación (I): Estafa. En Gómez Rivero, M.C. (Coord.) Nociones fundamentales de Derecho Penal: parte especial (adapto al EEES). Ed. Tecnos (2010)

The Gender Gap of Retirement Pensions in Spain, Causes and Improvements in the Legal Order

José Álvarez-García[1]([⊠]), Inmaculada Domínguez-Fabián[1], Francisco del Olmo-García[2], and Beatriz Rosado-Cebrián[1]

[1] Department of Financial Economics and Accounting, University of Extremadura, Avda. Universidad, n° 47, 10071 Cáceres, Extremadura, Spain
{pepealvarez, idomingu, brosadot}@unex.es
[2] Department of Economy and Business Management, University of Alcala-Madrid, Plaza de la Victoria, n° 2, 28802 Alcalá de Henares, Madrid, Spain
francisco.olmo@uah.es

Abstract. Despite the progress made, the gender gap in the labour market and in the Social Security System continues to be a pending issue of improvements for all the Spanish society. The objective of this paper is to analyse the part-time work situation in Spain and the gender gap that this may imply for pensions, as well as the regulatory changes and good practices developed in the Spanish legislation to reduce the gap. The analysis is developed from the Living Conditions Survey (LCS, 2013–2016) of the INE (National Institute of Statistics). The data show the gender inequalities in the labour market; unemployment, part-time work and inactivity are more present throughout women's working lives, which causes lower salaries and time contributed, and therefore, greater difficulties to access to contributory retirement benefits.

Keywords: Retirement pensions · Social security systems · Labour market Gender gap · Spain

1 Introduction

The concern for the financial health of the Social Security system in Spain is an important fact for all citizens. There is no doubt that the sustainability of the Social Security has been and will continue being one of the main concerns of the population in a context of changes in the labour market and in the demography.

The reforms that have been carried out in recent years have been aimed at improving the sustainability of the Social Security system in the face of demographic challenges. But beyond the ageing of the population, the system, of contributory type, is faced with situations such as an increase in the unemployment rate, temporary and part-time work, which puts its income at risk. An increase in the unemployment rate, in temporary and part-time work, in addition to aggravating the problem of sustainability,

J. Gil-Lafuente et al. (Eds.): AEDEM 2017, SSDC 180, pp. 349–364, 2019.
https://doi.org/10.1007/978-3-030-00677-8_28

also generates a problem of sufficiency, since the contributions made (both in time and amount) are determinants of the retirement pension.

Focusing on the challenge of part-time work and despite the fact that this type of contract is offered equally to men and women, it is noteworthy that 73.48% of total part-time contracts are performed by women. Many women prefer a part-time work contract in order to reconcile their professional and family life. But as indicated in the previous paragraph, in a pension system such as the Spanish one that is contributory, this implies that women are exposed to the difficulty of access to contributory benefits in the future. A significant fact is that in Spain there is a greater gender gap in pensions than in salaries, since although the salary difference between men and women is 16%, this difference is 39% in the pension case.

There are numerous underlying reasons for women's low income and increased risk of poverty: women participate less in the labour market, work fewer hours and receive, on average, a lower salary than men. Taking leave related to the family and the time dedicated to taking care of family members also have an impact on women's income. Women often face the "glass ceiling" when it comes to accessing to decision-making positions and this also has an effect on pensions; in fact only 1.4% of retired women receive the maximum pension. All this leads to the persistence of the gender wage gap and consequently the gender pension gap. The aim of this work is to examine some of the causes of this gender gap and the measures that have been taken to reduce them.

The Council of the European Union calls for awareness of the pension gender gap and to ensure that women and men "remain in quality jobs throughout their life cycle, which results in greater labour participation and greater professional careers and, therefore, an income record that will result in an adequate pension" (Council conclusions adopted in September 2015). This line is also included as a strategic commitment for gender equality 2015–2019 and the European Pact for Gender Equality (European Commission, 2015).

As stated in Gómez Saavedra (2016), the different regulatory developments, as well as the jurisprudential developments have created a series of mechanisms aimed at correcting the inequalities that were created in access to retirement for those who throughout their lives had worked part-time, and were and are mostly women. At first, the old regulation of the General Social Security Law 1994 offered a disproportionate treatment to those part-time workers who wanted to access to retirement pensions.

After the ECJ ruling on the *"Elbal Moreno case"*, and the subsequent ruling by the Constitutional Court, the regulations on the calculation of contributions for part-time workers were transformed, with the approval of the Royal Decree-Law 11/2013, which is included in the current TR-LGSS (Consolidated Text-General Law on Social Security), and where the contribution calculation is redefined by applying the correction mechanism of the Global Part-time Coefficient. For this, the applicable contributed periods in which the worker has been registered working part-time will be taken into account, applying the percentage of the working day that appears in the contract and that would correspond to the part-time coefficient, whose result will correspond to the effective number of days contributed by the worker. Then the so-called Global Part-time Coefficient will be calculated, which is the proportion between the days contributed and the number of days registered working throughout the worker's working life, which allows for a more accurate calculation to access to the different benefits for part-time workers.

But everything related to part-time work and pensions is discriminatory; in fact, the Constitutional Court in its judgment 156/2014 of September 25 considered the constitutionality of the legal precept that establishes the rule for the calculation of the regulatory base, including contribution gaps, of retirement and permanent disability benefits of part-time workers. On April 14, 2015, the Court of Justice of the European Union (2011) concluded that the way in which the Spanish Social Security regulations include the contribution gaps as of Law 27/2011, is not discriminatory for part-time workers.

The objective of this paper is to analyse the part-time work situation in Spain and the gender gap that this may imply for pensions, as well as the regulatory changes and good practices developed in the Spanish legislation to reduce the gap mentioned above. The analysis is developed from the Living Conditions Survey (LCS, 2013–2016) of the INE (National Institute of Statistics), from which data are obtained in relation to the evolution of employment and part-time work in recent years and if there are differences between men and women.

After this introduction, the structure of the work is as follows. In the second section, a descriptive analysis of the evolution of employment in Spain during the 2013–2016 period by gender is carried out based on the Living Conditions Survey. The third section analyses the regulatory changes related to eliminating indirect discrimination in Spanish retirement pensions, in the fourth section, the data of the pension system is shown by sex and in the fifth section, some good practices carried out in the Spanish regulations are indicated. Finally, the conclusions and recommendations extracted from the work and the bibliographic references used are collected.

2 Analysis of Part-Time Work in Spain Based on the Living Conditions Survey (2013–2016)

Despite the progress made, the gender gap in the labour market and in the Social Security system continues to be a pending issue of improvements for all the Spanish society. During 2017, 46.31% of Social Security affiliates were women, increasing up to five percentage points with respect to 2007, according to data from the Ministry of Employment and Social Security.

During the last few years, part-time recruitment has been promoted. However, 73.5% of such contracts are for women, with their average salary being 23.9% lower than men's (Ministry of Employment and Social Security 2017). As can be seen in Graph 1, of all the contracts in 2017, 85.70% are full-time and 14.30% are part-time. In addition, gender differences can be seen, since 93.10% of men's contracts are full-time, while in the case of women it is 76.80%.

With the objective of carrying out a descriptive analysis of the situation of women in the labour market, the Living Conditions Survey of the National Institute of Statistics (LCS) includes comparable and updated cross-sectional and longitudinal data on income, the level and composition of poverty and social exclusion, at national and European level.

Its design enables to obtain *longitudinal information*, i.e., referring to the same people at different moments in time. In this way, you can know the temporary evolution

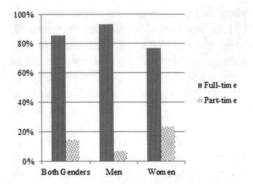

Graph. 1. Employment according to the day and gender in 2017. *Source:* Own elaboration based on the Ministry of Employment and Social Security (2017).

of the variables of interest of each individual. To carry out this work, the use of the latest version (2013–2016) of the Living Conditions Survey (LCS) was chosen.

Table 1 shows the evolution of part-time employment, and how it is distributed by gender. Based on the above results, we can see an increase in part-time employment in relation to the total of the Survey and to the employment between 2013 and 2016. Regarding the evolution of part-time employment by gender, it can be observed how part-time work has increased both for men and women in the last years and how it is considerably higher in the case of women.

Table 1. Evolution of part-time employment (2013–2016).

Years	Both genders		Men		Women	
	Partial/Total[a]	Partial/Employment	Partial/Total	Partial/Employment	Partial/Total	Partial/Employment
2013	7.01%	16.10%	3.87%	8.00%	10.00%	25.70%
2014	7.16%	16.10%	3.93%	7.60%	10.24%	26.10%
2015	7.56%	15.70%	4.43%	8.00%	10.59%	25.00%
2016	7.86%	15.30%	4.78%	7.70%	10.94%	24.40%

[a]It refers to the totality of situations that can occur in the labor market: full-time and part-time employment, unemployment and inactivity.
Source: Own elaboration based on the Living Conditions Survey (2013–2016).

In Graphs 2, 3, 4 and 5, there is a comparison of the employment situation in Spain for both men and women. In this way, and as can be seen, the percentage of the population in part-time employment has increased in the analysed period, it remains practically constant (about 72%) for women, and contributory unemployment decreases over the years, being much more significant in the case of men.

Regarding the gender comparison, we can see how part-time work is a phenomenon with a greater presence of women for all the years analysed, as well as inactivity. The inactive population includes those people who are not employed or receive contributory unemployment benefits (people who take care of their home, students, retired people and early retirees who receive a pension different from retirement and early retirement, those unable to work and those people who, without performing any economic activity, receive public or private aid).

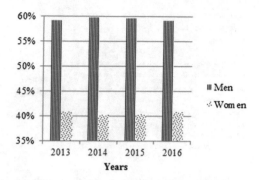

Graph. 2. Evolution of full-time contracts (2013–2016) in Spain according to gender. *Source:* Own elaboration based on the Living Conditions Survey (2013–2016).

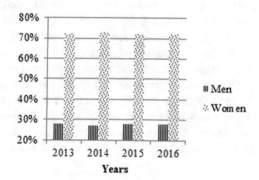

Graph. 3. Evolution of part-time contracts (2013–2016) in Spain according to gender. *Source:* Own elaboration based on the Living Conditions Survey (2013–2016).

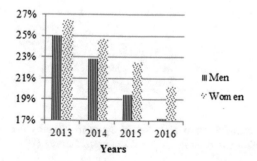

Graph. 4. Evolution of unemployment with benefit (2013–2016) in Spain according to gender. *Source:* Own elaboration based on the Living Conditions Survey (2013–2016).

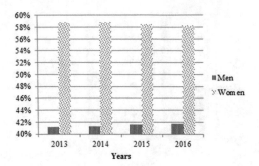

Graph. 5. Evolution of inactivity (2013–2016) in Spain according to gender. *Source:* Own elaboration based on the Living Conditions Survey (2013–2016).

The data shown, and compared between men and women, make us think that the contributions made to the Social Security by women will be lower.

Regarding the analysis performed by age groups (Graphs 6, 7, 8 and 9), it is observed that as the age of the individuals that make up the LCS increases in 2016, full-time and part-time employment increases, being the age groups between 35 and 44, the ones with a higher percentage of individuals with full-time and part-time jobs compared with the total. Such is the case that part-time work is predominant among younger people, in the age groups between 16–19 and 20–24, being part-time work much higher among women in all age groups observed. This is the case for inactivity, whose presence is noticeable in the case of women in all the observed age groups.

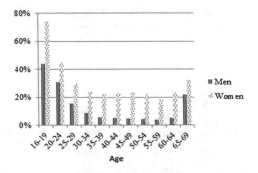

Graph. 6. Part-time contracts (in percentage) in Spain in 2016 by age and gender segments. *Source:* Own elaboration based on the Living Conditions Survey (2013–2016).

Despite the fact that part-time work, according to Gómez (2016), is an option that is presented equally to men and women, reality shows a differentiated behaviour by gender, as can be seen in Graph 6, and in Spain it is linked to younger ages and women.

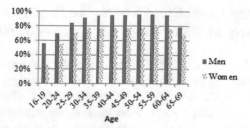

Graph. 7. Full-time contracts (in percentage) in Spain in 2016 by age and gender. *Source:* Own elaboration based on the Living Conditions Survey (2013–2016).

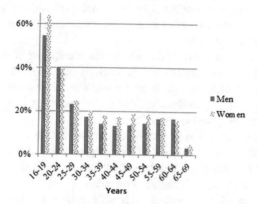

Graph. 8. Unemployment with benefit (in percentage) in Spain in 2016 by age and gender. *Source:* Own elaboration based on the Living Conditions Survey (2013–2016).

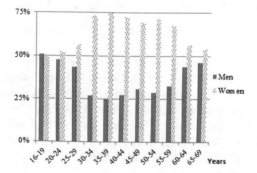

Graph. 9. Inactivity (in percentage) in Spain in 2016 by age and gender segments. *Source:* Own elaboration based on the Living Conditions Survey (2013–2016).

3   Changes in the Legal System of Social Protection of Part-Time Contracts to Compensate for Sex Discrimination

In order to verify the existence or not of a gender gap and the discrimination of workers with part-time contracts, following Alonso et al. (2015), the judgement of the European Court of Justice of April 14, 2015, is shown below, which determines non-discrimination by gender in the formula for calculating the regulatory base for disability and retirement pensions, in the case of using integration mechanisms of contribution gaps. And on the other hand, the sentence pronounced by the same Court, that considers discriminant the fact of requiring a proportionally larger contribution period in the case of part-time contracts to access to a contributory pension. This fact has been the subject of debates both at European and national level and has produced a series of legislative changes and modifications in the way of calculating the days contributed to calculate the retirement pension.

Thus, the social protection derived from part-time contracts will be governed by the rules of the Royal Legislative Decree 8/2015, of October 30, in relation to the requirement of the same effort for a full-time worker as for a part-time worker, avoiding disproportionate effects between the contributions actually made by the worker and the amount of the retirement pension.

In this way, the contribution to access to benefits will be:

- The contribution base to the Social Security and the contributions that are collected will always be monthly and will be constituted by the payments actually received based on the hours worked, both regular and extra.

- The contribution base determined this way cannot be lower than the amounts established in the regulations.

- The extra hours will contribute to the Social Security on the same bases and types as the regular hours.

In the accreditation of contribution periods the following rules will be applied:

(a)   The different periods during which the worker is registered working on a part-time contract will be taken into account.

(b)   Once the number of total days contributed has been determined, the Global Part-time Coefficient will be calculated, which is the percentage that represents the number of contribution days accredited out of the total number of registered working days throughout the employee's working life.

(c)   The minimum contribution period required for part-time workers for each of the economic benefits will be the result of applying to the general regulated period the global part-time coefficient referred to in section (b).

The regulatory base for retirement and permanent disability pensions will be calculated according to the general rule. For the amount of retirement and permanent disability pensions derived from a common illness, the number of resulting contributed days are increased by applying the 1.5 coefficient, but the resulting number of days must not exceed the registered part-time working period.

The way to calculate the days contributed in order to do the calculation corresponding to the waiting periods in order to obtain the right to a retirement pension for those workers recruited on a part-time basis has been subject to various reforms and, therefore, has evolved over time. Most of the changes undergone are due to the fact that the way of calculating the corresponding pension has been the subject of numerous debates by judicial doctrine, both at European and national level.

On November 22, 2012, the Court of Justice of the European Union ruled on the Elbal Moreno case. The main litigation was based on the claim made by the worker Elbal Moreno, 66 years old, who in October 2009 asked the National Institute of Social Security (INSS) for a retirement pension that was subsequently denied. The worker had worked for 18 years as a cleaner of a building of flats, with a part-time job of 4 h per week.

After an uphill legal battle, which is explained in detail in Alonso et al. (2015) and in Gómez (2016) a new calculation rule is added, based on the Global Part-time Coefficient. This new rule is more beneficial for part-time workers, especially in those cases where workers spend most of their professional career working part-time or also have short duration working days because by using the Global Part-time Coefficient the aim is to ensure access to Social Security benefits, respecting the principles of equality and proportionality that until now had not been achieved.

The calculation to obtain the Global Part-time Coefficient to request a retirement benefit is done in the following way:

1. The calendar days in which the worker has been registered will become accredited as days effectively contributed, applying the part-time coefficient to them, which represents the part-time working day.
2. Once we obtain the number of days effectively contributed, we will calculate the global part-time coefficient, which will be the result of dividing the accredited days as contributed by the total number of days in which the worker has been registered or recognised during his entire working life.

Finally, the global part-time coefficient will be applied to the minimum contribution period required, which in the case of a retirement pension is equivalent to 5,475 days (15 years). The coefficient will also be applied in those cases in which part or all of the contribution period is required to be included within a certain period of time. In this way, we will obtain the number of days of effective contribution that any part-time worker will need to access to the retirement benefit.

Paragraph (c) of the third rule of section 1 of the 7th Additional Provision of the LGSS-1994 was also subject to reform stating that in order to obtain the percentage applicable to the regulatory base, the correction coefficient of 1.5 must be applied to the days that have been accredited as effectively contributed. But the greatest novelty occurs in cases where the interested parties accredit a contribution period lower than the minimum period required, for these, once the correction coefficient has been applied to the days accredited as contributed, according to the provision: "*the percentage to be applied on the respective regulatory base will be the equivalent to that resulting from applying to 50 the percentage representing the contribution period accredited by the worker on fifteen years*".

This new form of calculation can be somewhat complex, by applying two coefficients: first the part-time coefficient and then the Global Part-time Coefficient. But these calculations offer a detailed adjustment of each worker, adjusting each of their needs to reality.

Therefore, this new regulation is able to ensure that part-time workers can access to a retirement pension respecting the criteria of proportionality established in the Constitution.

4   Men and Women in the Spanish Retirement Pension System

Once the regulatory changes made in Spain to ensure that part-time workers, mainly women, can access to a retirement pension on equal terms have been explained, some figures and data are analysed in this section that can help us assess the effects of these regulatory changes.

As we already know, in recent years, several key reforms have been approved to ensure the sustainability of our pension system, among which are highlighted those introduced by Law 27/2011, of August 1, on updating, adapting and modernizing the Social Security system, and Law 23/2013, of December 23, regulating the Sustainability Factor and the Revaluation Index of the Social Security Pension System.

The following are included in the main novelties introduced by the 2011 reform: delaying the legal retirement age to 67 years, increasing the calculation period of the regulatory base to 25 years, increasing the contribution period necessary to reach 100% of the regulatory base to 37 years. On the other hand, the necessary requirements for early retirement are tightened in order to stop an early exit from working life, and to approximate the real age and legal retirement age, and the incentives to prolong working life are improved by increasing the percentages that are applied for each year that individuals delay their retirement. All these measures tighten the conditions to access to a pension, especially in terms of contribution time.

Various authors analyse the sustainability of the Spanish pension system after the main reforms mentioned, taking into account real professional careers, Meneu et al. (2013), Rosado and Domínguez (2014), Rosado et al. (2015), and many others, that analyse the labour market in a more particular way through labour transitions–Cebrián et al. (2009), Malo and Toharia (2009), Kugler et al. (2002), Llorente et al. (2009), among others. All of them note a greater labour stability in men with respect to women, who show more cases of part-time work, inactivity and unemployment, and therefore lower contribution periods, which causes greater difficulties to access to a contributory pension and differences compared to men's pensions; even more after the last reforms of the system.

Analysing the average annual salary by sex (Table 2), there is a gender gap in salary that has practically increased in recent years, although it declined in 2015 to 22.86%. In addition, these differences by gender can be seen in terms of the type of working day, since most of the contracts and higher salaries correspond to men with full-time jobs. In relation to the age of the contributors, it can be observed how the average annual earnings are higher for men in all age groups analysed (Graph 10) (Table 3).

The gender gap shown for wages is also seen in the case of the pension, as can be seen in Table 4.

Table 2. Evolution of the average annual profit (in euros) and gender gap (2008–2015).

Years	Total	Men	Women	% Salaries women/mens	Gender gap
2008	21,883.42	24,203.33	18,910.62	78.13%	21.87%
2009	22,511.47	25,001.05	19,502.02	78.00%	22.00%
2010	22,790.20	25,479.74	19,735.22	77.45%	22.55%
2011	22,899.35	25,667.89	19,767.59	77.01%	22.99%
2012	22,726.44	25,682.05	19,537.33	76.07%	23.93%
2013	22,697.86	25,675.17	19,514.58	76.01%	23.99%
2014	22,858.17	25,727.24	19,744.82	76.75%	23.25%
2015	23,106.30	25,992.76	20,051.58	77.14%	22.86%

Source: Own elaboration based on INE (2017).
These are the most updated data a January 2018. For more detailed information consult: http://www.ine.es/dyngs/INEbase/es/operacion.htm?c=Estadistica_C&cid=1254736177025&menu=ultiDatos&idp=1254735976596

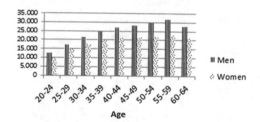

Graph. 10. Average annual profit (in euros) according to age and gender in 2015. *Source:* Own elaboration based on INE (2017).

As reflected in the previous table, there are significant differences both in the number and in the average amount to be received from retirement pensions for all ages analysed between both sexes. Thus, while male pensioners aged between 70 and 74 receive an average pension of 1,302.47 euros, women of the same age group receive 505.03 euros less.

However, in the last column of the table it can be seen how the gender gap is greater for the older age groups, which indicates that for younger retirees the gap is smaller. In addition, it can be seen how the gender gap in the case of contributory retirement benefits has fallen in the last nine years, although it remains quite high (36.30% in 2017), as can be seen in Table 5.

On the other hand, significant gender differences are also perceived in relation to the total time contributed to the Social Security system, as stated by Cebrián and Moreno (2015). Following these authors and from their analysis of the Continuous Work History Sample, it can be observed how the difference between the contributed

Table 3. Evolution of the average annual profit (in euros) and gender gap according to the type of working day (2008–2015).

Years	Men		Women		% Salaries women/mens		Gender gap	
	Full-time	Part-time	Full-time	Part-time	Full-time	Part-time	Full-time	Part-time
2008	25,415.12	11,392.62	21,936.20	9,662.44	86.31%	84.81%	13.69%	15.19%
2009	26,523.55	11,247.94	22,898.25	9,872.39	86.33%	87.77%	13.67%	12.23%
2010	27,335.23	10,960.91	23,932.01	10,133.18	87.55%	92.45%	12.45%	7.55%
2011	27,595.54	11,232.73	23,692.76	10,077.81	85.86%	89.72%	14.14%	10.28%
2012	27,898.06	11,032.10	23,674.19	9,988.41	84.86%	90.54%	15.14%	9.46%
2013	28,095.60	10,670.74	23,994.34	9,766.17	85.40%	91.52%	14.60%	8.48%
2014	28,318.14	10,028.85	25,041.55	9,690.50	88.43%	96.63%	11.57%	3.37%
2015	28,509.14	10,538.84	25,045.91	9,851.64	87.85%	93.48%	12.15%	6.52%

Source: Own elaboration based on INE (2017).

Table 4. Number of retirement pensions and average pension (in euros) by gender in 2017.

Age Groups	Retirement pension (December 1, 2017)					
	Men		Women		Pensions of women/pensions of men	Gender gap
	Number	Average pensions	Number	Average pensions		
60–64	2,55,335	1,567.91	96,129	1,417.42	90.40%	9.60%
65–69	9,42,823	1.379,44	5,69,190	993.83	72.05%	27.95%
70–74	8,57,811	1.302,47	5,05,941	797.45	61.23%	38.77%
75–79	6,00,867	1.161,89	3,52,911	672.31	57.86%	42.14%
80–84	5,28,109	1.046,13	3,20,061	610.66	58.37%	41.63%
85 y más	4,69,299	964.7	3,72,843	575.88	59.69%	40.30%

Source: Own elaboration based on data from the Ministry of Employment and Social Security (2017).

time of men and women increases with age, reaching a difference of more than 5 years for older ages. This indicates that in the ages closest to retirement, women have contributed less time than men of their generation. Therefore, there is a greater possibility that women have contributed less time than men, which will have a negative impact on the moment of access to retirement.

In this sense, and according to data from the Ministry of Employment and Social Security, the contribution careers of women amount to 30.61 years and that of men to 39.52 years, despite the fact that women retire, on average, at an older age than men (according to 2015 data). Following these authors and from their analysis of the Continuous Work History Sample, it can be observed how the difference between the contributed time of men and women increases with age, reaching a difference of more than 5 years in the most advanced ages. This indicates that in the ages closest to retirement, women have contributed less time than men of their generation.

Table 5. Evolution of the number of pensions, of the average pension (in euros) and of the gender gap in retirement pensions (2008–2017).

Years	Men		Women		Pensions of women/pensions of men	Gender gap
	Number	Average pensions	Number	Average pensions		
2008	3,220,736	954.15	1,770,132	569.29	59.66%	40.34%
2009	3,274,115	1,001.94	1,813,341	597.32	59.62%	40.38%
2010	3,333,320	1,038.75	1,859,663	619.91	59.68%	40.32%
2011	3,386,799	1,075.00	1,903,072	643.11	59.82%	40.18%
2012	3,441,062	1,117.01	1,950,351	670.31	60.01%	39.99%
2013	3,509,196	1,151.49	2,004,293	700.35	60.82%	39.18%
2014	3,557,309	1,173.53	2,053,715	720.44	61.39%	38.61%
2015	3,588,469	1,197.19	2,098,140	742.81	62.05%	37.95%
2016	3,628,587	1,219.44	2,156,098	767.06	62.90%	37.10%
2017	3,661,841	1,239.16	2,211,491	789.34	63.70%	36.30%

Source: Own elaboration based on data from the Ministry of Employment and Social Security (2017).

5 Measures Included in the Legal System to Reduce the Gender Gap

There are many studies that indicate the existence of a reduced pension motivated by the reduction of the working day and by leave for child or family care. In order to deal with these consequences, the Spanish legal system has devised a number of measures aimed at correcting these effects.

The General Social Security Law considers certain periods of time contributed in order to compensate for the periods of time dedicated to the care of children or dependent family members. Art. 237 TR-LGSS (Consolidated text of the General Social Security Law) establishes four different cases of fictitious contributions, indicating that they are considered as an effective contribution period for the purpose of retirement benefits, permanent disability, death and survival, maternity and paternity:

1. In the first place, the first 3 years of childcare or foster care leave in the case of permanent foster or custody for the purpose of adoption.
2. Secondly, the first year of leave for care of other family members up to the second degree of blood or family relationship due to age, accident, illness or disability reasons, that cannot take care of themselves and do not perform any paid activity. If the situations of leave indicated in the two previous sections have been preceded by a reduction of working hours in the terms provided for in article 37.5 TR-LET, for the purposes of considering the periods of leave that correspond as contributed, the contributions made during the reduced working day period will be calculated increased up to 100% of the amount that would have corresponded if the working day had been maintained without such a reduction.

3. The contributions made during the first two years of the reduced working day period for care of children under 12 years of age or the first year of the reduced working day period for family care, as established by article 37.6 TR-LET, are considered to be increased up to 100% of the amount that would have corresponded if they had been maintained without this reduction in working hours.

4. The contributions made in the case of reduced working days for the purposes of benefits of retirement, permanent disability, death and survival, maternity, paternity, risk in pregnancy or risk during breastfeeding: the parent, adopter, or foster parent will be entitled to a reduction of the working day, with the proportional reduction of the salary of, at least, half of its duration, for the care, during hospitalization and continued treatment, of the minor he/she is in charge of who is affected by cancer (malignant tumours, melanomas and carcinomas), or by any other serious illness that implies a long-term hospital stay and requires the need for direct and permanent care, accredited by a report from the public health service or health administrative body of the corresponding autonomous community and, maximum, until the child turns eighteen.

With regard to the recognition of retirement, permanent disability, death and survival, maternity and paternity benefits, the period considered as "effective contribution" will serve to accredit the minimum periods of contribution that give entitlement to such benefits.

In addition to the indicated measures and with the objective of eliminating or at least reducing the gender gap in pensions, in 2015, an allowance was approved for the pensions of women with natural or adopted children and who are beneficiaries of any regime of Social Security of contributory retirement, widowhood and permanent disability. This allowance, which came into force on January 1, 2016, has for all purposes the legal nature of a contributory public pension, and consists of an amount equivalent to the result of applying a certain percentage to the initial amount of the referred pensions, which depends on the number of children (5% of the pension if they have 2 children, 10% with 3 children and 15% for women who have 4 or more children).

According to data from the Ministry of Employment and Social Security (2017), more than 252,720 women received the maternity allowance in their pension (100,355 pensions correspond to those of retirement, 23,102 to those with permanent disability and 129,263 to widows). This allowance varies between 5 and 15% depending on the number of children they have, (of the 252,720 pensions with an allowance, 133,617 correspond to women with 2 children, 70,545 are paid to pensioners who had 3 children and 48,558 to pensions of women with 4 or more children), and can be considered as one of the causes for the pensions of women aged 65 in 2017 to have increased.

These types of allowances are to recognize the contribution of working mothers to the birth rate, in addition to correcting to some extent the differences between the average pensions of men and women, caused by more unstable working careers. Other European countries have already adopted measures that also incorporate not only the economic contribution but the concept of "demographic contribution" for the final value of the pension. For example, in France, female employees with more than three children are entitled to improvements in their pension. In Germany, on the other hand, all mothers are recognized-although the father can be recognized- between two and three years of contribution per child, as they are understood as children's periods of education.

6 Conclusions or Recommendations

The objective of this paper is to analyse the part-time work situation in Spain and the gender gap that this may involve for pensions, as well as the regulatory changes and good practices developed in the Spanish legislation to reduce the aforementioned gap. The data from the Living Conditions Survey (2013–2016), show the gender inequalities in the labour market. Unemployment, part-time work and inactivity are more present throughout women's working lives, which causes lower salaries and time contributed, and therefore, greater difficulties to access to contributory retirement benefits.

Based on this premise, a series of corrective mechanisms have been developed, with the aim of favouring the contributions of this type of workers. Corrections have been established in the legal system, in addition to compensation premises for the time spent caring for children. As a good practice we highlight the maternity allowance that has been positive to reduce the gender gap as a "post" measure, for those women who had already decided at the time to be mothers and have a certain number of children. The analysed data show the reduction in the gender gap in the Spanish Pension System, after the changes made, which is a good argument to continue working on and improving measures that favour the reduction of the gender gap of pensions.

References

Alonso, J.J., Devesa, J.E., Devesa, M., Domínguez, I., Encinas, B., Meneu, R.: El Tribunal de Justicia de la Unión Europea no aprecia discriminación en la fórmula de integración de lagunas para los contratos a tiempo parcial. Rev. Aranzadi Unión Eur. **5**, 49–59 (2015)

Cebrián, I., Moreno, G., Toharia, L.: ¿Por qué no reducen las bonificaciones la temporalidad? *Jornadas de Economía Laboral.* Universidad de Zaragoza, Julio (2009)

Cebrián, I., Moreno, G.: Tiempo cotizado, ingresos salariales y sus consecuencias para las pensiones: diferencias por género al final de la vida laboral. Cuad. Relac. Labor. **33**(2), 311–328 (2015)

European Union: Council conclusions of the 7 March 2011 of European Pact for Gender Equality (2011–2020). Official Journal of the European Union, (2011/C 155/02) (2011). http://eur-lex.europa.eu/LexUriServ/LexUriServ.do?uri=OJ:C:2011:155:0010:0013:EN:PDF

European Commission.: Strategic engagement for gender equality 2016–2019. Brussels, 3.12.2015, SWD 278 final (2015). http://ec.europa.eu/justice/gender-equality/files/documents/151203_strategic_engagement_en.pdf

Encuesta de Población Activa.: Instituto Nacional de Estadistica (2017). http://www.ine.es/dyngs/INEbase/es/operacion.htm?c=Estadistica_C&cid=1254736176918&menu=ultiDatos&idp=1254735976595

Gómez Saavedra, H.: La brecha de género en las pensiones: principales causas y mecanismos correctores. Universidad de la Laguna, Departamento de Derecho Público y Privado Especial y Derecho de Empresa (2016)

Instituto Nacional de Estadística.: Encuesta de Condiciones de Vida (2013–2016)

INE. Instituto Nacional de Estadística (2017). http://www.ine.es/

Kugler, A., Jimeno, J., Hernanz, V.: Employment consequences of restrictive permanent contracts: evidence from Spanish labour market reforms? IZA Discussion paper, 657 (2002)

Llorente, R., Sáez, F., Vera, J.: Dinámica de la inserción laboral: un análisis basado en la explotación de microdatos. Universidad de Zaragoza, Julio, Jornadas de Economía Laboral (2009)

Malo, M.A., Toharia, L.: ¿Qué se puede esperar de las reformas del mercado de trabajo? *Circunstancia*, 20. Instituto Universitario de Investigación Ortega y Gasset, Septiembre (2009)

Meneu, R., Devesa, J.E., Devesa, M., Nagore, A., Domínguez, I., Encinas, B.: El Factor de Sostenibilidad: diseños alternativos y valoración financiero-actuarial de sus efectos sobre los parámetros del sistema. *Economía Española y Protección Social*, V, 63–96 (2013)

Ministerio de Empleo and Seguridad Social.: (2017). http://www.empleo.gob.es/index.htm

Rosado, B., Domínguez, I.: Solvencia financiera y la equidad del sistema de pensiones español tras las reformas de 2011 y 2013. An.S Del Inst. Actuar. Esp., Terc. Época **20**, 122–163 (2014)

Rosado, B., Domínguez, I., Alonso, J.J.: Análisis empírico de la solvencia financiera y de la equidad del sistema de pensiones de jubilación español desde la perspectiva del empleo. *Ekonomiaz*, 88, 2° semestre, 344–366 (2015)

Author Index

© Springer Nature Switzerland AG 2019
J. Gil-Lafuente et al. (Eds.): AEDEM 2017, SSDC 180, pp. 365–366, 2019.
https://doi.org/10.1007/978-3-030-00677-8

Printed in the United States
By Bookmasters